5/0 ml

UNIVERSITY OF STRATHCLYDE

30125   00718188   5

are to be returned on or before

Catalysis in Application

Catalysis in Application

Edited by

S.D. Jackson
University of Glasgow, Glasgow, UK

J.S.J. Hargreaves
University of Glasgow, Glasgow, UK

D. Lennon
University of Glasgow, Glasgow, UK

RS•C

advancing the chemical sciences

The proceedings of the International Symposium on Applied Catalysis to be held at the University of Glasgow on 16–18 July 2003.

The cover artwork was based on an original drawing by Kirstin M. Jackson.

ISBN 0-85404-608-9

A catalogue record for this book is available from the British Library

Published by The Royal Society of Chemistry,
Thomas Graham House, Science Park, Milton Road,
Cambridge CB4 0WF, UK
Registered Charity No. 207890

For further information see our web site at www.rsc.org

Printed by Athenaeum Press Ltd, Gateshead, Tyne and Wear, UK

Preface

Catalysis in Application contains a selection of papers presented at the International Symposium on Applied Catalysis held at the University of Glasgow from 16–18 July 2003. The Symposium was a joint meeting of Surface Reactivity & Catalysis, Applied Catalysis and Process Technology subject groups of the Royal Society of Chemistry and the Institution of Chemical Engineers. The meeting also marked the retirement of Professor Geoff Webb after nearly 40 years active participation in catalysis research. The content of the meeting was focused around hydrogenation, deactivation, chiral catalysis and environmental catalysis, four areas that Geoff has made significant contributions to during his career. The meeting was attended by delegates from industrial and academic laboratories throughout the UK, Europe, USA and Asia.

Over the course of Geoff's career, catalysis has made significant advances, for example when he started in Glasgow the ICI Low Pressure Methanol process using Cu/ZnO/Al$_2$O$_3$ catalysts had not been invented. In the 60s and 70s new catalysts and catalytic processes changed the face of petrochemical and refinery processing. In the 80s and 90s the move for catalysis was into fine chemicals and pharmaceutical conversions. The application of catalysis has increased, such that catalysts are responsible for the manufacture or processing of a large number of products in daily use (from clothes to all plastic products), to preserving our environment and health (enabling a sustainable production, energy and mobility), to enabling the development of advanced and functional materials and devices. Even so, catalysis is still in its infancy, the simple problems have been solved, if not fully understood, but major challenges lie ahead. The development of highly selective catalysts for complex chemical transformations is a constant industrial driver. However advances in catalysis rarely come from a single discipline. The complexity of the catalytic process usually requires that chemistry and engineering intimately mix to deliver the desired effect. The need for a multi-disciplinary approach is reflected in this symposium.

The organisers would like to express their thanks to the participants and to the authors for their commitment to submitting camera-ready manuscripts on time. Finally we would like to thank all the people who have worked hard behind the scenes to enable this conference to take place and these proceedings to be published.

S. David Jackson
Justin S. J. Hargreaves
David Lennon

Contents

Contents   ix

MODIFICATION OF CATALYSIS AND SURFACE REACTIONS BY SURFACE CARBON

Michael Bowker*, Toseef Aslam, Chris Morgan*, Neil Perkins

Centre for Surface Science and Catalysis, Dept. Chemistry, University of Reading, Reading RG6 6AD

*Now at Chemistry Dept., Cardiff University, Cardiff CF10 3TB, Wales, UK

1   INTRODUCTION

This paper is devoted to considerations of the role of surface carbon in modifying surface reactivity, an area to which Geoff Webb has contributed significantly during his career [1,2]. It is generally considered that surface carbon is a poison for many reactions. Indeed, in the strict sense this is usually true (that is, as carbon builds up on a surface and total activity goes down). However, in this paper we give some examples of surface reactivity which show that carbon can have a very positive role to play in manipulating reaction selectivity, so much so that it can result in ***higher activity to desired products.***

Geoff Webb has been involved in this area during his years of contribution to the field of surface reactivity and catalysis. In particular he noted that the presence of carbon on metal surfaces may take a direct role in the catalysis of butene hydrogenation, by acting as a surface hydrogen exchange medium between hydrogen in the gas phase and the adsorbed olefin [2]. These kinds of ideas were extended by Somorjai [3] and others to hydrocarbon reactivity on surfaces by identifying the presence of certain intermediates on the surface (e.g. ethylidyne [4]). He also recognised that, although the metal surface can contain a very large amount of surface C, nevertheless hydrocarbon reactions can still proceed at a very high rate. In that case it was proposed that the reaction proceeds on the small amount of free surface still available [3].

Finally, a very nice example of the modification of surface reactivity by a surface poison is the case of methanol decomposition on Ni(100) studied by Johnson and Madix [5]. The clean Ni surface is a complete dehydrogenator, whereas the surface dosed with half a monolayer of S in an ordered structure results in a surface which is very selective to formaldehyde production, that is, the total dehydrogenation pathway is effectively blocked.

In what follows we show three very different examples of the influence of surface carbon on surface reactivity, namely, hydrocarbon reforming, the decomposition of a carboxylic acid and the decarbonylation of acrolein. In each case it can be argued that the adsorbed carbon layer plays a positive role in the catalytic reactions involved.

2   HEPTANE REFORMING ON Pt-Sn CATALYSTS

Coking is generally thought to be a problem in hydrocarbon reforming catalysis, but it is not so widely recognised that it is essential to the successful operation of modern catalysts for producing high octane fuel. Thus fig 1 shows data for the reforming of n-heptane on an alumina-supported Pt-Sn catalyst with 0.3 wt% of each of the latter components. Here it can be seen that, as the coke builds up on the catalyst, so the selectivity to toluene, a much desired reaction due to the high octane rating of toluene, increases significantly. The important coke layer is built up within a very short time on stream and our estimates indicate that it corresponds to about 1 monolayer of 'coke' spread over the whole catalyst, most of it therefore being located on the support. In fact, we believe that 'coke' is an inappropriate description of what is likely to be a well-defined, evenly spread layer. This layer on the support, then, appears essential to the good performance of these catalysts. This level of surface carbon is nearly constant for a significant time of the run (between 0.3 – 10 hours on stream). There is then evidence that, at much longer times, multilayer carbon builds up which is more properly described as 'coke'. Although we didn't carry out long-term tests, there was 2.2 wt% coke after 5 hrs on stream and others report an acceleration of coking after a long time on stream. This appears to be a second stage of detrimental carbon deposition and is part of the reason for recycling the catalysts for carbon removal in an oxidation step in industry. In summary then, carbon deposition is essential for the good performance of industrial naphtha reforming catalysts.

a)

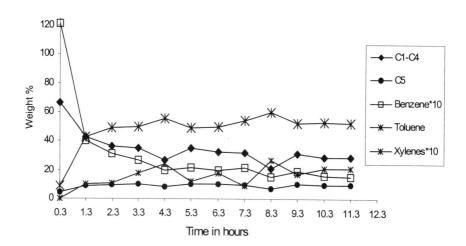

b)

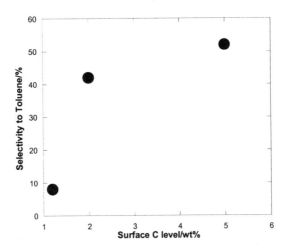

Figure 1   *a) Showing the change in reaction products with time for heptane reforming on a Pt-Sn catalyst which can be used for naphtha reforming. The 'fresh' catalyst predominantly exhibits hydrogenolysis to C1-C4 alkanes, whereas with time and coke build-up, it becomes very selective to aromatisation to toluene. b) The dependence of toluene selectivity upon coke level. Reaction conditions: temperature 515°C, H2:Heptane ratio = 3.7, heptane flow 4 mls liquid hr^{-1}*

3 THE ROLE OF SURFACE CARBON IN MODIFYING SURFACE REACTIVITY ON Pd(110)

3.1 Acetic Acid Decomposition on Pd(110)

Acetic acid decomposes on the clean surface at elevated temperature to produce gas phase CO_2 and hydrogen and leaves C (henceforth C_a for adsorbed carbon) on the surface [6,7]. However, this carbon has a surprising property, that is, it can modify the reaction pathway on the surface, yet does not affect the activity for adsorption very significantly. The C_a forms a well-ordered c(2x2) structure which is identified by LEED. As shown in fig 2 the carbon acts as a poison in one sense and in one regime of temperature, that is, it deactivates the surface for acetate decomposition in such a way that the acetate TPD peak is shifted from ~360-390K to 455K when the c(2x2) layer is preformed before dosing the acetic acid onto the surface. The overall reaction is –

$$CH_3COOH \rightarrow CO_2 + 2H_2 + C_a$$

Even though the C_a is there it does not poison acetate formation, which appears to occur with a similar adsorption probability, but it does stabilise it towards decomposition. In fact the desorption at 455K, occurring in the presence of C_a, is what is known as a 'surface explosion' [8], an autocatalytic decomposition, showing a very narrow half-width for the peak and anomalous desorption kinetics.

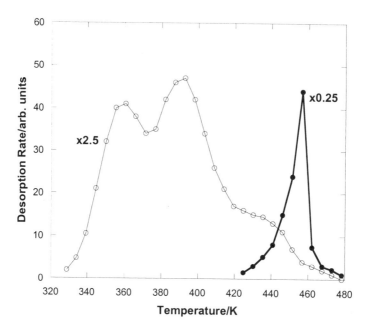

Figure 2   *Temperature programmed desorption experiment after acetic acid adsorption on Pd(110): a) on the clean surface and b) on the surface predosed with half a monolayer of C atoms in the c(2x2) structure*

When the reaction is carried out above 430K or so, then the reaction occurs at steady-state, notwithstanding the fact that a c(2x2) layer of carbon is present on the surface and that C_a is continually being deposited on the surface (fig. 3). The extra C_a appears to dissolve through the half monolayer of surface C into the bulk, presumably as a carbide. On the timescale of these experiments approximately 6 monolayers of C_a are deposited into the crystal with no apparent detriment to the reaction. In this regime the reaction does not appear to be limited by the surface C_a, that is, no net activation barrier is apparent and the rate is flux-limited. Presumably, if the pressure were much higher, then the surface would become populated by the stabilised acetate, which would then block sites and self-limit the reaction rate.

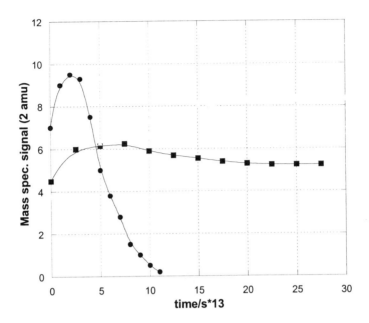

Figure 3 *Evolution of reaction products with time measured in the molecular beam reactor. At 423K(filled circles) hydrogen is evolved and stabilised acetate build-up on the surface, eventually blocking it to further reaction, whereas at 473K (squares) the acetate is unstable and the reaction proceeds at steady state on the c(2x2)-C layer*

3.2 Acrolein Decomposition on Pd(110)

This reaction shows a selectivity influence of adsorbed carbon on the decomposition reaction. If the acrolein is adsorbed at room temperature, then the molecule dehydrogenates to yield hydrogen in the gas phase and a mix of adsorbed CO and CHx species and then ceases due to blockage of the reaction sites by the latter. If the reaction is carried out at a slightly higher temperature then the reaction selectivity is changed. The surface is no longer a dehydrogenator and instead shows high activity and selectivity for the decarbonylation reaction (fig 4) , that is,

$$CH_2CHCHO \rightarrow C_2H_4 + CO$$

This reaction only occurs on the C-passivated surface and is selective in a limited temperature range; if the crystal temperature is > 350K, then dehydrogenation activity is seen again, presumably because extra C can be formed which can diffuse into the sub-surface region as for the acetic acid decomposition above. If the surface is pre-dosed with carbon, then the decarbonylation reaction begins immediately at high rate.

Thus C plays a very important role as a reaction modifier here. It reduces the dehydrogenation ability of the Pd and instead facilitates hydrogen intramolecular mobility. It must be noted that the decarbonylation reaction is thermodynamically well-favoured, but dehydrogenation is even more preferred on the clean surface.

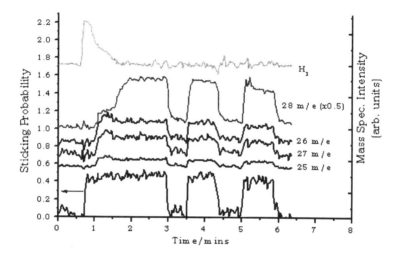

Figure 4   *A molecular beam reactor experiment in which the carbon-precovered Pd(110) surface is exposed to acrolein at 313K, beginning upon opening the beam shutter at 0.7 minutes into the experiment. The surface shows a high adsorptivity for the acrolein (s~0.4) and steady stare reactivity after 1.5minutes to the decarbonylation of the acrolein, producing ethene and CO. The beam is blocked and re-opened at 3-3.5 and 4.3 to 5 minutes to check background signals. The experiment is stopped at 5.8 mins*

4 CONCLUSIONS

We have given three examples of reactions where the nature of the surface is significantly changed by the presence of surface carbon. For hydrocarbon reforming the presence of a monolayer of carbon, mostly on the support, plays a very positive role in suppressing the hydrogenolysis reactions and enhances the rate of the desired aromatisation reactions. For acetic acid decomposition, there is little evidence of deactivation of the surface when a half monolayer of carbon is adsorbed, the reaction probability still being very high. In the case of acrolein, surface carbon changes the reaction from total cracking of hydrogen from the molecule to steady-state decarbonylation, occurring in a very clean fashion with very high reaction probability.

References

1. G. Webb, Specialist Periodical Reports: Catalysis, 1978, **2**, 145
2. S. Thomson and G. Webb, J. Chem. Soc., Chem. Comms., 1976,526.
3. G.A.Somorjai and F. Zaera, J. Phys. Chem., 1982, **86**, 3070.
4. G.A. Somorjai and B.E. Bent, Progr. Colloid Polym. Sci., 1985, **70**, 38.
5. S. Johnson and R.J. Madix, Surf. Sci., 1981, **103**, 361.
6. N. Aas and M. Bowker, J. Chem. Soc., Faraday Trans., 1993, **89**,1249.
7. C. Morgan and M. Bowker, Surf. Sci. submitted
8. J.L. Falconer, J. McCarty an R.J. Madix, Surf. Sci., 1974, **42**, 329.

CATALYTIC PROPERTIES OF THE PLATINUM–HYDROGEN–CARBON SYSTEM

Zoltán Paál and Attila Wootsch

Institute of Isotope and Surface Chemistry, Chem. Res. Center, Hungarian Academy of Sciences, P. O. Box 77, Budapest, H-1525 Hungary. Email: paal@iserv.iki.kfki.hu

1 INTRODUCTION

Pt catalysts are, as a rule, covered by "hydrocarbonaceous overlayers" during hydrocarbon reactions.[1] Their presence is necessary for steady-state activity in aromatization, C_5-cyclization, isomerization of alkanes.[2] Freshly regenerated catalysts (in a "Pt–H" state) exhibit high activity in hydrogenolysis. They become a *platinum-hydrogen-carbon* system, "Pt–C–H", after a short contact time with the reaction mixture.[2] The hydrocarbonaceous "Pt–C–H" entities correspond to the "reversible"[3] or to the "beneficial" carbon.[4] Catalyst after deactivation is transformed into "Pt–C".[2] Radiotracer methods can detect carbonaceous residues directly.[5–7] Hydrocarbonaceous deposits lose hydrogen and transform into "carbon" upon evacuation necessary for analysis by electron spectroscopy.[3] This may explain why relatively much C was detected by these methods.[8,9] Indirect methods involve carbon removal by oxidation[10,11] or hydrogenation.[2,12–14] Hydrogen treatment removed about 1 C atom per surface Pt.[13] Studies with ^{14}C radiotracer[6] showed ~0.7 C/Pt(surf.) after alkane exposures without hydrogen. This dropped to 0.15–0.20 C/Pt even in small H_2 excess. X-ray Photoelectron Spectroscopy (XPS) detected "massive" – graphitic and polymeric – carbon up to ~50% surface C after exposure to t,t-hexa-2,4-diene at 600 to 660 K.[15] Disordered C and ordered graphite layers on Pt were observed by lattice resolution transmission electron microscopy (TEM).[15] Exposure to hexane resulted also in similar amount and state of surface carbon.[16] "Regeneration" with O_2 and H_2 decreased the amount of "massive carbon", increasing the abundance of single C atoms or CH_x entities.[16]

The primary products of alkane reactions on Pt are dissociated alkyl radicals that give either reaction products or dehydrogenate further to form carbonaceous deposits.[17] They coexist with chemisorbed hydrogen, the abundance of which is, in turn, determined by the H_2 pressure, $p(H_2)$. Their competition results in maximum turnover rates[18,19] as a function of $p(H_2)$. Aromatization and dehydrogenation are preferred under small $p(H_2)$ values, together with "coking". A "polyene" route of coking[19,20] involves polymerization of *trans*-unsaturated intermediates whereas the "C_1 route"[11,21] would involve polymerization of the single C-atom entities. The deactivating effect of surface carbon depends on its amount and nature[15,16] and influences various reactions to a different extent. Catalysts representing platinum–hydrogen–carbon systems were obtained by intentional deactivating treatments of Pt. We report on their catalytic behaviour in hexane transformation, using this reaction itself as an indicator on the surface state of the catalyst.

2 EXPERIMENTAL

Pt-B is Pt black reduced by HCHO, dispersion D=1,36%[15,22b]; **EPT** is EUROPT-1, i.e. a "standard" 6.3% Pt on SiO_2, D=60 %[23]; and **Pt-CEO** is 2% Pt/CeO_2, D=35% (2Pt/HTR,[24] i.e. "high-temperature r educed", a t 7 73 K). T heir d eactivation a nd c atalytic t esting t ook place in a closed-loop reactor.[15,16,25] Two subsequent runs were monitored: 1) *pretreatment*, i.e., exposure to mixtures of 10 Torr (1 Torr=1.33 mbar) hexane (nH) or *t,t*-hexa-2,4-diene (HD) and 60 Torr H_2 for 20 minutes at temperatures between 483 and 663 K, followed, after evacuating the system, by 2) *test runs* at 603 K using a mixture of 10 Torr nH and 120 Torr H_2 for 50 minutes, with sampling also after 5 min.[15] Conversion (X) and selectivity (S) values of these *test runs* were compared to the well-reproducible[16,22a] results (X_0, S_0) of the regenerated samples, measured without deactivating *pretreatment*. Analysis after both runs used a gas chromatograph with a 50-m CP-Sil glass capillary column and FID detector. The same sample – 36 mg of **Pt-B**, 10.5 mg of **EPT** and 64 mg of **Pt-CEO** – was used and regenerated after each *test run* by exposure to 30 Torr O_2 for 2 minutes and then to 100 Torr H_2 for 3 min.[16,25]

3 RESULTS

Previous results showed that the presence of hydrogen hindered deactivation both by nH[16,26] and HD.[15,25] Another factor is the temperature: since more hydrogen would chemisorb at lower temperatures,[18,22] it would compete more efficiently with deactivating carbonaceous adsorbates. Table 1 shows transformations of nH and HD on **Pt-B** *during pretreatments*. Hexane conversion was negligible at 483 K and a few per cent at 543–663 K. The abundance of dehydrogenated products (benzene, hexenes) increased, those of saturated products (isomers, MCP) decreased at higher temperatures. The results obtained after exposure to nH alone at 603 K[26] are also shown. Different trends were observed after *pretreatment* with HD (Table 1). The HD conversion was very high, 100%, at 483 K, decreasing slightly at higher temperatures. The main reaction was saturation of both double bonds below 603 K. An abrupt change occurred at T≈600 K, where hexenes became the main products. At still higher temperature (T=663 K) double-bond isomerization prevailed. Highest benzene selectivity from both nH and HD appeared at 663 K (Table 1), where also thermodynamics favoured aromatization.

Similar trends appeared in the standard *test runs* after the treatments of Table 1. The residual activity (Figure 1) decreased in all cases with increasing pretreating temperature, parallel with increasing amount of deposited "coke".[10] Different patterns o f activity loss appeared, however, after deactivation with nH and HD. The residual activity was ~ 60% of that measured on regenerated Pt after pretreatments with nH at 483 K and 543 K but diminished dramatically after pretreatments at T=603 and 663 K. HD, in turn, caused hardly any deactivation at 483 K but an abrupt drop of the residual activity occurred above 543 K (Figure 1). The residual activities after deactivation with nH w ere slightly lower than after HD at 603 and 663 K. The sharp decrease in the residual activity can be attributed to the formation of more 3D deposits at higher temperatures. They predominated above 603 K after hexane treatment.[26] *t,t*-Hexa-2,4-diene, in turn, is a better precursor of poly-aromatic[18,20] as well as of 3D carbon of graphitic character. Their formation may have started at lower temperature, but HD apparently produced less "deactivating" disordered carbon[15] at 603 and 663 K than did nH. The strongest deactivation was caused by nH without H_2.

Table 1   *Conversion (X) and selectivity (S) values measured in pretreating runs of t=20 min at different temperatures. "No H_2" treatment took place at 603K, as reported earlier.[26] Catalyst: 36 mg **Pt-B**, p(nH or HD):p(H₂)=10:60 Torr.*

	Hexane(nH)/H₂ treatments					t,t-hexa-2,4-diene(HD)/H₂ treatments			
	483 K	543 K	603 K	663 K	No H₂		483 K	543 K	603 K	663 K
X (%)	0.2	2.5	5.3	4.5	0.3	X (%)	100.0	99.3	94.6	38.4
S (%):						S (%):				
<C₆	31.9	23.0	22.1	18.3	3.5	<C₆	0.4	0.6	1.0	2.8
C₆-iso.	48.2	31.2	23.5	9.0	4.9	Hexane	98.7	96.2	22.5	5.3
Hex.	6.0	1.8	3.2	34.6	60.9	C₆-iso.	0.4	0.6	0.2	0.0
MCP	13.9	31.4	35.3	14.8	0.2	Hexenes	0.4	0.2	68.7	23.3
Benz.	0.0	12.6	15.9	23.3	30.5	Dienes	0.0	0.1	2.7	59.9
						MCP	0.0	0.9	0.4	0.1
						Benzene	0.1	1.4	4.5	8.6

Since hydrogen was able to remove some carbon from Pt catalysts,[13,14] one can expect some "self-regeneration" during prolonged *test runs*. This was rather pronounced after nH treatments (Figure 1): the residual activity after *test runs* of 50 min (compared to the activity of 50-min runs over regenerated catalysts) was higher in all cases. Most conspicuous reactivation occurred after nH pretreatment at 603 K. The ability of self-reactivation was, as a rule, low after treatments by hexadiene (Figure 1).

The selectivity ratios measured on deactivated (S) and regenerated (S_0) catalyst are shown in Figure 2 after *pretreatments* with nH and HD at different temperatures. Deactivation was almost "non-selective" on Pt treated with nH at 483 K and 543 K where the activity was also hardly affected. The selectivity change was the highest at 663 K. Selectivities of isomer and fragment formation were most sensitive.[15,16,25] Benzene selectivity decreased almost proportionally to the overall activity loss (S/S_0 was around 0.8), it dropped markedly at 663 K only. MCP formation was *relatively* higher after pretreatments at lower temperatures (T≤543 K), but it deactivated after higher-temperature exposures. Hexene selectivity was nearly doubled at T≤543 K and increased dramatically at higher temperatures. The Figure includes points corresponding to the values measured after deactivation with nH without H_2.[26] These would correspond to a treatment with nH+H₂ at even higher temperature, giving additional evidence to the role of temperature in determining the hydrogen coverage under reaction conditions.

The selectivity ratio of deactivated to regenerated (S/S_0) catalyst (Figure 2) after exposure to 24HD at 483 K shows just more C_6 saturated products and less benzene. The selectivities of fragments, MCP and isomers decreased continuously after *pretreatments* at higher temperatures, opposite to the abrupt change in the activity. The S/S_0 value for benzene, in turn, was around 1 from 543 to 663 K.

"Self-regeneration" during *test runs* of 50 min brought the selectivities closer to the values of regenerated samples. For example at 603 K, the restoration of activity was more marked after *pretreatment* with nH as compared to HD. The selectivity differences were lower at the end of the *test run*. The S/S_0 values after 5 min are shown in Figure 2. After 50 min., the S/S_0 values of fragments and isomers increased to 0.75–0.8, S/S_0 for benzene was ~1 in both cases, S/S_0 for MCP was around 1 after HD and 1.2 after nH *pretreatment*. The stronger "self-reactivation" after nH treatment manifested itself mainly in hexene formation. S/S_0 for hexenes dropped by a factor of 2 (from ~7 to ~3) after nH whereas this change was much less (from ~7 to ~5) after HD *pretreatment*.

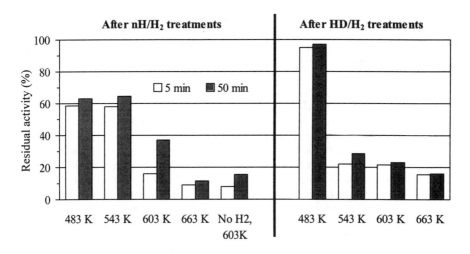

Figure 1 *Residual activity of the test reaction after the different pretreating exposures presented in Table 1. Test reaction: p(nH):p(H₂)=10:120 Torr, T=603 K, t=5 to 50 min.*

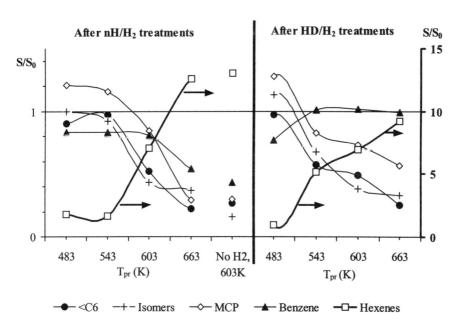

Figure 2 *Selectivity ratio of individual product classes of the test reaction at t_{tr}=5 min as a function of pretreatment temperature. Conditions see in Table 1, Figure 1.*

Catalysis in Application

XPS and UPS spectra of Pt black after similar treatments showed carbon accumulation up to ~50 atom% of the surface, consisting mainly of "graphitic" and "polymeric, C_xH_y" carbon.[15,16,25] Figure 3 compares C 1s spectra measured after exposure to 24HD and nH at 543 K. The difference spectrum indicates a more marked carbonization after HD treatment, the excess corresponding to "massive" deposits. The UPS difference spectrum also showed a more pronounced carbonaceous overlayer after HD treatment (Figure 3). Note that – in spite of a considerable amount of surface C – the well-developed Fermi-edge showed much metallic Pt, indicating relatively little chemical interaction between Pt and C.[16] Mostly two-dimensional carbonaceous deposits were reported on Pt single-crystals under 550 K, with three-dimensional (3D) structures[8] developing gradually and becoming predominant above 600 K. The higher abundance of 3D carbonaceous deposits – including graphitic species – can be responsible for the overall activity drop. Thus, the difference in activities observed in *test runs* (Figure 1) should be attributed not only to the relatively small difference in the carbon content but also to its different chemical state.

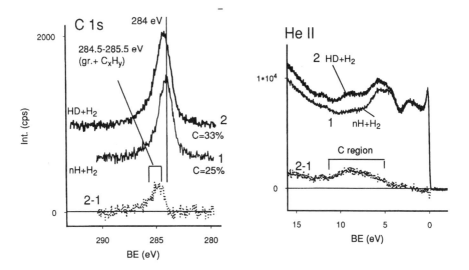

Figure 3   *XPS C 1s (left) and He II UPS (right) spectra measured on **Pt-B** after pretreating with the mixture of hexane-H_2 (1) and t,t-hexa-2,4-diene-H_2 (2) and their difference. T=543 K, t=20 min, p(nH, or HD):p(H_2)=10:60 Torr.*

Similar deactivation treatments were carried out with the two supported catalysts. Results obtained in *test runs* have been summarized in Table 2. Pronounced differences are seen in the selectivities of regenerated samples: **EPT** producing much more C_6 saturated products and less fragments than **Pt-B**.[19] Higher fragmentation ability is characteristic of **Pt-CEO**, at the expense of aromatization while the selectivities for producing MCP and isomers were similar to **Pt-B**. The residual activity of both supported Pt was much higher than that of **Pt-B**. The hexene selectivity on **EPT** increased less dramatically than on **Pt-B**. No hexene was produced on **Pt-CEO**. The S/S_0 values for benzene were almost the same for all three catalysts. Isomer selectivity decreased to a smaller extent and MCP selectivity even increased after deactivation. Details of temperature and hydrogen pressure effects will be reported elsewhere.

Table 2 *Activity and selectivity changes over different catalysts during the test reaction – p(nH):p(H$_2$)=10:120 Torr, T=603 K, t=5 min – upon treatments with hexane (nH) or to t,t-hexa-2,4-diene (HD) - hydrogen mixtures at 603 K.*

	Regenerated samples				After nH treatment			After HD treatment		
	Pt-B	EU-Pt	Pt-Ce		Pt-B	EU-Pt	Pt-Ce	Pt-B	EU-Pt	Pt-Ce
X$_0$ (%)	6.2	13.4	4.2	X/X$_0$	0.18	0.73	0.74	0.16	0.70	0.69
S$_0$ (%)				S/S$_0$						
<C$_6$	31.4	8.8	44.3	<C$_6$	0.52	0.94	0.99	0.49	0.84	0.76
Iso	18.1	36.3	19.4	Iso	0.43	0.94	0.81	0.38	0.88	0.89
Hex	5.3	0.8	0	Hex	6.95	2.20	–	7.04	2.34	–
MCP	17.7	32.7	18.9	MCP	0.84	1.09	1.40	0.73	1.12	1.67
Benz	27.5	21.4	17.4	Benz	0.81	0.94	0.81	1.02	1.02	0.99

3 DISCUSSION

The catalytic propensities (activity and selectivity) are obviously determined by the amount and availability of hydrogen and carbon on the surface. Hexane chemisorbed initially a s C $_6$H$_{13}$–* c an r eact i n t wo d irections: (i) b y further d ehydrogenation l eads t o hexenes and to benzene via dienes and trienes or (ii) to give isomers via C$_5$-cyclic surface intermediate.[18,19] This latter is less dissociated and can also be desorbed as MCP. The ratio of the two intermediates – thus the ratio of saturated and aromatic products – is regulated by the available hydrogen.[27] Hexane reactions[2,23] and also ethene hydrogenation[7] exhibited their maximum rates after the carbon coverage of Pt reached an optimum value. Three different sites and three different ways of chemisorption of the C$_2$ molecule were assumed, leading to ethane and surface carbon, respectively.[7,28]

The slight activation (including higher C$_6$ saturated selectivity) after 24HD treatment at 483 K (Figures 1 and 2) can be interpreted that this treatment produced a "Pt–C–H" state favourable for these reactions. Hexadiene at 483 K adsorbed, likely, by π-adsorption[29] on a hydrogen-covered Pt surface, likely on "Type I" sites,[7] "picked up 4 hydrogen" from "Pt–H" and left the surface, as a fully hydrogenated hexane (Table 1). At this low temperature (483 K) hexane was not reactive (Table 1) it was chemisorbed only, most probably by σ-adsorption having split one or two C–H bonds as demonstrated by the H–D exchange activity in this temperature range.[30] Carbon detected by XPS after treatments at 483 K must have formed by dehydrogenation of the "hydrocarbonaceous overlayer" during evacuation.[1,3] In spite of its pronounced hydrogenation at 543 K, HD started to produce σ-bonded "deactivating" carbon. Still, much π-bonded intermediates of the prevailing reaction: hydrogenation o r double-bond shift caused less deactivation at 603 and 663 K than observed after *pretreatments* with nH.

Surface deposits of various degree of dehydrogenation could be formed from hexane either by its fragmentation or by polymerization. Large contiguous Pt surfaces would be more favourable for this latter process than small supported particles. That is why much "massive carbon" was present on **Pt-B**, causing its more pronounced deactivation, first of all at higher temperatures, in the presence of less surface hydrogen. The formation of those products, which requires multiple adsorption and the presence of surface hydrogen, namely fragments (uptake of 2 H atoms) and isomers (H-neutral), were always poisoned selectively. The changes of MCP (loss of 2 H atoms) selectivity upon deactivating treatments were, however, a more complicated case. It was usually poisoned selectively on

Pt-B, but increased after moderate deactivation on **Pt-B** (Figure 1 and 2), and always over supported catalysts (Table 2). A "common surface C_5-cyclic intermediate" was reported to be the precursor of both isohexanes and MCP.[31] Their ratio was determined by the availability of surface hydrogen.[19,22] The formation of isohexanes requiring hydrogen deactivated first. The selectivity of MCP increased in the initial stages of deactivation at the expense of the isomer selectivity – see low temperature treatments on **Pt-B** (Figure 2) and over supported catalysts (Table 2). When surface carbon was likely not deeply dehydrogenated (nH at T ≤ 603 K, HD at T < 543 K), reactivation during *test run* was also more pronounced. More severe treatments, in turn, deactivated C_5-cyclization, too. This reaction was shown to be almost structure-insensitive,[32] that is, two neighbouring Pt atoms in any position could catalyze it. Its deactivation could have been caused by randomly accumulating "single C_1 entities" and 3D deposits decreasing the number of free doublets. The "proportional" deactivation for aromatization can be interpreted by their active sites consisting of 3 Pt atoms in triangular arrangement.[33] These sites must have deactivated proportionally to the overall activity loss. Hexenes could be produced on single-atom Pt sites or on Pt-C ensembles.[29,34] With progressive carbonization, the number of these sites changed rather slowly. That is the reason why the yield of hexenes remained almost constant up to the loss of ~85% of the original activity,[35] this yield, representing, of course, higher and higher selectivities on deactivated catalysts.

The lower degree of deactivation of **EPT** can be explained by the relatively small Pt particles (D=60%), facilitating hydrogen spillover from Pt sites to the support.[22a,23] Its re-migration to the metal would hinder carbonization. **PT-CEO** showed exceptional behaviour. CeO_2 is readily reduced to Ce(III) and the Pt metal on this support can undergo surface oxidation.[36] This oxygen was able to oxidize CH_x entities generated from C_4–C_8 alkanes on metal (Pd, Pt, Rh), producing hydrogen and liberating surface centres.[37] Cyclization of larger alkanes was observed under the conditions of steam reforming.[38] CeO_2 may have oxidized surface C_1 entities in our case, too. This way, the "Pt–C" sites producing hexenes in other cases would disappear and the liberated free platinum atom could participate in the active doublet required for C_5 ring closure. This is a possible explanation for the absence of hexenes and the increase of methylcyclopentane selectivity on **Pt-CEO** (Table 2).

Acknowledgements

The authors thank Prof. R. Schlögl and Ms U. Wild for the XPS and UPS experiment. We thank the Hungarian National Science Foundation OTKA Grant T037241 for support.

References

1   S. J. Thomson and G. Webb, *J. Chem Soc. Chem. Comm.* 1976, 526; G. Webb, *Catal. Today,* 1990, **7**, 139
2   A. Sárkány, *Catal. Today*, 1989, **5**, 173; A. Sárkány, *J. Chem. Soc. Faraday Trans. 1,* 1989, **85**, 1523.
3   F. Garin, G. Maire, S. Zyade, M. Zauwen, A. Frennet and P. Zielinski, *J. Mol. Catal.,* 1990, .**58**, 185.
4   P. G. Menon, *J. Mol. Catal,.* 1990, **59**, 207.
5   Z. Paál, S. J. Thomson, G. Webb and N. J. McCorkindale, *Acta Chim. Hung.* 1975, **84**, 445.
6   Z. Paál, M. Dobrovolszky and P. Tétényi, *J. Catal.,* 1977, **46**, 65.
7   E. A. Arafa, G. Webb, *Catal. Today*, 1993, **17**, 411.

8   G. A. Somorjai and F. Zaera, *J. Phys. Chem.,* 1982, **86**, 3070.
9   Z. Paál, R. Schlögl and G. Ertl, *J. Chem. Soc. Faraday Trans.,* 1992, **88**, 1179.
10 J. Barbier, *Appl. Catal.,* 1986, **23**, 225.
11 S. D. Jackson, J. Greenfell, I. M. Matheson, S. Munro, R. Raval and G. Webb, *Catalyst Deactivation 1997,* (Ed. C. H. Bartholomew and G. A. Fuentes), Elsevier, Amsterdam, 1997, p. 167.
12 K. Foger and H. J. Gruber, *J. Catal,* 1990, **122**, 307.
13 K. Matusek, A. Wootsch, H. Zimmer and Z. Paál, *Appl. Catal.,* 2000, **191**, 141.
14 A. Wootsch, C. Descorme, Z. Paál and D. Duprez, *J. Catal,* 2002, **208**, 273.
15 N. M. Rodriguez, P. E. Anderson, A. Wootsch, U. Wild, R. Schlögl and Z. Paál, *J. Catal.,* 2001, **197**, 365.
16 Z. Paál, U. Wild, A. Wootsch, J. Find and R. Schlögl, *Phys. Chem. Chem. Phys,* 2001, **3**, 2148.
17 G. C. Bond, *Appl. Catal. A,* 2000, **191**, 23.
18 Z. Paál, *Adv. Catal.,* 1980, **29**, 273
19 A. Wootsch and Z. Paál, *J. Catal.,* 2002, **205**, 86.
20 Z. Paál and P. Tétényi, *J. Catal.,* 1973, **30**, 350.
21 A. Sárkány, H. Lieske, T. Szilágyi and L. Tóth, *8th Internat. Congr. Catalysis, Berlin, 1984,* Verlag Chemie, Weinheim, 1984, Vol. 2, p. 613.
22 **a**. Z. Paál, H. Groeneweg, H. and J. Paál-Lukács, *J. Chem. Soc. Faraday Trans.* 1990, **86**, 3159; **b**. A. Wootsch and Z. Paál, *J. Catal.,* 1999, **185**, 192.
23 G. C. Bond and Z. Paál, *Appl. Catal. A,* 1992, **86**, 1.
24 D. Teschner, A. Wootsch, K. Matusek, T. Röder and Z. Paál, *Z. Solid State Ionics,* 2001, **141**, 709.
25 Z. Paál, A. Wootsch, K. Matusek, U. Wild and R. Schlögl, *Catal. Today,* 2001, **65**, 13.
26 J. Find, Z. Paál, H. Sauer, R. Schlögl, U. Wild and A. Wootsch, *Proc. 12th Intern. Congr. Catal., Granada, 2000,* Elsevier, Amsterdam, Part C, p. 2291.
27 Paál, Z., Székely, G., and Tétényi, P., *J. Catal.* 1979, **58**, 108.
28 A. S. Al-Ammar and G. Webb, *J. Chem. Soc. Faraday Trans.,* 1978, **74**, 195; A. S. Al-Ammar and G. Webb, *J. Chem. Soc. Faraday Trans.,* 1978, **74**, 657.
29 G. A. Somorjai and G. Rupprechter, *J. Phys. Chem.,* 1999, **103**, 1623.
30 V. Ponec, G. C. Bond, *Catalysis by Metals and Alloys,* Elsevier, Amsterdam, 1995, pp. 464-475.
31 F. G. Gault, *Adv. Catal.,* 1981, **30**, 1.
32 F. Zaera, D. Godbey and G. A. Somorjai, *J. Catal.,* 1986, **101**, 73.
33 P. Biloen, J. N. Helle, H. Verbeek, F. M. Dautzenberg and W. M. H. Sachtler, *J. Catal,* 1980, **63**, 112.
34 V. Ponec, *Adv. Catal.,* 1983, **32**, 189.
35 Z. Paál and A. Wootsch, *React. Kinet. Catal. Lett.,* 2002, **77**, 355.
36 L. Yang, O. Kresnawahjuesa and R. J. Gorte, *Catal. Lett.,* 2001, **72**, 33.
37 X. Wang and R. J. Gorte, *Catal. Lett.,* 2001, **73**, 15.
38 X. Wang and R. J. Gorte, *Appl. Catal. A,* 2002, **224**, 209.

DEACTIVATION KINETICS OF COBALT-NICKEL CATALYSTS IN A FLUDISED BED REFORMER

K.M. Hardiman, M.M. Mohammed and A.A. Adesina[*]

Reactor Engineering and Technology Group, School of Chemical Engineering and Industrial Chemistry, University of New South Wales, Sydney, New South Wales 2052, Australia.
E-mail: a.adesina@unsw.edu.au

1   INTRODUCTION

Catalyst deactivation is a common pathological phenomenon in many industrial reactions. In the case of hydrocarbon steam reforming to produce synthesis gas, catalyst activity loss may be due to coke arising from carbon deposition[1]. Carbon lay-down usually occurs via undesired side reaction, namely;

$$C_nH_{2n+2} = nC + (n+1)H_2 \qquad (1)$$

$$2CO = C + CO_2 \qquad (2)$$

$$CO + H_2 = C + H_2O \qquad (3)$$

$$C_nH_m \rightarrow coke + H_2 \qquad (4)$$

Coke build-up manifests itself in pore blockage and hence denying access of reactant hydrocarbon and steam to surface active sites[2]. Left unabated, further interparticle coke growth leads to excessive pressure drop in fixed bed reactors and ultimately cessation of fluid flow requiring reactor shut-down. It has therefore become necessary to develop catalysts with strong anticoking attributes. Kepinski et al.[3] reported the beneficial effects of Mo addition to Ni catalysts during steam reforming. Doping of Ni catalysts with rare-earth oxides and noble metals also causes reduction in coke formation[4-5]. Even so, catalyst mortality in inevitable and from an industrial perspective it is important to carry out reactor operation with a view to minimizing coke accumulation and hence optimal catalyst regeneration. Since coking and steam reforming takes place concurrently, deactivation kinetics will reflect the contributions of various reaction steps describing both processes and thus, the resulting parameter estimates can be meaningfully employed for reactor optimisation which admits on-line catalyst activity decay[6-7].

In a previous study[8], we reported the superior performance of a Co-Ni bimetallic catalyst for methane reforming. Higher hydrocarbon substrates, however, tend to coke more easily, and hence the present study explores the deactivation behaviour of the Co-Ni catalyst in a fluidised bed reactor for propane reforming.

$$C_3H_8 + 3H_2O = 3CO + 7H_2 \qquad \Delta H_{298} = 498.6 \text{ kJ mol}^{-1} \qquad (5)$$

Although deactivation kinetics is often investigated as if it were occurring by a mechanism independent of the desirable reaction, this decoupling denies the inseparability of reaction-deactivation in coke-induced catalyst activity loss during steam reforming. A transient model has been developed from the coupling of a phenomenological activity decay rate to power law steam reforming kinetics and was used to estimate deactivation and reaction rate constants[9] from the conversion-time profiles under various steam:propane ratios and temperatures. Spent and fresh catalysts were further subjected to temperature-programmed reduction and oxidation runs to understand the nature of deposited carbon.

2   EXPERIMENTAL DETAILS

2.1   Catalyst

The bimetallic catalyst was prepared as 5Co-15Ni on 80γ-Al_2O_3 support via multiple impregnation. The support was pre-treated at 1073 K for 6 hours to ensure thermal stability of the alumina phase. It was then mixed with an aqueous solution of $Co(NO_3)_2$ and stirred at 303 K for 3 hours. The slurry was left to dry overnight in an oven at 393 K. The dried solution was subsequently mixed with calculated amount of $Ni(NO_3)_2$ solution. Each metal nitrate impregnation was carried out at a constant pH of 2. The final slurry was further dried at 393 K for 12 hours and calcined for 5 hours at a heating rate of 5 K min^{-1}. The calcined solid was then crushed and sieved to 212-250 μm for further use.

2.2   Deactivation Experiments

The experimental set-up is illustrated in Figure 1. Propane (BOC Gases, Sydney) was used as the hydrocarbon feed, while diluent helium and other gases were supplied by Linde (Sydney). High purity grade (>99.99%) was used in all cases. Gas flows were regulated by Brooks electronic mass flow controllers and mixed in a 110 mL stainless steel chamber before passing into preheater where it was further mixed with steam. The steam itself was generated via controlled injection of ultra-pure water (using a Razel A-99 injector) into the preheater maintained at 533 K. The feed line to the reactor was insulated and heated to 533 K to avoid any condensation of steam. The reactor was made from a quartz tube (ID=20 mm, H=490 mm) fitted with a 3 mm thick sintered quartz plate (50 μm holes) as gas distributor and an axially-movable 1/8-inch stainless steel thermocouple. The reactor was loaded with typically 1 gram of catalyst (5.5 mm bed thickness) and placed vertically in a temperature-controlled furnace. Product gases from the reactor were sent to an ice-cooled condenser and subsequently passed over drierite bed ($CaSO_4$) to remove non-condensable water. The moisture-free gas was then analysed isothermally (393 K) by Shimadzu gas chromatograph (TCD) model 8A equipped with an Alltech Haysep DB column with argon as the carrier gas.

Although steam reforming is normally carried out at steam-to-carbon ratio greater than 2, in order to investigate coke-induced deactivation, experiments were performed with 4 different steam-to-carbon ratios of 0.4, 0.8, 1.2 and 1.6 at 4 temperatures over the range of 773-923 K using a fluidising gas flow rate of 300 ml min^{-1} (ambient conditions). Prior to each run, the catalyst was reduced using H_2 flow at 873 K for 2 hours. Each deactivation experiment in the fluidised bed reactor was conducted for 10 hours with composition sampling every 40 minutes.

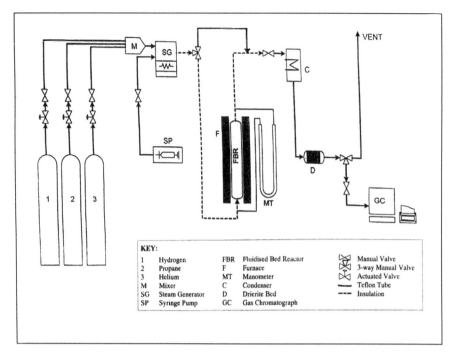

Figure 1   *Schematic diagram of experimental set-up*

2.3   Carbon Deposition Analysis

Morphological changes and nature of carbon on spent catalysts were characterised by different methods. N_2 adsorption at 77 K was used to measure the total (BET) surface areas, pore volume and pore size distribution on a Micromeritics TriStar 3000 system. The total carbon content was measured by a Shimadzu Total Organic Carbon (TOC) Analyzer 5000(A) integrated with a Solid Sample Module SSM-5000A. About 10 mg of the catalyst sample was combusted in a chamber at 1173K chamber using zero (99.99%) oxygen. X-ray diffractograms were obtained from Philips X'pert system using a Ni-filtered CuKα radiation (λ = 1.542 Å) at 40 kV and 30 mA. Temperature programmed experiment (oxidation and reduction) were performed on the coked catalysts on a ThermoCahn TGA 2121. Air at 55 mL min^{-1} was used for TPO at 5 K min^{-1} to 923 K and held constant for 5 hours, before cooling down to room temperature in air to at the same rate. TPR runs where conducted with 50%H_2/N_2 mixture at 55 mL min^{-1}.

3   RESULTS AND DISCUSSION

3.1   Deactivation Experiments

Transient concentration data obtained from 16 runs (4 steam-to-carbon ratios × 4 temperatures) were used to probe the deactivation of the Co-Ni catalyst. Typically, propane consumption was rapid within the first hour and after reaching a minimum, propane composition in the exit gas gradually rose towards a steady-state value, as seen in Figure 2.

Temporal product (H_2, CO, CO_2 and CH_4) distribution was, however, dependent on the feed composition and temperature. Stoichiometrically, propane steam reforming requires a steam-to-carbon (S:C) of 1, therefore catalysts used with S:C<1 are referred to 'severely' coked specimens, while those obtained under conditions with S:C>1 are considered 'lightly' coked. As may be seen in Figure 2, CH_4 was the only hydrocarbon produced peaking within the 1st hour before gradually decreasing and levelling off to a non-zero steady production. CO production was somewhat smaller and slower, attaining a maximum after 2 hours. CO_2 was practically non-existent while H_2 was produced in relatively higher concentration albeit falling steadily with time-on-stream.

Figure 3 shows the transient profile with time-on-stream for 'lightly' coking conditions. It is evident that H_2 was still the most abundant product with maximum concentration attained after 1 hour of operation. However, production stabilized after 4 hours on stream. Interestingly, the behaviours of other species are practically invariant with time. In particular, time-averaged value for H_2/CO ratio was 3.81, while the CO_2/CO ratio is 0.45. These data suggest that propane dehydrogenation to CH_4 and carbon was probably the dominant reaction at S:C<1 with CO being produced for partial steam gasification of the surface carbon. H_2 produced from C_3H_8 dehydrogenation was possibly continuously reincorporated into the deposited carbon to form CH_x (x<2) species, which subsequently polymerise. At S:C ratio about the stoichiometric requirement, H_2 was apparently produced from the steam reforming reaction although a CO_2/CO ratio of 0.45 indicates that carbon deposition may be due to the Boudouard reaction. It would therefore seem that site blockage due to carbon deposition was the primary cause of catalyst activity decay.

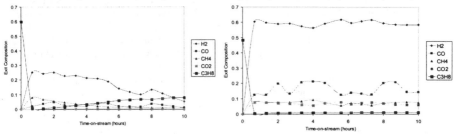

Figure 2 Transient concentration profile for 'severely' coking conditions

Figure 3 Transient concentration profile for 'lightly' coking conditions

Whilst linear, exponential and reciprocal decay laws have been used to describe deactivation kinetics, the reciprocal decay law has been recommended[10] for most coke-induced activity deterioration. Thus, time-dependent activity, a, may be given as

$$a = A_0 t^{-b} \qquad (6)$$

Following Levenspiel[9], we obtain the transient propane concentration profile in the well-mixed fluidised bed reactor as

$$\ln\left(\frac{C_{A_0}}{C_A} - 1\right) = \ln A_0 k' \tau' - b \ln t \qquad (7)$$

where k' is a pseudo 1st order rate constant in the power law kinetics for steam reforming and τ' is a modified gas residence time ($\tau' = W \cdot C_{A0}/F_{A0}$) and b is a deactivation coefficient. Data from the 16 runs were used to obtain k' and b for each run. Estimates are provided in Table 1.

Table 1   Parameter estimates of transient deactivation-reaction model

Temperature (K)		S:C ratio			
		0.4	0.8	1.2	1.6
773	b	1.7773	0.4377	0.4143	0.4240
	k'/b	0.8842	0.3184	0.6251	0.6593
823	b	2.1943	0.4356	0.1856	0.1355
	k'/b	0.6589	0.3618	0.7823	1.8077
872	b	1.7051	0.2081	0.1198	0.1326
	k'/b	0.5248	0.7052	1.1935	3.0594
923	b	1.8549	0.7888	0.6732	0.1061
	k'/b	0.7246	0.5670	1.6232	12.1805

Figure 4 shows that the deactivation coefficient, b, decreased rather sharply with increased S:C ratio and then approaches a constant value at around S:C=1. In this situation, pyrolitic carbon accumulation via propane dehydrogenation may be more prominent in steam-deficient conditions leading to high b values. However, at S:C>1, the reforming reaction is more favourable although limited coke deposition may persist via other routes such as CO disproportionation. This explains why b has a non-zero value even at high S:C ratio. The reforming kinetic constant, k', increases with relatively to b with S:C ratio as shown in Figure 5. Clearly, temperature also has a strong influence on these parameters. Intriguingly, the dependency does not seem to follow Arrhenius prediction. Inspection of the data in Table 1 (k'/b data) suggests a reasonable fit to Arrhenius expression at S:C>1, however below this value propane consumption is controlled by both chemical reaction and other physical processes. This interaction between temperature and composition (S:C ratio) was statistically substantiated via a 2-way Analysis of Variances (ANOVA) treatment of the data[11]. Error variance was computed from replicated runs. Table 2 presents the summary of pertinent statistics. This analysis shows that calculated F-values are greater than the expected values from standard tables at 95% confidence level[11], thus confirming that not only are temperature and S:C ratio significant factors, the interaction between temperature and S:C ratio also has a statistically-meaningful effect on the k' and b.

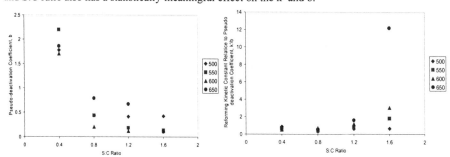

Figure 4   Profiles of deactivation coefficient, b, as a function of S:C ratio for various temperatures

Figure 5   Profiles of reforming kinetic relative to deactivation, k'/b, as a function of S:C for various temperatures

Table 2   *Calculated F-values from 2-way ANOVA with error variance as denominator*

	Temperature	S:C Ratio	Temperature-S:C Ratio Interaction	Tables [Box et al, 1978]
k'	648.9 (3, 16)	233.3 (3, 16)	49.49 (9, 16)	F = 3.24 (3, 16) at 95% confidence
B	40.64 (3, 16)	1412.5 (3, 16)	23.14 (9, 16)	F = 2.54 (9, 16) at 95% confidence

∗ The number in brackets is the degree of freedom for numerator and denominator respectively.

3.2   Carbon Deposition Analysis

The 2 coked specimens (Figures 2 & 3) were then placed in a furnace at 723 K for 3 hours with flowing air to remove the carbon deposits. BET surface area of fresh, 'lightly' and 'severely' coked specimens were obtained as 108.7 m^2 g^{-1}, 102.8 m^2 g^{-1} and 97.8 m^2 g^{-1} respectively. The associated pore volumes were 0.435, 0.466 and 0.414 cm^3 g^{-1}. These relatively small changes in physical properties suggest that carbon deposition rather than sintering was responsible for the loss in catalyst activity. Carbon analysis gave 14.04% and 76.83% for the 'lightly' and 'severely' cokes specimens. The carbon content was totally organic in nature as expected.

Bulk phase changes in the fresh, 'lightly' and 'severely' coked catalysts were determined from XRD analysis. The diffractograms shown in Figure 6 indicate substantial changes in both composition and metallic phases. Unreduced fresh catalyst shows a strong signal for spinel-type NiCo$_2$O$_4$ (2θ=36.4°), Co$_3$O$_4$ (36.8°) and NiO (43.2°). The 'lightly' coked catalyst showed the presence of carbonaceous matter at 2θ=26°. At least 5 different forms of carbon, namely, adsorbed atomic carbon (C$_\alpha$), amorphous carbon (C$_\beta$), vermicular carbon (C$_v$) and crystalline graphitic carbon (C$_c$) have been reported[12]. As may be seen from XRD, bulk Ni carbide was not detected and neither was graphite present, which should have appeared at 2θ=30[13]. However, pyrolitic carbon is thermodynamically favourable during C$_3$H$_8$ dehydrogenation at temperatures greater than 870 K while below 770 K carbon deposition occurs primarily by slow polymerisation of CH$_x$ (x<2) radicals on the metal surface. In view of the temperature used and the XRD results, we believe 2 types of carbon may be present - C$_\alpha$ (pyrolitic) and C$_\beta$ (amorphous containing highly unsaturated C$_x$H$_x$). Signals for the metal oxide phases present in the unreduced sample were somewhat attenuated but still evident. However, the strong peak for Co (44.2°) is evidence of the reducing environment under which the reaction was conducted. As expected, a large carbon peak was recorded from the 'severely' coked sample.

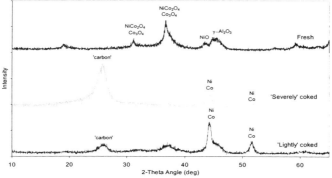

Figure 6   *XRD diffractograms of fresh, 'severely' coked and 'lightly' coked catalysts*

Temperature programmed oxidation-reduction (TPO-TPR) results are presented in Figures 7 & 8. Figures 7a shows that the air readily oxidises the carbon deposit on the 'severely' coked catalyst. A weight drop of about 79.6% at 650°C reveals the presence of 2 types of carbonaceous compounds as seen from the weight percent derivative curves – a sharp peak and a shoulder. This agrees with interpretation from XRD data. We believe the first (sharp) peak is the C_β at 702 K while C_α corresponds to the shoulder and has a higher oxidation temperature (800 K). Since C_β is essentially polymerised CH_x species and C_α is adsorbed atomic carbon, C_α would be primarily responsible for site blockage and would therefore be responsible for coking control in view of its higher resistance to react. A similar pattern, but smaller weight drop (13.8%) was also observed in Figure 7b for the 'lightly' coked sample. It would appear that the 2 types of carbon have different reactivities. The % weight drop for both 'lightly' and 'severely' coked catalysts are 13.8% and 79.6% respectively compared with 14.04% and 76.83% from TOC analysis.

TPR data obtained immediately after TPO are shown in Figures 8a and b. It is apparent from these plots that the % weight loss after reduction was reversibly gained during the air cooling portion of the cycle. Interestingly, the weight % change for the 'lightly' coked catalyst, valued at 5.01% is essentially the same as that seen in the 'severely' coked specimen (5.56%) even though the original carbon content were vastly different (13.8% and 79.6%). This shows that the catalysts were fully oxidised during TPO, such that both were in the same oxidation state prior to TPR. This is noteworthy since XRD shows the catalyst for multiple oxidation states.

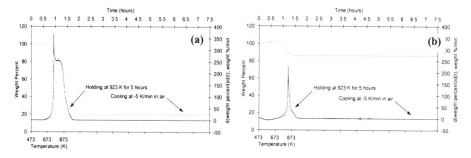

Figure 7   TPO profiles for (a) 'severely' coked catalyst, and (b) 'lightly' coke catalyst

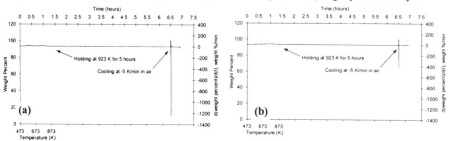

Figure 8   TPR profiles for (a) 'severely' coked catalyst, and (b) 'lightly' coke catalyst

4   CONCLUSIONS

This study has examined the effect of both steam-to-carbon (S:C) ratio and temperature on deactivation. Conversion-time data during steam reforming were fitted to a transient reactor-model incorporating both reaction and deactivation kinetics in a well-mixed fluidised bed. Estimates of the pseudo-rate constant and deactivation coefficient exhibited strong dependency on both S:C ratio and temperature. The lack of fit to an Arrhenius expression signalled possible interaction between S:C ratio and temperature. This was subsequently confirmed via a 2-way ANOVA and implicates the possibility of other physical processes in the deactivation mechanism, e.g. interphase carbon transport. Independent XRD and thermal analysis spectra (TPO-TPR) also revealed the existence of 2 types of carbonaceous deposits (C_α and C_β) on the catalyst. On the streangth of their reactivities, it seems that C_α (atomically-adsorbed carbon) is the culprit for coking control.

References

1   R. Hughes, *Deactivation of Catalysts*, Academic Press, New York, 1984.
2   J.C. Rodriguez, E. Romeo, J.L.G. Fierro, J. Santamaria and A. Monzon, *Catal. Today*, 1997, **37**, 255.
3   L. Kepinski, B. Stasinska and T. Borowiecki, *Carbon*, 2000, **38**, 1845.
4   J.C. Summers and S.A. Ausen, *J. Catal.*, 1979, **131**, 58.
5   B. Harrison, A.F. Diwell and C. Hallet, *Plat. Met. Rev.*, 1988, **32(2)**, 73.
6   G.F. Froment, *Appl. Catal. A: General.*, 2001, **212**, 117.
7   A.F. Ogunye and W.H. Ray, *AIChE J.*, 1971, **17**, 365.
8   K. Opoku-Gyamfi and A.A. Adesina, *Chem. Eng. Sci.*, 1999, **54**, 2575.
9   O. Levenspiel, *Ind. Eng. Chem. Res.*, 1999, **38**, 4140.
10   F. Shadman-Yazdi and E.E. Petersen, *Chem. Eng. Sci.,* 1972, **27**, 227.
11   G.E.P. Box, W.G. Hunter and J.S. Hunter, *Statistics for Experimenters*, Wiley, New York, 1978.
12   J. R. Rostrup-Nielsen, *Catal. Today*, 1997, **37**, 225.
13   S. Wang and G.Q.M. Lu, *Appl. Catal. B: Environ.*, 1998, **16**, 269.

DEACTIVATION BEHAVIOUR OF Zn/ZSM-5 WITH A FISCHER-TROPSCH DERIVED FEEDSTOCK

Arno de Klerk

Fischer-Tropsch Refinery Catalysis, Sasol Technology Research and Development, PO Box 1, Sasolburg 1947, South Africa

1 INTRODUCTION

ZSM-5 based catalysts are widely used commercially, either in their unpromoted acidic form, or as metal promoted catalysts. The aromatisation of light paraffins is an example of a refining process where ZSM-5 catalysts are often encountered. With the global shift towards higher octane unleaded petrol, aromatisation is one of the processes that can be used to produce high octane blending stock. In a Fischer-Tropsch environment, where the unrefined Fischer-Tropsch product contains very little aromatics, this is an especially attractive option. However, Fischer-Tropsch derived material does not only contain paraffins, but also contains olefins and oxygenates.

Despite some debate on the details of the aromatisation mechanism,[1-3] there is consensus that the mechanism entails olefin formation at some stage during the process. The presence of olefins in the feed should therefore not present problems. The effect of oxygenates on ZSM-5 is known from the work done on the methanol to gasoline (MTG) process[4] and from hydrothermal dealumination studies.[5-9] Since water is invariably produced during regenerative coke burn-off subsequent to aromatisation, hydrothermal dealumination is a familiar deactivation mechanism for such catalysts. What is not well described, is the combined effect of olefins and oxygenates on the short term and long term deactivation behaviour of a metal promoted ZSM-5 aromatisation catalyst. Coke formation, hydrothermal dealumination, metal loss and sintering could all potentially contribute to deactivation. The aim of this study was to determine the main cause of reversible deactivation and permanent deactivation respectively.

The deactivation behaviour was studied by monitoring the change in product spectra with time on stream and over multiple reaction-regeneration cycles. The findings were based on macroscopic observations only.

2 EXPERIMENTAL

All test work was done with a commercial Zn/ZSM-5 catalyst and unrefined Fischer-Tropsch derived feedstock taken from Sasol Synfuels West in Secunda, South Africa. Two different feed batches were prepared from the unrefined material and characterised (see table 1).

Table 1 *Feed characteristics of C₇/C₈-cut Fischer-Tropsch material.*

Property	Feed for cycle 1-10	Feed for cycle 11-19
Density (kg.m^{-3})	731	734
Oxygen content (mass % O)	3.1	2.6
Olefin content (%)	52	45

The aromatisation test work was done in a fixed bed reactor with an internal diameter of 28 mm. A single batch of catalyst was loaded in its commercial uncrushed form and the catalyst bed was supported on a bed of low surface area alpha-alumina (corundum) balls. The test work consisted of multiple cycles, each consisting of a production run followed by in situ regeneration of the catalyst. Each production run lasted 48 hours. The following operating conditions were used: catalyst bed temperature of 480°C, pressure of 500 kPa gauge (atmospheric pressure 85-86 kPa) and feed rate of 2.8 g.g^{-1}.h^{-1} (weight hourly space velocity). Catalyst regeneration was done by increasing the temperature to 520-530°C under nitrogen and then bleeding in air at 0.75 % oxygen for the first 6 hours, followed by an increase to 1.7 % oxygen until the carbon monoxide content and the carbon dioxide content in the off-gas were both less than 10 μg.g^{-1}.

The liquid products were quantified by GC-FID (gas chromatograph with flame ionisation detector) and characterised with GC-MS (gas chromatograph with mass spectrometer). The aromatics content was monitored on-line with a near infrared spectrometer that was calibrated by proton nuclear magnetic resonance spectrometry.

The hydrocarbon content of the gaseous product was analysed by GC-FID and the hydrogen, carbon monoxide content and carbon dioxide content were determined by mass spectrometry. A correction was made for water vapour based on the conditions in the off-gas system.

3 RESULTS AND DISCUSSION

The processes leading to reversible deactivation take place in parallel with those causing permanent deactivation. Both reversible and permanent deactivation processes cause a steady deterioration in catalyst performance, although the activity lost due to reversible deactivation can be regained by catalyst regeneration. Since ZSM-5 based catalysts are used commercially, it stands to reason that the permanent deactivation processes should be slow. However, the reversible deactivation processes found during aromatisation are comparatively fast and catalyst regeneration is necessary after only days on stream. The difference in the rate of reversible and permanent deactivation enables its decoupling.

3.1 Reversible deactivation

The catalyst deactivated with time on stream and this was observed in all runs. Despite differences in catalyst performance between runs, there were common trends characterising the deactivation with time on stream. The following general trends were observed:
  a) The conversion, expressed in terms of the mass of feed converted, decreased gradually with time on stream (see figure 1).
  b) Aromatic selectivity decreased with time on stream (see figure 2).
  c) The C₁-C₅ aliphatic selectivity increased with time on stream.

d) The paraffin to olefin ratio decreased with time on stream (see figure 3).
e) The benzene to C_7/C_8-aromatics ratio generally decreased with time on stream, although some mid-run to end-run increases (local maxima) were sometimes observed.

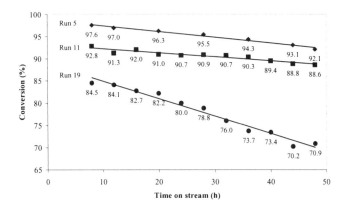

Figure 1   *Change in conversion of C_7/C_8 feed with time on stream.*

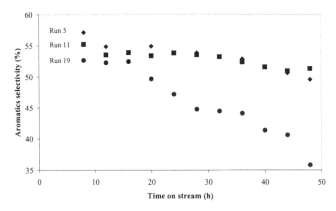

Figure 2   *Change in overall aromatics selectivity with time on stream.*

The dominant deactivation process is coke formation, which explains why the catalyst can be regenerated by controlled coke combustion. The work done by Viswanadham, et al.[10] and Echevsky, et al.[11] on H/ZSM-5 catalysts indicated that coke tended to accumulate on the binder and external surface of the zeolite and occurred in such a way as to leave the active sites accessible. Deactivation occurred mainly via pore blockage. A decrease in acid strength was caused by the shifting of electron density from the condensed aromatic structures onto the active proton sites. Little loss of mean free pathlength was observed due to coke formation on H/ZSM-5 and coke did not increase diffusion resistance.[12] Yet, the location of the coke deposits led Viswanadam, et al.[10] to postulate that a molecular traffic control mechanism may be operative due to the position of the coke that forms in the zeolite itself.

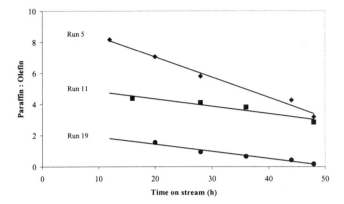

Figure 3 *Change in paraffin to olefin ratio with time on stream.*

The high olefin content of the feed resulted in high conversions over the commercial Zn/ZSM-5 catalyst. The conversion decreased gradually in almost linear fashion with time on stream and even at the end of run, the conversion was still quite high. If deactivation by coke formation prevented access to the active sites in general, non-selective deactivation behaviour would be expected. This was not the case and aromatic selectivity decreased with time on stream. Since aromatisation is primarily promoted by the metal sites and cracking conversion by the acid sites, these observations tend to indicate a preferential deactivation of the metal sites in agreement with the observations by Cumming and Wojciechowski[13] for reforming catalysts.

The decrease in paraffin to olefin ratio can partly be explained by noting that Zn/ZSM-5 is sensitive to the hydrogen partial pressure with respect to selectivity and conversion.[2,14] Consequently, a decrease in hydrogen partial pressure due to a decrease in aromatics yield may lead to a decrease in the paraffin to olefin ratio. However, the decrease in hydrogen partial pressure is not sufficient to explain the magnitude of the change in the paraffin to olefin ratio for the initial runs, although it might explain the decrease observed in the later runs (for example run 19).

From comparative studies of H/ZSM-5 and Zn/ZSM-5 it is known that the acid sites are poor at removing hydrogen after C-H bond activation.[2] However, metal sites not only increase the rate of dehydrogenation, but also the rate of hydrogenation. Since the feed is already olefinic, it is expected that the metal sites should promote both hydrogenation and aromatisation. The decrease in paraffin to olefin ratio therefore suggests a preferential deactivation of the metal sites.

This interpretation is further supported by the decrease in benzene to C_7/C_8-aromatics with time on stream. The formation of benzene from a C_6 molecule on an acid site would probably involve a primary carbenium ion intermediate, which explains the low propensity for benzene formation over H/ZSM-5.[14] The ability of metal sites to aromatise C_6 molecules from the olefin pool led to an increase in benzene selectivity over Zn/ZSM-5 when compared to H/ZSM-5.[2,3,15,16] The observed decrease in benzene formation relative to C_7/C_8-aromatics therefore points toward deactivation of the metal sites.

This leaves only the increase in C_1-C_5 aliphatic selectivity unexplained. Since acid sites regulate the carbon number distribution, the increase in light material would indicate unimpaired acid site activity. The increase in C_1-C_5 selectivity can also be explained in terms of their molecular sizes and not necessarily in terms of a shift in catalyst selectivity favouring cracking over aromatisation. If the coke only constrained access to the active sites rather than prevented access, there would be some shift toward lighter products due to some form of molecular traffic control mechanism. The increase in diffusion resistance of larger molecules would also explain the decrease in aromatics production, spurious local selectivity maximum for benzene compared to other aromatics and the decrease in conversion.

Coke formation therefore leads to selective deactivation in two ways: it affects the metal sites more than it affects the acid sites and it constrains access to the zeolite by increasing the diffusion resistance for larger molecules. This results in a selectivity shift to lighter products and a decrease in the aromatics selectivity, paraffin to olefin ratio and conversion.

3.2 Permanent deactivation

The time on stream data presented for runs 5, 11 and 19 (see figures 1-3) already gave some indication that the Zn/ZSM-5 catalyst was also subject to permanent deactivation. It also underlined the importance of doing comparisons between runs after the same time on stream, or alternatively, on composite samples after the same time on stream. The latter approach was followed.

The differences in composition between composite samples of each run gave an indication of the permanent change that took place and which could not be reversed by catalyst regeneration. The following general trends were observed:

a) The conversion of C_7/C_8 aliphatic material decreased slightly (see figure 1).

b) The C_6+ liquid yield did not change apart from small run to run fluctuations.

c) The methane selectivity did not change apart from small run to run fluctuations.

d) The aromatics yield decreased slightly during the first eleven runs, followed by a marked decrease before stabilising again at a lower aromatics yield (see figure 4).

e) The benzene to C_7/C_8-aromatics ratio decreased (see figure 5).

f) The paraffin to olefin ratio of the light material remained constant during the first nine runs, apart from some run to run fluctuations, followed by a marked decrease before stabilising at a lower ratio (see figure 6).

g) The paraffin to olefin ratio of the C_6-C_8 aliphatic material fluctuated, but did not change much.

Deactivation of zeolites by dealumination is well known. Although the catalyst was not analysed, a fair degree of dealumination was suspected to be caused by the water vapour generated during reaction and regeneration. Dealumination can destroy activity in three ways: a) a decrease in the number of Brønsted acid sites; b) destruction of the zeolite pore structure; and c) a decrease in metal sites through the loss of the active zinc species attached to the alumina.

Although alumina is lost from the zeolite framework, this loss is not always accompanied by a serious loss of activity.[4,7] This can partly be explained by the formation of very strong acid sites during dealumination,[6,9] but more likely this is caused by the high rate of the C-C bond formation and scission reactions on the acid sites. The cracking-oligomerisation reactions, which are much faster than dehydrocyclisation, quickly establish a pseudo-equilibrium hydrocarbon mixture that is independent of the size of the feed.[3,17] Even a substantial loss of aluminium would therefore not cause a serious decrease in conversion, but product selectivity will be influenced.

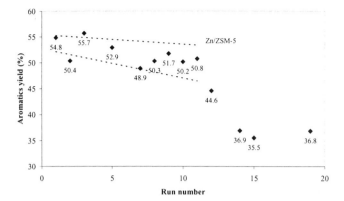

Figure 4   *Run to run change in average aromatics yield of composite samples collected over a 48 hour period.*

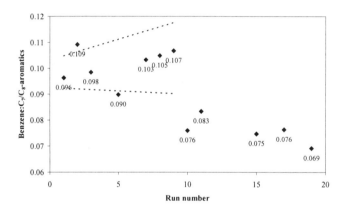

Figure 5   *Run to run change in benzene to C_7/C_8-aromatics selectivity.*

It was therefore not surprising that the observed C_6+ liquid yield and the carbon number distribution were not significantly affected by the run to run deactivation. The slight decrease in conversion in the absence of a shift in the carbon number distribution was due to the decrease in aromatisation of the C_7/C_8 material. (All aliphatic C_7/C_8 material was considered as unconverted, irrespective of whether it was converted to naphthenes or skeletal isomers).

The decrease in the aromatics yield, decrease in the benzene to C_7/C_8-aromatics ratio and the decrease in paraffin to olefin ratio, are all indicative of metal site deactivation. What is of special interest is the form of the deactivation profiles (see figures 4-6). At first it seemed as if the change in feed to a less olefinic feed precipitated the changes. However, the benzene to C_7/C_8-aromatics ratio and paraffin to olefin ratio started to decrease by run 10 already, while the feed was only changed after run 10. Yet, the aromatics yield only started to decrease in run 12. The change in feed may have accentuated these changes, but these changes were not caused by the change in feed.

It has already been noted that Zn/ZSM-5 catalysts have higher benzene selectivity than H/ZSM-5 catalysts and that the decrease in the benzene to C_7/C_8-aromatics ratio can be seen as indication of permanent metal site loss. The simultaneous decrease of the paraffin to olefin ratio is probably not a coincidence and lends further support to such an interpretation. It is not clear why the decrease in aromatics yield was offset with two runs, but the data suggests that there is a difference in the metal site density threshold for the reactions.

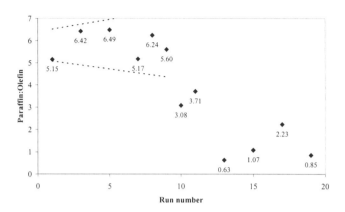

Figure 6   *Run to run change in paraffin to olefin ratio.*

As the metal site density decreases below a certain threshold, there are sufficient metal sites to aid in the dissociative chemisorption of hydrogen molecules, but it is not sufficient to have a noticeable contribution towards hydrogenation and dehydrogenation reactions. Aromatisation is therefore not yet impaired, although only the formation of acid catalysed aromatic molecules is aided in this way. The aromatic product distribution therefore resembles that of H/ZSM-5, notwithstanding that the rate of aromatics formation still resembles that of Zn/ZSM-5. As more active metal sites are lost, the character of the catalyst becomes that of H/ZSM-5 and the aromatics yield also drops to a level similar to H/ZSM-5.

Dealumination consequently led to permanent deactivation that is mainly characterised by a loss of metal sites. This caused the catalyst to change character from being a Zn/ZSM-5 catalyst to an H/ZSM-5 catalyst. The loss of acid sites and structural damage to the catalyst may also have contributed, but these contributions were not especially noticeable.

The rate of permanent deactivation was surprisingly fast even considering that dealumination could take place during both reaction and regeneration. Calculations based on the water partial pressure[5] indicated that a catalyst lifetime in the order of 100 runs could be expected, but in practice, the catalyst lasted only about 10 runs. Under the reducing conditions prevalent during reaction, with olefins, hydrogen and carbon monoxide present, some of the active zinc species might have been reduced and lost as zinc vapour.[18] Such a mechanism was not suspected to be of importance and to what extent such a mechanism of deactivation contributed to the loss of metal sites, could not be ascertained.

4 CONCLUSION

Reversible deactivation by coke formation led to selective deactivation of the Zn/ZSM-5 catalyst with time on stream. The metal sites were more affected than the acid sites and access to the zeolite was constrained, which increased the diffusion resistance for larger molecules. This resulted in a selectivity shift to lighter products and a decrease in the aromatics selectivity, paraffin to olefin ratio and conversion.

Permanent deactivation was caused mainly by a loss of metal sites, which probably occurred due to hydrothermal dealumination, but possibly to some other mechanism too. This caused the catalyst to change character from being a Zn/ZSM-5 catalyst to an H/ZSM-5 catalyst. The loss of acid sites and structural damage to the catalyst may also have contributed in some way, but were not especially noticeable.

Although the deactivation mechanisms for reversible and permanent deactivation are different, their main effect was common, namely a loss of active metal sites.

Acknowledgements

Sasol Technology (Pty) Ltd, Research and Development is gratefully acknowledged for the use of their facilities and permission to publish the results.

References

1   Kanai, J. and Kawata, N., *J. Catal.*, 1988, **114**, 284.
2   Biscardi, J.A. and Iglesia, E., *J. Catal.*, 1999, **182**, 117.
3   Urdă, A. Tel'biz, G. and Săndulescu, I., *Stud. Surf. Sci. Catal.*, 2001, **135**, 4017.
4   Bauer, F. Ernst, H. Giedel, E. and Schödel, R., *J. Catal.*, 1996, **164**, 146.
5   Sano, T. Suzuki, K. Shoji, H. Ikai, S. Okabe, K. Murakami, T. Shin, S. Hagiwara, H. and Takaya, H., *Chem. Lett. – Japan*, 1987, 1421.
6   Topsøe, N.Y. Joensen,A. and Derouane, E.G., *J. Catal.*, 1988, **110**, 404.
7   Smirniotis, P.G. and Zhang, W., *Ind. Eng. Chem. Res.*, 1996, **35**, 3055.
8   De Lucas, A. Canizares, P. Durán, A. and Carrero, A., *Appl. Catal. A*, 1997, **154**, 221.
9   Sahoo, S.K. Viswanadham, N. Ray, N. Gupta, J.K. and Singh, I.D., *Appl. Catal. A*, 2001, **205**, 1.
10 Viswanadham, N. Murali, D.G. and Rao, T.S.R.P., *J. Mol. Catal. A: Chem.*, 1997, **125**, L87.
11 Echevsky, G.V. Ayupov, A.B. Paukshtis, E.A. O'Rear, D.J. and Kibby, C.L., *Stud. Surf. Sci. Cat.*, 2001, **139**, 77.
12 Pradhan, A.R. Wu, J.F. Jong, S.J. Chen, W.H. Tsai, T.C. and Liu, S.B., *Appl. Catal. A*, 1997, **159**, 187.
13 Cumming, K.A. and Wojciechowski, B.W., *Catal. Rev. - Sci. Eng.*, 1996, **38**, 101.
14 Mole, T. Anderson, J.R. and Creer, G., *Appl. Catal.*, 1985, **17**, 141.
15 Bhattacharya, D. and Sivasanker, S., *Appl. Catal. A*, 1996, **141**, 105.
16 Nakamura, I. and Fujimoto, K., *Catal. Today*, 1996, **31**, 335.
17 Biscardi, J.A. and Iglesia, E., *Catal. Today*, 1996, **31**, 207.
18 Berndt, H. Lietz, G. Lücke, B. and Völter, J., *Appl. Catal. A*, 1996, **146**, 365.

IN-SITU ULTRAVIOLET RAMAN SPECTROSCOPY OF SUPPORTED CHROMIA/ALUMINA CATALYSTS FOR PROPANE DEHYDROGENATION

V. S. Sullivan[1], P. C. Stair[2] and S. D. Jackson[3]

[1]Argonne National Laboratory, Argonne, IL 60439, USA
[2]Department of Chemistry and Centre for Catalysis and Surface Science, Northwestern University, Evanston, IL 60208, USA
[3]Department of Chemistry, Joseph Black Building, The University, Glasgow G12 8QQ, Scotland, UK

1 INTRODUCTION

The characterization of the surface of supported metal oxide catalysts is vital to the understanding of many catalytic reactions. Supported chromium oxide catalysts are used for many industrial catalytic processes.[1] Chromium oxide supported on alumina is used as a catalyst for propane and butane dehydrogenation.[2-11] Determination of the surface structure under reaction conditions is important for a complete understanding of the catalyst system.

The focus of this paper is the application of Raman spectroscopy to study coke formation during the dehydrogenation of propane to propene over 1 % chromium oxide catalysts supported on alumina. Catalytic dehydrogenation of light alkanes with catalysts based on platinum or chromium oxides on alumina is an established technology.[12] Catalyst deactivation by carbon deposition is a significant problem. Regular regeneration of the catalysts is required to sustain the activity of the catalyst.[12] Better understanding of coke formation chemistry and the role of the catalyst properties in coke formation could facilitate optimization of catalysts and possible reduction of side reactions leading to coke.

Raman spectroscopy is a powerful technique for characterization of solids and surfaces, and is well suited for examining oxides and supported oxide catalysts.[13-19] Over the past few years, our group successfully examined a number of catalytic systems using UV Raman spectroscopy.[2, 20-25] The use of UV excitation prevents fluorescence from the Raman spectra by exciting the sample at a frequency where fluorescence does not occur.[26, 27] In this study, Raman spectra were obtained from the 1% chromium on alumina catalyst during exposure to propane or propene under reaction conditions. Calcined catalysts were compared to catalysts activated in hydrogen.

2 EXPERIMENTAL

Catalyst samples were prepared and provided by Synetix, a division of Johnson-Matthey. A solution of ammonium dichromate or potassium hydroxide and ammonium dichromate was added to the support (Engelhard Al-3992E, S.A. 200 m^2g^{-1}) and the resulting suspension evaporated to dryness. The solid was then calcined in air at 823 K for 3 hours. The catalyst, with 1% Cr_2O_3 by weight, is yellow. The catalyst was pressed into pellets, lightly ground, and sieved to 90 µ diameter spheres. The pre-coked sample was reduced in 5%H_2/N_2 by heating to 600°C at 10°C/min. and held for 0.5 hours at 600°C. Helium was used to purge the system at 600°C; then propane at 50 cc/min. was run for 10 minutes. Finally, the sample was purged in He while cooling to room temperature.

Raman spectra were measured in a fluidized bed sample cell described previously.[25] Briefly, a stainless steel porous disk with 40-µm-diameter pores is secured near the top of a vertical, stainless steel tube. The sample, as a loose powder, is placed on the porous disk, and gas is introduced into the cell and flows up the tube and through the sample bed. An outer quartz tube, surrounding the stainless steel tube and capped by a quartz optical window, provides a gas-tight seal for gases exiting the catalyst bed. A cylindrical furnace surrounding the cell is used to produce elevated temperatures during collection of the Raman spectra. An electromagnetic shaker is attached to the base of the cell to facilitate tumbling of the sample particles.

Raman scattering was excited by a 244 nm wavelength UV laser beam produced by a Lexel 95 SHG Argon ion laser. The laser was focused down onto the top surface of the sample bed. The scattered light was collected by an ellipsoidal mirror and focused into a Spex 1877 triple grating spectrometer with an imaging multichannel photomultiplier tube (IMPT) detector. The power delivered to the samples was between 0.8 and 30 mW.

Helium was used as the fluidizing gas for measurements on the pre-coked catalyst sample. To study the reactions of propane and propene on the catalyst, propane and propene were used as the fluidizing gas. The gas flow rate was greater than 100 ml/min during Raman spectra collection. Raman spectra were collected for 900-1800 seconds at the temperature indicated for each experiment below, with approximately 30 mW laser power to the sample. Pre-reduced samples of catalyst were heated to 500°C in hydrogen, held at that temperature for 1 hour and cooled in hydrogen before use. All gases were high purity and used without additional purification.

3 RESULTS AND DISCUSSION

3.1 Pre-coked Catalyst

The Raman spectrum from the pre-coked catalyst is shown in Figure 1. The main peak appears at 1592 cm^{-1} with a shoulder at 1550 cm^{-1}. Broad features at 1170 cm^{-1} and 1380 cm^{-1} are observable. With the exception of the shoulder at 1550 cm^{-1} the Raman spectrum is characteristic of polynuclear, aromatic carbon. Espinat et al.[28] characterized a wide variety of carbon compounds including coke on a series of mono- and bi-metallic catalysts and assigned the Raman bands found from 1300 to 1700 cm^{-1}. The D (1350 cm^{-1}) and G (1600 cm^{-1}) bands can be used to identify the type of coke deposited on the catalyst. For the spectrum in Figure 1, the D band is half the intensity of the G band. This ratio indicates pre-graphitic coke, which is composed of graphite with crystal defects or pre-graphitic paracrystals.[28] The shoulder at 1550 cm^{-1} is characteristic of conjugated olefins or polyenes.

3.2 Pre-reduced Catalyst

After calcination, Cr/Al_2O_3 catalysts are typically activated by reduction in hydrogen. The Raman spectra in flowing propane at 400°C and 500°C from the catalyst pre-reduced in

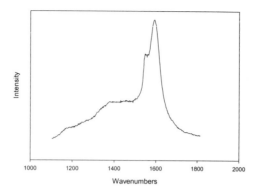

Figure 1   *UV Raman spectrum of coke deposited on 1% Cr/Al$_2$O$_3$ catalyst used for propane dehydrogenation.*

hydrogen are shown in Figure 2. At 400°C the equilibrium for dehydrogenation of propane to propene is 3%. The spectrum, measured at 400°C, has two features at 870 cm^{-1} and 1560 cm^{-1}. The band at 870 cm^{-1} is assigned to the CC stretch of molecular propane in the gas phase and adsorbed on the catalyst. The broad band centred at1560 cm^{-1} is assigned to the C=C stretch in conjugated olefins deposited on the catalyst.[29] Baruya et al.[30] have shown that the C=C stretching frequency in Raman spectra from a conjugated olefin d epends on the l ength of c onjugation and that the frequency of the C=C s tretch decreases from approximately 1 600 c m^{-1} t o 1 461 c m^{-1} w ith i ncreasing c onjugation. A n empirical formula in reference 30 indicates that the polyenes deposited at 400°C have 5-8 double bonds.

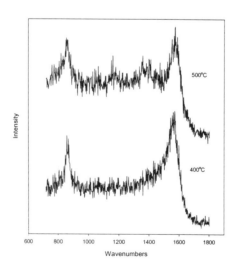

Figure 2   *Raman spectra from reactant propane over the pre-reduced Cr/Al$_2$O$_3$ catalyst.*

At 500°C there is a slight shift of the peak center (to 1574 cm⁻¹) and broadening of the most intense feature to higher wavenumbers and the appearance of new peaks at 1170, 1350, and 1400 cm⁻¹. These changes are consistent with the formation of polynuclear aromatic carbon on the catalyst with ring stretches in the regions 1150-1400 cm⁻¹ and 1600-1615 cm⁻¹ in addition to the continued presence of deposited polyenes.[31] Previous studies using (2-^{13}C)propane[32] have shown that as the temperature is increased the extent of molecular fracture of the propane into C-1 fragments increases suggesting that the polynuclear aromatic carbon is built up from C-1 species.

The results for flowing propene over the pre-reduced catalyst at 400°C and 500°C are shown in Figure 3. At both temperatures the features due to deposited carbon have significantly lower intensities than observed for propane. The sharp band at 1300 cm⁻¹, present at 400°C is due to molecular propene adsorbed on the catalyst. Note that this band is not detectable at 500°C. The broad features between 1350 cm⁻¹ and 1625 cm⁻¹ are due to polynuclear aromatic carbon.

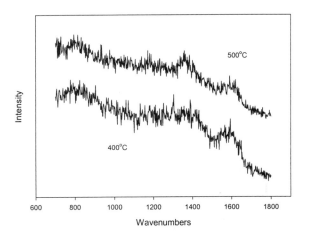

Figure 3   *Raman spectra of reactant propene over the pre-reduced Cr/Al₂O₃ catalyst.*

The differences between propane and propene reactants are quite intriguing. Polyenes are readily formed by propane, even at a reaction temperature where dehydrogenation is very sluggish. They are not formed to any detectable extent by propene. Clearly their formation does not proceed through propene as an intermediate. It is generally accepted that alkenes are more active in coke formation than alkanes, but this does not appear to be the case for propane and propene Cr/Al₂O₃ at low reaction temperatures when the coke is composed primarily of conjugated olefins.

3.3   Calcined Catalyst

The amount and chemical nature of the carbon deposited by both propane and propene reactants are remarkably different on calcined Cr/Al₂O₃ from the pre-reduced catalyst. The spectra in flowing propane and propene are shown in Figures 4 and 5, respectively. For propane (Figure 4) at 400°C the strong peaks at 1400 cm⁻¹ and ca. 1600 cm⁻¹ indicate the presence of polynuclear aromatic hydrocarbons with a chainlike topology (phenanthrene, anthracene, and pentacene).[31] The presence of polyenes is indicated by a shoulder at 1550 cm[-1]. After increasing the reaction temperature to 500°C the main peak shifts to 1580 and the spectral features near 1400 cm[-1] disappear. These changes are consistent with a transformation of the adsorbed species from polynuclear aromatic hydrocarbons to conjugated olefins. While the fate of the polynuclear aromatic hydrocarbons is unknown, the most likely possibility is that they desorb from the catalyst surface at the higher reaction temperature.

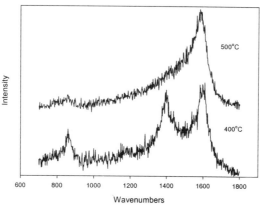

Figure 4   *Raman spectra from reactant propane over the calcined Cr/Al₂O₃ catalyst.*

The amount of deposited hydrocarbons from reactant propene is qualitatively larger on the calcined catalyst compared to the pre-reduced catalyst (compare Figures 3 and 5). The spectra from the calcined catalyst are consistent with the formation of a mixture of polynuclear aromatic hydrocarbons and conjugated olefins at both reaction temperatures. Comparison of the spectral features at 400°C and 500°C suggests that the chemical c omponents c ontributing t o t he t wo spectra are similar with p ossibly a l arger contribution from polyenes at 500°C.

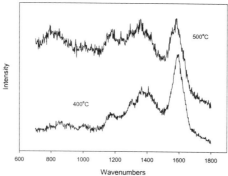

Figure 5   *Raman spectra from reactant propene over the calcined Cr/Al₂O₃ catalyst.*

The difference in the quantity and chemical nature of the deposits on the calcined and reduced catalysts is striking. Changes in the catalyst colour (yellow to green) suggest that Cr^{3+} are formed during hydrogen reduction. The present results indicate that the deposited hydrocarbons produced by reduced chromium are smaller in amount but potentially more potent for catalyst deactivation (polynuclear aromatics vs. polyenes). The most interesting question, left unanswered by the present study, is why the deposition of carbon is so facile on the calcined, unreduced catalyst by comparison to the catalyst activated in hydrogen.

4 CONCLUSIONS

The quantity and chemical nature of deposited hydrocarbons from reaction of propane and propene over 1% Cr/Al_2O_3 catalysts are substantially different and depend strongly on reaction t emperature a s well a s t he p resence o r absence o f C r^{3+} p roduced b y h ydrogen reduction. Hydrocarbon deposits are readily formed at temperatures below where propane dehydrogenation is appreciable. Propene does not appear to be a required intermediate in the formation of deposited polyenes under propane dehydrogenation conditions.

Acknowledgements

Financial support of this work was provided, in part, by the Department of Energy, Office of Science, Office of Basic Energy Science, Division of Chemical Sciences, under Contract DE-FG02-97ER14789.

References

1   B. M. Weckhuysen and I. E. Wachs, *J. Phys. Chem.*, 1996, **100**, 14437.

2   S. D. Jackson, I. M. Matheson, M. L. Naeye, P. C. Stair, V. S. Sullivan, S. R. Watson, and G. Webb, *Stud. Surf. Sci. Catal.*, 2000, **130C**, 2213.

3   S. D. Jackson, J. Grenfell, I. M. Matheson, and G. Webb, *Stud. Surf. Sci. Catal.*, 1999, **122**, 149.

4   S. Udomsak and R. G. Anthony, *Ind. Eng. Chem. Res.*, 1996, **35**, 47.

5   L. R. Mentasty, O. F. Gorriz, and L. E. Cadus, *Ind. Eng. Chem. Res.*, 1999, **38**, 396.

6   A. Brito, R. Arvelo, R. Villarroel, and M. T. Garcia, *React. Kinet. Catal. Lett.*, 1995, **55**, 85.

7   A. Brito, R. Arvelo, R. Villarroel, F. J. Garcia, and M. T. Garcia, *Chem. Eng. Sci.*, 1996, **51**, 4385.

8   A. Brito, R. Arvelo, R. Villarroel, and M. T. Garcia, *React. Kinet. Catal. Lett.*, 1995, **55**, 77.

9   F. Cavani, M. Koutyrev, F. Trifiro, A. Bartolini, D. Ghisletti, R. Iezzi, A. Santucci, and G. D. Piero, *J. Catal.*, 1996, **158**, 236.

10   S. De Rossi, G. Ferraris, S. Fremiotti, E. Garrone, G. Ghiotti, M. C. Campa, and V. Indovina, *J. Catal.*, 1994, **148**, 36.

11   O. F. Gorriz, V. Cortes Corberan, and J. L. G. Fierro, *Ind. Eng. Chem. Res.*, 1992, **31**, 2670.

12   S. D. Jackson, J. Grenfell, I. M. Matheson, S. Munro, R. Raval, and G. Webb, *Stud. Surf. Sci. Catal.*, 1997, **111**, 167.

13   L. Dixit, D. L. Gerrard, and H. J. Bowley, *Appl. Spectrosc. Rev.*, 1986, **22**, 189.

14   J. M. Stencel, 'Raman Spectroscopy for Catalysis', Von Nostrand Reinhold, 1990.
15   R. M. Pittman and A. T. Bell, *Catal. Lett.*, 1994, **24**, 1.
16   J. Miciukiewicz, T. Mang, and H. Knoezinger, *Appl. Catal., A*, 1995, **122**, 151.
17   G. J. Hutchings, A. Desmartin-Chomel, R. Oller, and J. C. Volta, *Nature (London)*, 1994, **368**, 41.
18   M. Scheithauer, H. Knozinger, and M. A. Vannice, *J. Catal.*, 1998, **178**, 701.
19   J. P. Dunn, H. G. Stenger, Jr., and I. E. Wachs, *J. Catal.*, 1999, **181**, 233.
20   C. Li and P. C. Stair, *Stud. Surf. Sci. Catal.*, 1996, **101**, 881.
21   C. Li and P. C. Stair, *Catal. Lett.*, 1996, **36**, 119.
22   C. Li and P. C. Stair, *Stud. Surf. Sci. Catal.*, 1997, **105A**, 599.
23   C. Li and P. C. Stair, *Catal. Today*, 1997, **33**, 353.
24   P. C. Stair and C. Li, *J. Vac. Sci. Technol., A*, 1997, **15**, 1679.
25   Y. T. Chua and P. C. Stair, *J. Catal.*, 2000, **196**, 66.
26   S. A. Asher, *Anal. Chem.*, 1984, **56**, 720.
27   S. A. Asher and C. R. Johnson, *Science (Washington, D. C., 1883-)*, 1984, **225**, 311.
28   D. Espinat, H. Dexpert, E. Freund, G. Martino, M. Couzi, P. Lespade, and F. Cruege, *Appl. Catal.*, 1985, **16**, 343.
29   F. R. Dollish, W. G. Fateley, and F. F. Bentley, 'Characteristic Raman Frequencies of Organic Compounds', Wiley-Interscience, 1973.
30   A. Baruya, D. L. Gerrard, and W. F. Maddams, *Macromolecules*, 1983, **16**, 578.
31   Y. T. Chua and P. C. Stair, *Journal of Catalysis*, 2003, **213**, 39.
32   S. D. Jackson, J. Grenfell, I. M. Matheson, and G. Webb, *Studies in Surface Science and Catalysis*, 1999, **122**, 149.

BUTANE DEHYDROGENATION OVER A Pt/ALUMINA CATALYST.

S David Jackson, David Lennon, and John M McNamara

Department of Chemistry, The University, Glasgow G12 8QQ, Scotland.

1 INTRODUCTION

The relative demand of propene, butene and isobutene present a need and an opportunity for on-purpose manufacture of these chemicals, as opposed to their traditional sourcing as by-products from steam cracking and catalytic cracking. While various schemes have been proposed for the on-purpose production, the most feasible on a commercial scale at present is catalytic dehydrogenation of the relevant alkane. The dehydrogenation of light alkanes has been known as a catalytic process for a significant number of years[1,2]. Hence catalytic dehydrogenation has now developed into a major route for the valorisation of alkanes and for the production of active, functionalised molecules. For simple catalytic dehydrogenation reactions, such as the dehydrogenation of light alkanes, the industrial catalytic processes fall into two categories, those based on platinum as the active phase[3,4] and those based on chromia as the active phase[5]. However in both cases the catalytic process is complex with a series of competing reactions occurring simultaneously, for example, the basic alkane dehydrogenation reaction is accompanied by various side reactions, comprising essentially the cracking of the feed hydrocarbon to lower hydrocarbons and the formation of dehydrogenated carbonaceous species on the catalyst surface. The formation of this "coke" leads to the deactivation of the catalyst, and the catalyst requires regular regeneration to restore its activity. None of the current industrial processes are ideal and ways of improving the process, including the use of oxidative dehydrogenation, are always being sought[2, 6-9]. To achieve this we have established a multi-disciplinary group to investigate all aspects of dehydrogenation from the underlying surface physics to the applied chemical engineering of the reactors[10].

In this paper we will examine the dehydrogenation of butane to butene, $C_4H_{10} \Leftrightarrow C_4H_8 + H_2$. The reaction is highly endothermic ($\Delta H = +126$ kJ mol^{-1}) and is equilibrium limited (at 1 atm. 6% conversion at 673 K, 58% conversion at 873 K). Typical operating conditions for commercial reactors are 0.3 - 3 atm and 753 – 923 K. We have studied the reaction under continuous flow conditions looking at a range of temperatures and space velocities.

2   EXPERIMENTAL

The catalyst used throughout this study was Pt/alumina and was prepared by impregnation. H_2PtCl_6 (Johnson-Matthey) was dissolved in de-ionised water and the solution added to γ-alumina (Engelhard Al-3992E, S.A. 180 m^2g^{-1}). The resulting suspension was dried and heated to 313 K for 16 h and calcined at 823 K for 3 h. The Pt loading of the sample was 0.66 % w/w. The dispersion of the catalyst, measured by CO chemisorption, was 78 % and the particle size from TEM was measured at <2 nm[11].

Continuous flow reaction studies were performed in a 0.101 MPa, continuous flow microreactor with the gas stream exit the reactor being sampled by on-line GC. Using this system the catalysts (typically 0.25 g) could be reduced *in situ* in flowing 6% hydrogen in nitrogen (30 cm^3min^{-1}) by heating to 573 K at 5 K min^{-1} and then holding at this temperature for 3 h. Whilst still at temperature the flow was switched to helium and held for 2 h. The temperature of the catalyst bed was then altered to that required in the experiment. Once at temperature the flow was switched to butane and samples of the exit gas were analysed by on-line GC.

3   RESULTS

The extent of dehydrogenation was measured in the absence of a catalyst and with the support in the absence of the metal. The reactor was found to be inert at all temperatures investigated. Only at 873 K was there any detectable reaction over the support and this was over an order of magnitude less than the reaction detected in the presence of the catalyst.

The catalyst was tested at 673 K, 773 K, and 873 K, a fresh sample was used at each temperature. The reaction was run in pure butane with a GHSV of 9677 h^{-1}. The results are shown in Table 1.

At 873 K, methane, ethene/ethane, and isobutane were also formed; typical yields were 0.4 %, 1.3 %, and 1.5 % respectively.

The effect of varying space velocity was also examined. The results are shown in Table 2.

4   DISCUSSION

The effect of increasing the temperature of reaction is shown in Table 1 and, as might be expected, the activity of the catalyst is seen to increase with temperature but decrease with time. At 673 K the equilibrium conversion of butane to butenes is 6 %, at 773 K 23 % and at 873 K 58 %. Therefore from Table 1 the system is at equilibrium at 673 K and 773 K but not at 873 K. The reason for this is likely to be two-fold. Firstly the reaction is endothermic and as the conversion increases the extent of the temperature decrease increases, therefore the temperature may be lower in the catalyst bed than that detected in the thermocouple pocket. Secondly there is far more extensive carbon laydown, so that the catalyst will have suffered a far more rapid deactivation than at the lower temperatures.

Given that the dehydrogenation reaction is at equilibrium then it is informative to examine the isomer distribution of the butenes. At 673 K the equilibrium ratio of 1-butene:trans-2-butene:cis-2-butene is 1:2:1.7. If we examine the experimentally observed ratio at 673 K we find that 1-butene has the lowest yield throughout and at 0 min the ratio

Table 1 *Dehydrogenation of butane over Pt/alumina.*

Temperature (K)	Time on stream (min)	% Conversion	Yield (%)			
			1-butene	Trans-2-butene	Cis-2-butene	1,3-butadiene
673	0	12.0	1.8	3.3	2.8	
	30	4.9	0.8	1.8	1.7	
	50	5.5	0.6	1.4	1.2	
	73	3.5	0.5	1.3	1.1	
	105	3.0	0.4	1.1	0.8	
	141	2.9	0.3	0.9	0.7	
773	0	37.1	10.2	7.2	6.5	1.0
	25	11.7	2.5	2.2	2.0	0.8
	69	4.8	0.9	0.9	0.8	0.3
	109	3.1	0.8	0.8	0.7	0.3
	132	4.2	0.7	0.7	0.6	0.2
873	0	55.2	10.8	5.3	4.8	2.5
	23	23.3	4.0	1.9	1.5	1.4
	46	20.0	3.3	1.4	1.2	1.1
	78	15.1	4.1	1.1	0.8	0.7
	145	12.5	0.8	0.5	0.5	0.5
	210	10.0	0.6	0.4	0.4	0.6

is 1:1.9:1.6, showing that the system is at equilibrium. However as the time on stream increases the system moves away from equilibrium, e.g. at 141 min on stream the ratio of 1b:t2b:c2b is 1:2.7:2.3. However the cis and trans isomers are still in equilibrium with a ratio of 1.2:1 trans:cis. Therefore it would appear that the reaction chemistry of the system to 1-butene and cis- and trans-2-butene is not identical. The data at 773 K and 873 K show different behaviour in that 1-butene is not produced in the lowest yield. Indeed 1-butene is initially the major product, cis-2-butene is now the butene produced in the smallest yield. Hence the ratio of 1-butene:trans-2-butene:cis-2-butene is well removed from equilibrium, as from equilibrium calculations 1-butene should be the minor product. However the trans:cis ratio is at its equilibrium position; a position that is maintained throughout the test. Therefore once again we have the situation that 1-butene and cis- and trans-2-butene reaction chemistries are not the same. If we examine the literature we find that equilibration of the butenes is not expected to be fast over platinum[12, 13, 14]. Under hydrogenation conditions[12] the isomerisation of the butenes is slow while in the absence of hydrogen no isomerisation was observed at 373 K over a platinum powder[14]. However at 408 K isomerisation is observed over a platinum black[14].

In Table 2 it can be seen that, as expected, an increase in space velocity results in a decrease in the overall conversion. However the yield of butenes remains approximately constant at the equilibrium conversion. Therefore the main reaction affected by the increase in space velocity is not the dehydrogenation reaction but the carbon deposition reactions. As carbon laydown is one of the major other reactions occurring we analysed

Table 2   *Effect of Space Velocity on Conversion and Yield.*

GHSV (h⁻¹)	Time on stream (min)	% Conversion	Yield (%)		
			1-butene	Trans-2-butene	Cis-2-butene
5806	0	21.1	1.22	2.7	2.13
	25	6.4	0.98	2.45	2.16
	47	4.9	0.73	1.97	1.72
	77	5.5	0.69	1.86	1.59
	102	4.1	0.66	1.8	1.51
9677	0	12	1.75	3.29	2.84
	30	4.9	0.76	1.81	1.65
	50	5.5	0.59	1.42	1.17
	73	3.5	0.49	1.27	1.06
	105	3	0.4	1.08	0.82
	141	2.9	0.32	0.87	0.74
19354	0	11.2	1.29	2.79	2.5
	19	8.9	0.25	0.83	0.58
	43	3.5	0.16	0.58	0.49
	61	2.1	0.08	0.42	0.34
	117	2.9	0.02	0.33	0.33
25161	0	7.4	1.52	2.99	2.66
	52	3	0.11	0.44	0.39
	77	2.4	0.06	0.46	0.12
	107	1.2	0.04	0.28	0.23
	127	1.3	0	0.34	0

the behaviour of each of isomers to carbon deposition. It has been recognised for over 30 years that the deactivation due to the deposition of carbonaceous species can be modelled can be modelled as follows[15]:
The rate of reaction "i" at time "*t*" is given by

$$R_{it} = \phi \cdot R_{i0} \qquad \text{(Eqn 1)}$$

where R_{i0} is the initial rate (fresh catalyst) and φ is an activity factor given by :

$$\phi = f(C_c) \qquad \text{(Eqn 2)}$$

where φ is the activity relative to fresh catalyst and C_C is the accumulated coke concentration on the catalyst. The same paper presents a number of alternative expressions relating φ to C_C . The paper concludes that an expression of the form below is generally preferred

$$\phi = exp(-\alpha \cdot C_c) \qquad \text{(Eqn 3)}$$

where α is an adjustable parameter, frequently assumed to be a constant.

Rearranging and integrating Equations (1) and (3) gives:

$$\frac{1}{R_{it}} = \frac{1}{R_{i0}} + \alpha \cdot t \qquad \text{(Eqn 4)}$$

Hence α and R_{i0} can be evaluated for each reaction by the relatively simple plot of $1/R_{it}$ versus time where α is the gradient of the resulting straight line and R_{i0} its intercept on the y-axis. This approach has subsequently been successfully applied to dehydrogenation of butene-1[16]. A number of detailed mechanistic studies have followed on from this utilising more sophisticated approaches, but confirming the validity of the general approach, see for example Acharya and Hughes[17] or various papers presented at the 7th International Symposium on Catalyst Deactivation[18]. In the studies with butene-1[16, 17] both papers note that α is a constant; independent of the reaction, temperature and partial pressure. It was concluded from this that all the reactions occur on the same active sites.

Analysis of the butene data at 673 K, 773 K and 873 K gave α values as shown in Table 3.

Table 3   *α Values for the Butenes as a Function of Temperature.*

Temperature (K)	α values ($\times 10^{-3}$)		
	1-butene	Trans-2-butene	Cis-2-butene
673	17.4±0.9	5.7±0.4	7.2±0.6
773	10.3±1.6	9.7±1.2	11.2±1.5
873	8.1±1.1	12.5±0.8	10.9±0.7

It can be clearly seen that the 1-butene values are decreasing with temperature while the cis- and trans-2-butene values are increasing with temperature. Hence it is unlikely that the carbon laydown reaction from 1-butene occurs on the same site as that for cis- and trans-2-butene.

In conclusion the dehydrogenation of butane has been studied over a Pt/alumina catalyst under continuous flow conditions. The system is operating at equilibrium in terms of total butene yield, however the 1-butene is does not undergo fast equilibration with the 2-butenes. The 2-butenes are usually in thermodynamic equilibrium. Analysis of the deactivation profiles of the butenes confirms that the carbon deposition reaction takes place on different sites for 1-butene and the 2-butenes.

Acknowledgements

The authors would like to acknowledge the support of EPRSC and Synetix/Johnson Matthey through the ATHENA project for part of this work.

References

1 K. Kearby, in *Catalysis*, ed., P.H. Emmet, Reinhold, New York, 1955, Vol. 3, p.453.
2 M.M. Bhasin, J.H. McCain, B.V. Vora, T. Imai, and P.R. Pujado, *Appl. Catal. A*, 2001, **221**, 397.
3 T. Hutson, Jr., and W.C. McCarthy, in *Handbook of Petroleum Refining Processes*, ed., R.A. Meyers, McGraw-Hill, London, 1986.
4 P.R. Pujado and B.V. Vora, *Hydrocarbon Process.*, March, 1990, 65.
5 G.F. Hornaday, F.M. Ferrell, and G.A. Mills, *Advan. Petrol. Chem. Refining*, 1961, **4**, 451.
6 R.J. Rennard and J. Freel, *J. Catal.*, 1986, **98**, 235.

7   L. Jinxiang, G. Xiuyiang, Z. Tao, and L. Liwu, *Thermochimica Acta*, 1991, **179** 9.
8   S. De Rossi, G. Ferraris, S. Fremiotti, E. Garrone, G. Ghiotti, M.C. Campa, and V. Indovina, *J. Catal.*, 1994, **148,** 36.
9   S.D. Jackson, P. Leeming and J. Grenfell, *J. Catal.*, 1994, **150,** 170.
10 ATHENA project.   An EPSRC/Johnson Matthey funded multi-centre project examining selective hydrogenation, selective dehydrogenation, and selective oxidation.
11 J. M. McNamara, Ph.D. Thesis, University of Glasgow (2000).
12 G.C. Bond, J.J. Phillipson, P.B. Wells and J.M. Winterbottom, *Trans Faraday Soc.*, 1964, **60**, 1847.
13 P.B. Wells and G.R. Wilson, *J. Catal.*, 1967, **9**, 70.
14 V. Ragaini, *J.Catal.*, 1968, **10**, 230.
15 G.F. Froment and K.B. Bischoff, *Chem.Eng.Sci.*, 1961, **16**, 189.
16 F.J. Dumez and G.F. Froment, *Ind.Eng.Chem. Process Res.Dev*, 1976, **15**, 291.
17 D.R. Acharya and R. Hughes, *Can.J.Chem.Eng.*, 1990, **68**(2), 89.
18 C.H. Bartholomew and G.A. Fuentes, *Catalyst Deactivation 1997*, Studies in Surface Science and Catalysis, Vol 111, Elsevier, Amsterdam 1997.

SELECTIVE HYDROGENATION OF CINNAMALDEHYDE TO CINNAMYL ALCOHOL USING AN Ir/C CATALYST: INFLUENCE OF REACTION CONDITIONS

J.P. Breen[1], R. Burch[1], J. Gomez-Lopez[1], K. Griffin[2], M. Hayes[2]

[1] School of Chemistry, Queen's University Belfast, David Keir Building, Stranmillis Road, Belfast BT9 5AG, N. Ireland
[2] Johnson Matthey plc, Orchard Road, Royston, Hertfordshire, SG8 5HE, UK

1 INTRODUCTION

The selective hydrogenation of α, β-unsaturated carbonyls is a key step in the manufacture of pharmaceuticals and flavours and fragrances. The hydrogenation of the C=C bond to yield saturated carbonyls is thermodynamically favoured and can be readily achieved with high selectivity. The selective hydrogenation of the C=O bond to yield unsaturated alcohols is much more difficult to achieve and as a result has been the focus of much research. Giroir-Fendler et al. [1] studied the hydrogenation of cinnamaldehyde with various noble metal catalysts supported on carbon and graphite and found that the selectivity towards the unsaturated alcohol follows the series 0 = Pd < Rh < Ru < Pt < Ir. The best results were obtained with a carbon supported iridium catalyst. Ruthenium and rhodium catalysts are active for the reduction of the carbonyl group but also for the reduction of the C=C bond and palladium catalysts give very high selectivity for the production of the saturated alcohol. Platinum tends to be less selective than iridium for hydrogenation to the unsaturated alcohol but can be improved by the addition of a metallic salt such as FeCl₃ [1-5].

The majority of published research has concentrated on the preparation of the catalyst the effect of different supports and different metals, the addition of second metals and the effect of different preparation methods on the selectivity of the catalysts for selective hydrogenation [2,3,5,6-10]. The effects of reaction conditions on selectivity have received considerably less attention. Gallezot and Richard [4] commented on the scarcity of systematic studies on the influence of reaction parameters such as pre-reduction of the catalyst, temperature, pressure, concentration of reactant and nature of the solvent for a given catalyst and reaction. Since then Singh et al. [11] have obtained quantitative kinetic data on the liquid phase hydrogenation of citral over Pt/SiO₂ catalysts and have used this information to present a kinetic model which fits their data.

This paper aims to systematically study the effect of reaction parameters such as pre-reduction of the catalyst, temperature, pressure and concentration of reactant for an Ir/C catalyst for the hydrogenation of cinnamaldehyde. The results of earlier studies [1,12] have shown that Ir/C is both selective and active for the hydrogenation of cinnamaldehyde.

2   EXPERIMENTAL

The experiments were carried out in a 300 cm³ stainless steel autoclave equipped with a glass liner, magnetic stirrer, pressure gauge and thermocouple. A high pressure pipette for liquid degassing and reactants introduction under inert atmosphere was designed and connected. The gasses were introduced into the autoclave by means of a constant pressure delivery system. A pressure transducer was also connected to the autoclave in order to monitor the pressure inside the reaction vessel. The catalyst used, 5 % iridium on graphite, was supplied by Johnson Matthey. All the reactants were used as received without further purification.

For a typical experiment, 1 g of catalyst was placed in the autoclave; the autoclave was flushed 3 times with 5 bar nitrogen in order to remove any traces of oxygen. Prior to the reaction, the catalyst was pre-reduced in the autoclave in 40 cm³ of degassed isopropanol (Riedel de Haen) at 150 °C under 20 bar H_2 for 2 hours. After the reduction stage, and once the autoclave was cooled, the hydrogen was vented. The autoclave was then flushed 3 times with 5 bar nitrogen and 20 cm³ of degassed cinnamaldehyde were introduced under inert atmosphere. The temperature was then raised to the desired value and 20 bar of hydrogen introduced. The reaction was carried out at maximum stirring speed and hydrogen consumption was monitored. Samples were withdrawn at regular intervals, filtered and analysed by gas chromatography.

3   RESULTS

3.1 Mass Transfer

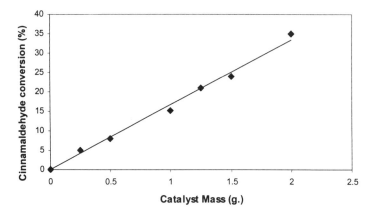

Figure 1   *Cinnamaldehyde (CAM) conversion as a function of catalyst mass. Reaction conditions: Catalyst: 5% Ir/graphite, T: 85 °C, P: 20 bar H_2, reaction time: 2 hours*

Mass transfer can play an important role in liquid phase hydrogenations using porous catalysts. Therefore, it is critically important that mass transfer effects are ruled out before attempting to obtain reliable kinetic data. In order to reduce the chances of mass transfer limitations the smallest catalyst particle sizes and highest stirring speed were used (1100 rpm). These two precautions served to increase the rate of mass transfer from the gas to the liquid phase, to increase the rate of reactant transfer from the bulk liquid to the catalyst surface and minimise internal diffusion resistance. To check that mass transfer was not rate controlling the conversion of cinnamaldehyde was measured over a range of different catalyst masses, the absence of mass transfer limitations was confirmed by a straight-line plot of conversion against catalyst mass (Fig. 1).

3.2 Effect of catalyst pre-reduction

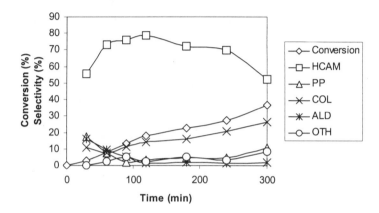

Figure 2 *Conversion of CAM and selectivity to products for the hydrogenation of CAM over an un-reduced 5% Ir/graphite catalyst. Reaction conditions: P: 20 bar H_2, T: 100 °C, reaction time 5 hours*

The hydrogenation of cinnamaldehyde (CAM) typically produces a mixture of the desired unsaturated alcohol - cinnamyl alcohol (COL), the undesired saturated aldehyde - hydrocinnamaldehyde (HCAM) and the saturated alcohol - phenyl propanol (PP). Aldol condensation (ALD) and other unidentified products (OTH) may also be produced in small amounts over particularly unselective catalysts. A plot of conversion and selectivity as a function of time (Fig. 2) illustrates the range of products that can be produced using an Ir/C catalyst at 100 °C. The selectivity to COL increases with increasing conversion but clearly HCAM is the predominant product at all conversions, thus showing that the hydrogenation of the C = C bond is favoured under these conditions. However, pre-treatment of the catalyst in hydrogen at 150 °C for a period of time results in a significant improvement in the conversion of CAL and a dramatic increase in the selectivity towards COL (Fig. 3). The duration of the pre-treatment is also important, both conversion and selectivity to COL being increased by increasing the pre-reduction time from one to two hours. These results are in agreement with those of Cordier et al. [13], who showed that pre-treatment of Os, Ir and Pt carbon-supported catalysts could improve selectivity towards C = O bond activation. The effect that the pre-treatment has on the catalyst is complex and is dependent on the reduction time, temperature, the presence or absence of solvent, type of solvent and hydrogen partial pressure [12]. All results of kinetic tests presented hereafter are based on catalysts that have been pre-reduced for 2 hours with hydrogen in the presence of isopropanol (IPA) at 150 °C.

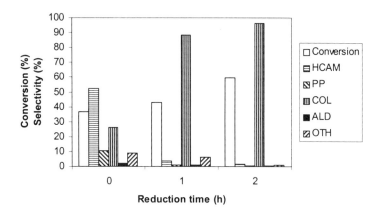

Figure 3   *The effect of pre-reduction time on the conversion of CAM and selectivity to products for the hydrogenation of CAM over a 5% Ir/C catalyst. Reaction conditions: P: 20 bar H₂, T: 100 °C reaction time: 5 h.*

3.3   Effect of reaction temperature

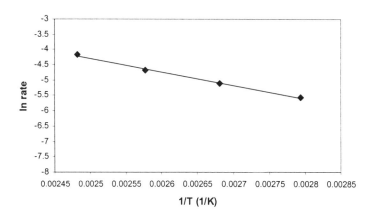

Figure 4   *The rate of reaction as a function of temperature. Reaction conditions: 1 g 5% Ir/graphite, P: 20 bar H₂.*

The influence of temperature on the kinetics of the hydrogenation reaction was studied at 85, 100, 115 and 130 °C. There are a number of trends evident from the data; the conversion of CAM increases with time of reaction for each temperature, selectivity to COL increases with increasing conversion and the rate of conversion increases with increasing reaction temperature. The selectivity to COL is > 90% throughout, approaching 100% at high temperatures and conversions. Fig. 4 demonstrates that the reaction exhibits conventional Arrhenius behaviour over the temperature range of 85 to 130 °C, giving an apparent activation energy for the reaction of 37 kJ mol^{-1}.

3.4   The effect of hydrogen pressure and CAM concentration

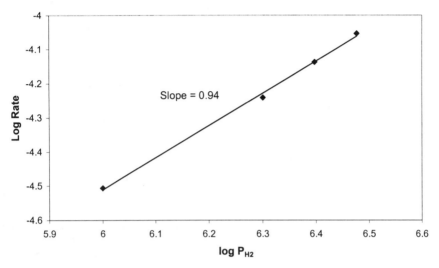

Figure 5   *The rate of reaction as a function of hydrogen partial pressure.   Reaction conditions: 1g 5% Ir/graphite, T= 85 °C*

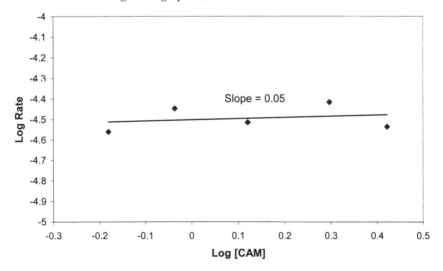

Figure 6   *The rate of reaction as a function of cinnamaldehyde concentration.   Reaction conditions: 1g 5% Ir/graphite, 20 bar H$_2$, T = 85 °C.*

    The hydrogenation reaction was carried out at hydrogen pressures ranging from 1 to 4 MPa while maintaining a constant temperature and initial CAM concentration. The results showed that increasing the hydrogen pressure resulted in an increase in the rate of conversion of CAM. Fig. 5 indicates that the rate of reaction is directly proportional to the hydrogen pressure, effectively exhibiting a first order dependency on hydrogen. The selectivity to COL remained high (> 90%) across the range of pressures.

    The effect of varying CAM concentration on the rate of reaction was studied by loading the reactor with different concentrations of CAM in IPA and monitoring the conversion and selectivity as a function of time at constant temperature and initial hydrogen pressure. The results of these experiments showed that conversion decreased with increasing CAM concentration but selectivity remained high. The initial rates of CAM hydrogenation were found to be independent of CAM concentration, this is illustrated by the log(rate) versus log([CAM]) plot shown in Fig. 6.

4 DISCUSSION

The CAM hydrogenation reaction showed conventional Arrhenius type behaviour over the Ir catalyst, giving an apparent activation energy of 37 kJ mol^{-1}. There are no comparable studies of the effect of temperature on the hydrogenation of CAM over iridium catalysts. Singh et al. [11] have studied the liquid phase hydrogenation of citral over Pt/SiO$_2$ and Pt/TiO$_2$ catalysts. They found that these reactions did not exhibit conventional Arrhenius behaviour, instead the rates of these reactions decreased when the temperature was increased from 25 to 100 °C. This unusual behaviour was explained by a reaction scheme incorporating decarbonylation of the organic compound and subsequent inhibition of the hydrogenation reaction by adsorbed CO. Pt/TiO$_2$ showed conventional Arrhenius behaviour for citral hydrogenation with an activation energy of 75 kJ mol^{-1} but only after high temperature pre-reduction (at 500 °C). Low temperature pre-reduction (at 200 °C) gave a catalyst that again showed an activity minimum with respect to temperature, thus emphasising the importance of the pre-treatment step in liquid phase hydrogenations. Cobalt catalysts have been found to exhibit Arrhenius type behaviour for the hydrogenation of CAM with activation energies ranging from 18.0 kJ mol^{-1} for a Co-B amorphous alloy catalyst to 35 kJ mol^{-1} for a Raney cobalt catalyst [7].

    From the kinetic data a power law rate expression was derived:

$$v = k[\mathrm{H_2}][\mathrm{CAM}]^0$$

    The zero order with respect to CAM implies a surface largely covered with hydrocarbon intermediates. There is no deactivation with time on stream, indicating that the hydrocarbon intermediates are reactive. Dissociatively adsorbed hydrogen is frequently invoked for modelling hydrogenation reactions over Pt catalysts [14]. Considering that iridium can also dissociatively adsorb hydrogen it is also reasonable to invoke a mechanism involving competitive adsorption between hydrogen and the organic compound (as in the case of Pt). In mechanistic terms the main difference between Pt and Ir appears to lie in the differences in selectivity for the hydrogenation reaction, Ir being more selective for the reaction than Pt. Delbecq and Sautet [15] showed that differences in selectivities between various metals could be attributed to differences in the radial expansion of their *d* bands. The larger the band, the stronger the four-electron repulsive interaction with the C = C bond and the lower the probability of its adsorption. The *d* band width has been found to increase in the order Os, Ir > Pt > Pd, an order which closely matches the selectivity of these metals for the hydrogenation of the C = O bond.

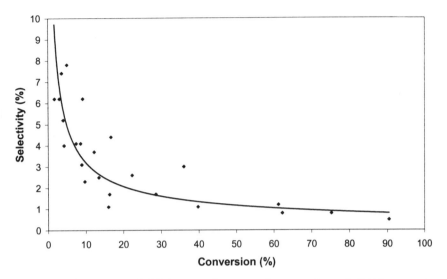

Figure 7 *Selectivity to HCAM as a function of conversion of CAM at various temperatures, pressures and initial concentrations of CAM*

One of the most interesting aspects of the results presented here is the relationship between the selectivity to HCAM and conversion of CAM. The selectivity to HCAM remained low (<10%) throughout the various kinetic experiments detailed above. However, there was a noticeable trend of decreasing selectivity to HCAM with increasing conversion of CAM. The plot of conversion against selectivity (Fig. 7) illustrates this trend. It is apparent from the data in Fig. 7 that most of the HCAM is formed in the initial stages of the reaction and that very little CAM is converted to HCAM thereafter. The implication of this is that the catalyst has some initial activity for the hydrogenation of the C = C bond but is modified during the reaction so that it becomes almost 100% selective for the hydrogenation of the C = O bond. This result and the results of further experiments to be reported elsewhere point to a steric ligand effect which further enhances the selectivity of Ir for the hydrogenation of CAM.

5 CONCLUSIONS

The selective reduction of cinnamaldehyde was carried out over a graphite supported iridium catalyst. The effects of catalyst pre-reduction, reaction temperature, hydrogen pressure and cinnamaldehyde concentration were studied. The pre-reduction of the catalyst plays a fundamental role in the reaction: Unreduced catalysts are less active and much less selective to the desired products. A treatment with hydrogen at high temperature for a short period of time dramatically increased the rate of reaction and the selectivity to cinnamyl alcohol.

Reactions carried out in the range between 85 and 130 °C showed selectivities over 90% and increasing selectivity with cinnamaldehyde conversion. Arrhenius behaviour was observed in the range of temperatures studied with an apparent activation energy of 37 kJ mol^{-1}. The reaction was found to be first order in hydrogen and zero order in cinnamaldehyde.

There is an indication that a steric ligand effect may be responsible for the increasing selectivity to cinnamyl alcohol with increasing reaction time.

Acknowledgements

The authors gratefully acknowledge the support of Johnson Matthey in funding this project.

References

1. A. Giroir-Fendler, D. Richard, and P. Gazellot, *Het. Catal. and Fine Chem.*, (1988) 171-178.
2. L. Mercadante, G. Neri, C. Milone, A. Donato, and S. Galvagno, *J. Mol. Catal. A*, **105** (1996) 93-101.
3. U. Shröder and L. de Verdier, *J. Catal.*, **142** (1993) 490-498.
4. P. Gallezot and D. Richard, *Catal. Rev. Sci. Eng.*, **40** (1998) 81-126.
5. A.B. da Silva, E. Jordão, M.J. Mendes, and P. Fouilloux, *App. Catal. A*, **148** (1997) 253-264.
6. G.J. Hutchings, F. King, I.P. Okoye, M.B. Padley, and C.H. Rochester, *J. Catal.*, **148** (1994) 464-469.
7. H. Li, X. Chen, M. Wang, and Y. Xu, *Appl. Catal. A*, **225** (2002) 117-130.
8. P. Reyes, M.C. Aguirre, J.L.G. Fierro, G. Santori, and O. Ferreti, *J. Mol. Catal.*, **184** (2002) 431-441.
9. W. Yu, Y. Wang, H. Liu, and W. Zheng, *J. Mol. Catal. A*, **112** (1996) 105-113.
10. M. Arai, H. Takahashi, M. Shirai, Y. Nishiyama, and T. Ebina, *App. Catal. A*, **176** (1999) 229-237.
11. U.K. Singh and M.A.Vannice, *J. Catal.*, **191** (2000) 165-180.
12. L. Theodoulou, *PhD Thesis*, University of Reading (2001).
13. T. Birchem, C. M. Pradier, Y. Berthier, Y. Berthier, and G. Cordier, *J. Catal.*, **161** (1996) 68-77.
14. U.K. Singh and M.A. Vannice, *Appl. Catal. A*, **213** (2001) 1-24.
15. F. Delbecq and P. Sautet, *J. Catal.*, **152** (1995) 217-236.

STUDY OF CATALYZED WALL-FLOW AND FOAM-TYPE FINE PARTICULATE FILTERS

Athanasios G. Konstandopoulos[1], Dimitrios Zarvalis[1]
John M. McNamara[2], Stephen Poulston[2], Raj R Rajaram[2]

[1]Aerosol & Particle Technology Laboratory, CERTH/CPERI, PO Box 361, 57001 Thermi Thessaloniki, Greece
[2]Johnson Matthey Technology Centre, Blount's Court, Sonning Common, Reading RG4 9NH. UK

1 INTRODUCTION

There is widespread concern about the threat posed to human health by the emission of particulates from internal combustion engines. This concern focuses not only on the mass of particulate emitted, which is already legislated, but also on the number and size of these particles, which may be the object of future legislation. In particular there is concern that small particles may cause respiratory disease[1] and so require the use of particulate control systems. For example, gasoline vehicles produce far less particulate than diesel in terms of mass, but it is reported that under certain conditions the number of small particles they emit can be comparable[2], therefore particulate emission standards for all vehicles could become effective in the not-too distant future. Compliance with such future standards is likely to require the deployment of particulate control devices. While several technologies could be employed particulate filters appear a promising option, as in the case of heavy duty diesel vehicles. One of the key requirements of a particulate filter system is effective regeneration, preferably during normal engine operation. In the environment of stoichiometric engine exhaust, particulate filter regeneration needs to be effected almost in the absence of oxygen, and this represents the motivation for the present research. In this study a number of catalysts were compared for their promotion of the reactions of oxygen, water and carbon dioxide with soot expressed by reactions (1)-(5).

$$C + O_2 \rightarrow CO_2 \qquad \text{... (1)}$$
$$C + CO_2 \rightarrow 2CO \qquad \text{... (2)}$$
$$C + H_2O \rightarrow CO + H_2 \qquad \text{... (3)}$$
$$C + 2H_2O \rightarrow CO_2 + 2H_2 \qquad \text{... (4)}$$
$$2C + 2H_2O \rightarrow CO_2 + CH_4 \qquad \text{... (5)}$$

The reactions of carbon dioxide and water with carbonaceous species are well known reactions and have been employed in various chemical processes to control or remove unwanted carbonaceous material[3]. The most important application is found in the gasification of coal[4], forming a mixture of hydrogen and carbon monoxide (syngas), which is also used in important industrial processes such as methanol synthesis and in the production of hydrocarbons (Fischer-Tropsch synthesis). In these reactions, alkali metal catalysts and promoters have been used extensively for their ability to activate water and carbon dioxide species, thus speeding up their reaction with carbon. Alkali metals are therefore a clear candidate for this study. A number of mechanisms have been suggested for the catalytic effect of alkali metals on these reactions:

1. Contact between the soot particles and the active phase is achieved through the mobility of alkali-metal ions. These ions act as oxygen transfer agents, which convert soot to carbon dioxide, before themselves being re-oxidised by gas phase oxidants (oxygen, water and carbon dioxide).
2. In the presence of a reactive metal oxide support, the alkali metal can be re-oxidised by oxide ions from the support. The gas phase oxidants can then replenish the oxide ions within the support, as has been demonstrated for reduced ceria[5].
3. A reactive metal oxide support may directly oxidise the soot to carbon dioxide. The alkali-metal ions again act as oxygen transfer agents, by activating the gas phase oxidants and re-oxidising the support.

For this study catalytic materials were prepared and tested in two forms. First, model studies using the powder form of the catalyst mixed with carbon black with micro-reactor testing. Second the catalysts were coated on porous filters and they were tested in an engine test cell. Previous studies have shown that the nature of the contact between the catalyst/coated filter and the soot is of crucial importance in the kinetics of soot oxidation reactions[5,6]. The most convincing test of the true activity of these materials with exhaust soot therefore requires that combustion-generated soot aggregates are deposited in the filter structure under realistic conditions employing an engine based test cell. Both the precise catalysts formulations and the method of coating are proprietary information.

2 METHOD AND RESULTS

2.1 Catalyst powders

The catalyst/soot samples for testing were prepared by lightly mixing the catalysts with soot using a mortar and pestle in order to achieve loose contact between the catalyst and the soot. Tight contact, obtained by heavy grinding, would not provide a suitable simulation. The catalyst/carbon mixtures were composed of 100mg of catalyst and 30mg of BP2000 carbon (high surface area graphite). BP2000 was chosen as it was found to combust at a temperature closer to that of particulate collected on a diesel filter than some other lower surface area more graphitic synthetic carbons.

Sample testing was carried out using thermogravimetric analysis (TGA). This technique was used in the reaction of carbon with oxygen and carbon dioxide. Temperature programmed oxidation (TPO) experiments were carried out to measure the effect of water as a source of oxygen for combustion. TPO involves passing helium through a water saturator at room temperature and then over the catalyst. The production of carbon dioxide and carbon monoxide is then monitored as a function of temperature using a mass spectrometer (VG Gaslab 300).

Table 1 *Conditions used for combustion reactions*

Technique	Gases	Flow rate (ml min^{-1})	Final temperature (°C)	Temperature ramp rate (K min^{-1})
TGA	Air	100	700	15
TGA	10% CO_2/Ar	100	1200	15
TGA	Ar	100	1200	15
TPO	H_2O/He	75-80	900	10

In order to obtain more fundamental catalytic activity data of the catalytic materials of interest a number of model catalysts consisting of alkali metal and precious metal were prepared and tested for their ability to promote the reactions of water and carbon dioxide with solid carbon. These tests provide basic information about the ability of the catalysts to catalyse soot combustion with CO_2, H_2O and O_2. Results are summarized in Table 2. Both alkali metal and precious metal (PM) doped supports were used. Two supports were used which can be categorised as an inert and a reducible oxide support. Clearly the presence of the alkali metal has a significant effect on catalysing the soot combustion as anticipated. The effect of the reducible oxide support is not significant. In addition to the experiments summarised in Table 2 two further samples of alkali metal supported on an alumina foam and cordierite wall flow filter were prepared and coated with soot in a similar manner to that described above. Measurement of the soot combustion characteristics of these samples in O_2, CO_2 and H_2O were very similar to the powder samples.

Table 2 *Combustion characteristics of fresh alkali-metal catalysts in oxygen, H_2O and CO_2. T_{ig} is the temperature at which soot oxidation begins.*

Catalyst support	O_2 T_{ign} /°C	H_2O T_{ign} /°C	CO_2 T_{ign}/°C
No catalyst	580	850-900	980
Alkali metal / Inert support	410	505	731
Reducible oxide	573	>850	-
Reducible oxide + alkali metal	407	550	712
Reducible oxide + alkali metal + PM	378	-	754

2.2 Coated Filters

A number of catalyst formulations based on reducible oxides, promoted with alkalis and precious metals were prepared using cordierite wall flow and foam type 25 mm diameter x 50 mm long filter samples.

Seven samples were prepared for evaluation. Six of the samples under consideration (No. 1 to No. 6) were small cordierite (Corning EX-80 type) monoliths (diameter: 24.8 mm, length: 52 mm, cell density 200 cpi) and one sample was an alumina foam monolith (diameter: 25.2 mm, length: 53 mm). Table 3 shows the catalyst applied to each sample tested. The catalyst formulations were similar to those used in 2.1 above in terms of the active components and their relative loading.

Table 3 *Catalyzed samples tested*

Sample No.	Catalyst
1	Blank cordierite monolith
2	Alkali metal
3	Reducible oxide
4	Reducible oxide / alkali metal
5	Reducible oxide / PM
6	Reducible oxide / PM / alkali metal
7	Alkali metal

2.3 Permeability measurements

The testing programme initially involved measurements of the permeability and inertial loss coefficients of the catalytic filters and their pressure drop behaviour to determine the influence of the catalytic coating on the filter flow resistance, according to the methodology in reference [7]. The filter sample was placed with the aid of a special holder (reactor) into the testing unit for permeability measurement and the pressure drop across it was measured at various flow rates employing air as the fluid. The test is done at ambient temperature. Figure 1 shows the experimental setup.

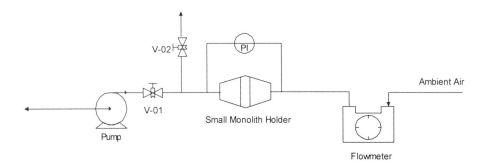

Figure 1   *Flow diagram of the experimental set-up used for permeability*

measurements.

The data are interpreted using published theoretical models[8,9] and the *permeability*, k (m^2) and *contraction/expansion inertial loss coefficient* ζ, (dimensionless) of the monolith sample are determined:

$$\Delta P = \frac{\mu Q}{2V_{trap}}(\alpha + w_s)^2\left[\frac{w_s}{k\alpha} + \frac{8FL^2}{3\alpha^4}\right] + \frac{\rho Q^2(\alpha + w_s)^4}{V_{trap}^2\alpha^2}\left[\frac{\beta w_s}{4} + 2\zeta\left(\frac{L}{\alpha}\right)^2\right] \quad (1)$$

The permeability of the foam monolith was calculated using the Darcy-Forchheimer law and the *permeability and Forchheimer coefficient* of the filter are determined by regression using equation (2). Table 4 summarizes the results for each filter sample tested.

$$\Delta P = \frac{1}{k}\frac{Ql\mu}{A} + \beta\rho l\left(\frac{Q}{A}\right)^2 \quad (2)$$

Figure 2 presents the evolution of pressure drop versus flow rate for each sample under consideration. The catalytic coating had a measurable effect on the permeability of most samples (a maximum reduction of about 40%) but this will not affect adversely their pressure drop under soot loading, since most of the pressure drop of the loaded filter comes from the soot cake built up on the wall.

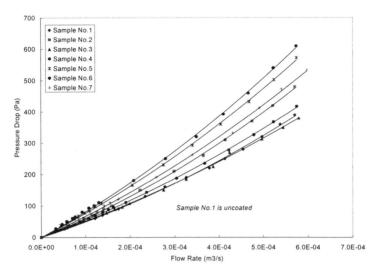

Figure 2   *Effect of catalytic coating on pressure drop of clean monolith*

Table 4   *Permeabilities and inertial loss coefficients*

Sample No.	k [m²]	ζ	β [m⁻¹]
1	8.50e-13	8	-
2	8.41e-13	27.8	-
3	8.40e-13	4.5	-
4	7.34e-13	9.2	-
5	5.82e-13	11.7	-
6	5.11e-13	10.7	-
7	2.8e-10	-	5.4e5

2.4 Filter Collection Efficiencies Measurements

Filter collection efficiency was evaluated by comparing filter upstream and downstream particle concentrations measured by an SMPS (Scanning Mobility Particle Sizer) consisting of a long DMA (Differential Mobility Analyzer – Model 3080) and a U-CPC (Ultrafine Condensation Particle Counter – Model 3025). The cordierite wall-flow filters exhibited their well-known soot collection efficiency while the foam sample exhibited a lower collection efficiency.

In Figure 3 a typical result of the transient evolution of size specific collection efficiency for the cordierite sample No. 2 is presented. The decrease of the filter permeability due to the coating is beneficial for the collection efficiency of the coated filters and makes them approach the cake filtration regime more quickly.

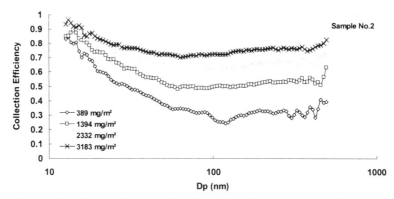

Figure 3   *Transient evolution of number—based size specific collection efficiency versus soot mass challenge for sample No. 2.*

2.5 Catalyst Screening

Since soot emission rates from gasoline engines are not high enough to provide fast loading in 1-2 hours of filters for screening tests, soot particles emitted from a passenger car turbo-charged direct injection diesel engine with a 1.9L displacement (rated at 66 kW/4000 rpm) operated on a dynamometer under well-defined conditions were employed to load the catalyzed filter samples in a specially-built filter testing unit.   Filter regeneration was effected by switching the inlet stream to the filter samples to the exhaust of a stoichiometric 4-stroke gasoline engine (4 kW) operated at full load (oxygen content <0.1 %). All samples of Table 3 with the exception of sample No.2 were screened for their loading and regeneration behaviour. The filters were exposed to a part of the exhaust flow of the engine which was operated at 2400 rpm and 50% load.

For the regeneration, the monoliths were exposed to a part of the exhaust flow from a 4-stroke gasoline engine with a 163 cm^3 displacement (rated at 4 kW/4000 rpm) operated at 80% load. The O_2 content in the exhaust was kept for all experiments at 0.1 %.

The alternate exit channels of each filter sample (with the exception of the foam sample) were plugged and packed inside stainless steel cartridges, wrapped with heat expanding fibrous mat (3M Interam). The filter cartridge was placed inside a tube furnace that allows the control of its temperature. The soot reactor unit is instrumented with thermocouples and incorporates a by-pass valve that permits isolation of the filter from the exhaust flow under an inert (N_2) atmosphere during switching from diesel to gasoline exhaust exposure.   A heated tube before the reactor heats the exhaust gases to the desired temperature.   The unit sits on a bench that incorporates sampling pumps and ancillary controllers and transducers to maintain constant volume flow through the filter. A dedicated data acquisition system collects temperature, pressure and exhaust gas analysis data from the unit. In Table 5 the soot loading conditions for each filter are summarized.

Figure 4 represents the filter samples loading behavior. The dips in pressure drop are due to pressure disturbances during sampling of the filter soot loading rate. In all wall-flow filter samples a well-defined soot cake build-up region is setting in (linear part of the pressure drop curve) after a relatively fast 'deep-bed' loading transition during the first 500 sec or so of each the experiments.   The monolithic filters exhibit a loading behavior that scales inversely with their clean permeability and have more or less the same pressure drop slopes, as expected for cake filtration under equivalent exposure conditions.

Table 5 *Loading conditions for each of the tested samples*

Sample No.	Filtration Velocity(cm/s)	Loading Temperature (°C)	Filter loading rate (mg soot /min)
3	3.34	363	1.56
4	4.05	364	1.52
5	4.14	357	1.46
6	4.47	361	1.43
7	128	363	1.43

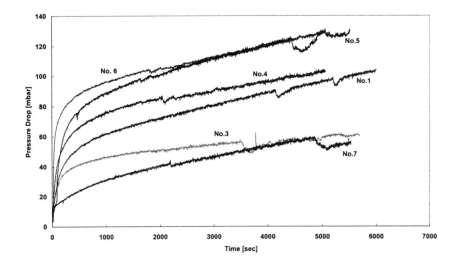

Figure 4 *Soot loading behavior of filter samples.*

For the study of regeneration, the soot loaded filters were exposed to the flow of gasoline engine exhaust and the temperature was ramped from about 410°C (exhaust temperature of gasoline exhaust) to 820°C at a rate of 3°C/min. Experimental conditions are shown in Table 6 while the regeneration results are presented as temporal and thermal traces of the pressure drop in Figures 5 and 6. Soot conversion cannot be directly determined since the production of CO_2 and CO during regeneration is not measurable against the gasoline CO_2 and CO emissions background. However, soot oxidation can be inferred by the measured pressure drop as a function of temperature. With a constant soot loading the pressure drop would be expected to increase gradually with temperature due to the increase of viscosity and the decrease of exhaust density, a falling pressure drop therefore represents soot oxidation. Through an independent experiment with oxygen content of 0.5% it was established that the exhaust HC and CO at the particular regeneration operation point consume 0.1% of oxygen content by the time the filter temperature reaches 650°C. Therefore performing the regeneration experiments with oxygen content of 0.1% guarantees that insignificant amount of oxygen remains in the exhaust above 650°C.

Table 6 *Regeneration conditions*

Sample No.	CO_2 (%)	CO (%)	O_2 (%)	NO (ppm)	Total HC (ppm)	Exhaust Flow (std lpm)
3	13.1	4.6	0.1	849	444	15.5
4	12.8	4.7	0.09	776	467	15.9
5	12.9	4.8	0.1	760	476	16.4
6	12.9	4.7	0.09	798	480	16.5
7	12.7	4.8	0.09	709	478	16.2

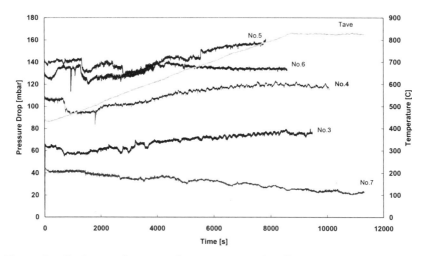

Figure 5   *Evolution of pressure drop versus time for all samples during regeneration*

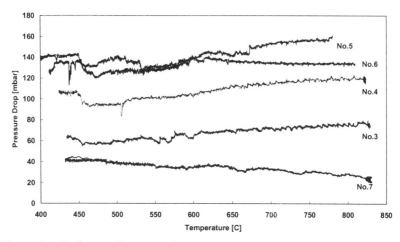

Figure 6   *Evolution of pressure drop versus temperature for all during regeneration*

The drop in pressure seen around 450° C in samples No 4-6 is most likely due to localized oxidation of soot particles that are deposited in good contact with the catalyst inside the filter wall, and can oxidize by the catalyst lattice oxygen.

Due to space limitations it is not possible to show the evolution of gaseous concentrations during all regeneration experiments. A representative evolution of gaseous concentrations versus temperature, measured downstream of the filter samples (Sample No. 6) is presented in Figure 7. As explained above all available oxygen under the conditions of the experiment is consumed by CO and THC oxidation below 650°C.

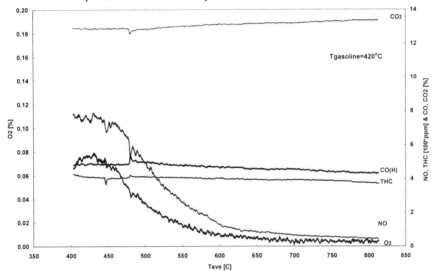

Figure 7 *Evolution of gaseous emissions versus temperature during regeneration of Sample No. 6.*

3 DISCUSSION-CONCLUSIONS

Of the different catalytic coatings tested sample No. 3 did not show any evidence of soot oxidation. Incipient soot oxidation was evident on samples No. 4, 5 (a less steep pressure drop trace increase than sample No. 3). Samples No. 6, 7 (foam) exhibited clearly soot oxidation as their pressure drop trace was decreasing with temperature. All pressure drop decreases with the catalytic filters have been observed above 650°C under oxygen deficient conditions and therefore it can be stated that the soot is oxidized catalytically by other species than oxygen in these experiments. The best catalyst formulation appears to be the combination of reducible oxide/alkali metal and precious metal.

In conclusion, some of the catalysts tested exhibit a definite soot oxidation activity in the absence of oxygen, although expectedly its rate is substantially less than the rate of soot oxidation by oxygen. The mass of particulate generated by stoichiometric internal combustion engines is however small. In addition though the temperatures required for soot oxidation in the absence of oxygen are high they are within the limits of achievable temperatures on catalysts in stoichiometric gasoline engine powered vehicles. The feasibility of soot oxidation in oxygen deficient gasoline exhaust by other than oxygen species (e.g. CO_2, H_2O) under realistic conditions is thus demonstrated. Future research should continue towards increasing the activity and durability of the developed catalysts under oxygen deficient conditions. Methods for combining these soot oxidation catalysts with novel particulate trapping devices have also begun and are continuing[10].

　　　　　　　　　　　　　　　　　　　　　　　　　　　　Catalysis in Application

References

1. D.E. Hall, D.J. King, T.B.D. Morgan, S.J. Baverstock, P. Heinze, B.J. Simpson, *SAE Tech. Paper*, 982602, 1998
2. D.E. Hall, C.L. Goodfellow, P. Heinze, D.J. Rickeard, G. Nancekievill, G. Martini, J. Hevesi, L. Rantanen, P.M. Merino, T.B.D. Morgan, P.J. Zemroch, *SAE Tech. Paper*, 982600, 1998.
3. S.R. Vatcha, Catalytica, Studies Division, California, 1993
4. K. Nagase, T. Shimodaira, M. Itoh, Y. Zheng, *Phys. Chem. Chem. Phys.*, 1999 **1**, 5659
5. A.G. Konstandopoulos, M. Kostoglou, *Combustion and Flame* 2000 Paper No. 121, pp 488-500.
6. J.P.A. Neeft, M. Makkee, J.A. Moulijn, *Chem. Eng. J.*, 1996 **64**, 295.
7. A.G. Konstandopoulos, *SAE Tech. Paper* 2003-01-0846, 2003.
8. A.G. Konstandopoulos, J.H. Johnson, *SAE Trans.* **98** sec. 3 (*J. Engines*) Paper No. 890405, 1989, pp625-647.
9. A.G. Konstandopoulos, M. Kostoglou, E. Skaperdas, E. Papaioannou, D. Zarvalis, E. Kladopoulou, *SAE Tech. Paper* 2000-01-1016, 2000.
10. J.M. McNamara, R.I. Crane, L. Rubino, C. Arcoumanis, S.E. Golunski, S. Poulston, R.R. Rajaram, *SAE Tech. Paper*, 2002-01-1894, 2002.

Abbreviations

A (m^2) : exposed surface to the flow and

ΔP (Pa) : pressure drop across the filter

Q (m^3/s) : ambient air flow through the monolith,

L (m) : length of the monolith,

μ : viscosity of the gas flow (Pa·s),

w_s (m) : filter wall thickness,

α : monolith cell size,

V_{trap} (m^3) : monolith effective volume,

F : factor equal to 28.454,

k (m^2) : permeability of the monolith,

ζ (-) : contraction/expansion inertial loss coefficient (not applicable to foam sample)

ρ (kg/m^3) : exhaust gas density

β (m^{-1}) : Forchheimer coefficient of the porous wall (insignificant for EX-80 monoliths)

A NOVEL "THRIFTED" PALLADIUM-ZINC CATALYST SUPPORTED ON CERIA STABILISED ZIRCONIA FOR USE IN THREE-WAY VEHICLE EXHAUST CATALYSIS

J.Thomson[*], P.J.C.Anstice, and R.D.Price.
Division of Physical and Inorganic Chemistry,University of Dundee,
Dundee, DD1 4HN

([*]correspondence e-mail address: j.z.thomson@dundee.ac.uk)

1 INTRODUCTION

The catalytic removal of vehicle exhaust gases (hydrocarbons, carbon monoxide and oxides of nitrogen) is an important process in the control of atmospheric pollution in the urban environment. The catalytic processes operating in a conventional system normally requires two noble metals i) platinum or platinum/palladium to perform the oxidation of hydrocarbons and carbon monoxide to carbon dioxide and steam and ii) rhodium to perform the reduction of the oxides of nitrogen by hydrocarbons or carbon monoxide to dinitrogen, carbon dioxide and steam [1-9]. The conventional loading of these metals is typically *ca* 120 g (cu ft of monolith)$^{-1}$ of platinum or a mixture of platinum and palladium and *ca* 80 g (cu ft of monolith)$^{-1}$ of rhodium. These metals are normally supported on a lanthanide oxide such as zirconia doped ceria which acts as an oxygen storage phase, mobilises anionic oxygen to the combustion sites and also stabilises the metal components against sintering [10]. The entire catalytic system is coated onto a cordierite or metal monolith, containing a γ-alumina coating. The catalyst formula operating in the temperature range of 400-700^{0}C gives *ca* a 98% reduction of the polluting eluent.

Vehicle exhaust emission standards and conversion requirements are progressively becoming more stringent, owing to the respective environmental legislation imposing lower emissions criteria. Hence, greater demands are placed on the catalyst manufacturers to increase performance. The imminent introduction of Euro IV legislation halves the emissions threshold for hydrocarbons (HCs) to 0.1 g km^{-1}, for carbon monoxide to 1.0g km^{-1} and for oxides of nitrogen (NO$_x$) to 0.08 g km^{-1}. This increased demand in catalyst performance is compounded by a constant increase in the cost of the noble metal component of the vehicle exhaust catalyst, resulting in a desire by the manufacturers to "thrift" the metal loading of any potential catalyst system, while continuing to deliver ever better performance in conjunction with lower noble metal loadings.

This paper describes the development and performance of a 16 mol% ceria stabilised zirconia supported palladium-zinc catalyst in vehicle exhaust after treatment. The main aspects of the novel catalytic system are i) the low temperature preparation of a thermally stable solid solution of tetragonal phase ceria stabilised zirconia, ii) the rôle of zinc in the dispersion of the palladium component and its synergetic interaction with palladium in an oxygen ion transfer mechanism, iii) the *in situ* perfomance of the catalyst under "close-coupled" engine conditions in producing ammonia after "light-off" and iv) a thermodynamic chemical process which results in a reduction in temperature of the catalyst eluent gas by ca 150^0C relative to that observed for a commercial system.

2 METHODS AND RESULTS

2.1 Low Temperature Preparation of the Ceria Stabilised Zirconia Phase

Zirconia exhibits three polymorphs namely the monoclinic, tetragonal and cubic phases [11]. The conventional preparation of a dopant stabilised zirconia for the desired tetragonal phase can involve i) ball milling the respective dopant oxide in the presence of the zirconia phase followed by a high temperature anneal, or ii) the co-precipitation of the dopant cation in the presence of zirconia or iii) by the thermal decomposition of a mixture of lanthanide and zirconium salts [12-14]. The doped zirconias are then fired to *ca* 1000^0C to allow the diffusion of the dopant into the zirconia crystal lattice. The purpose of the dopant is to stabilise the surface area of the zirconia against sintering and to stabilise a metastable tetragonal phase of zirconia at the temperatures between $400\text{-}1100^0C$ [15]. The co-impregnation method described here involves the application of sol technology to yield a solid solution of ceria doped zirconia.

An inorganic sol of ceria stabilised zirconia is prepared by the addition of a stoichiometric quantity of zirconium carbonate (MEL Chemicals plc, Manchester, England U.K.) to a solution of 50% nitric acid (BDH Chemicals, England) to give an acid solution of zirconyl nitrate. Cerium nitrate hexahydrate (Aldrich Chemical Co, England, UK) was added to give a 16 mol% ratio w.r.t. the zirconium concentration. The sol was digested for 2h at 80^0C ensuring that no cloudiness or a precipitative phase being formed. Samples of this mother sol were collected for spectroscopic analysis with the rest of the liquor being co-impregnated with palladium nitrate hexahydrate (Aldrich Chemical Co., 5wt% w.r.t. zirconia content) and zinc nitrate hexahydrate (Aldrich Chemical Co., to give Pd:Zn=0.5).

The XRD of the calcined ceria stabilised zirconia are given below (Figure 1).

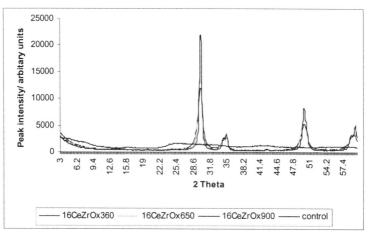

Figure 1   *The XRD diffraction pattern for 16 mol% ceria stabilised zirconia over the temperature range 360-900⁰C.*

The solid solution produced by a low temperature evaporation of the sol gives initially an amorphous state. Calcination at 360^0C produces reflections consistent with the tetragonal phase of $Ce_{0.16}Zr_{0.84}O_2$ [15]. The diffraction pattern shows that the inorganic sol procedure exclusively yields the metastable tetragonal phase of ceria stabilised zirconia ($Ce_{0.16}Zr_{0.84}O_2$). The diffraction pattern also confirms that the solid solution is stable to thermal stress over the temperature range 350-900^{0}C with no separation of the ceria dopant to give discreet regions of ceria during the calcinations procedure.

The addition of the zinc doped palladium, as the nitrates, to the co-impregnated 16 mol% ceria stabilised zirconia sol results in a highly dispersed form of palladium (Figure 2).

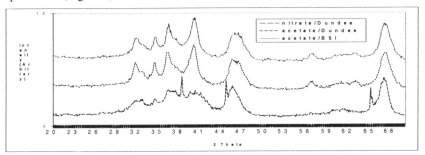

Figure 2   *XRD diffraction pattern for γ-alumina supported palladium/zinc (mole atom ratio=1:2). Each sample prepared by co-impregnation methods and calcined in air at 250^0C followed by reduction in 5% dihydrogen at 350^0C, with 30 min anneal steps at 50^0C intervals.*

The XRD diffraction pattern produced from the acetate precursors shows peaks at the 2σ value of 39^0 accords with that of the palladium (111) plane [7]. This reflection plane is not observed in the sample prepared from the nitrate precursor. The sharp peaks at 2σ values of 64.9^0 and 45^0 confirms the presence of zinc(II) oxide and are notably absent in the spectra produced from the acetate precursors. The reflections at 35^0 and 57.5^0 confirm the presence of palladium(II) oxide in the sample, and are of greater intensity in the samples prepared from the acetate precursors. The XRD analysis shows that using the nitrate precursors and calcining in air followed by a reduction in dihydrogen, results in the dispersion of the palladium component. The XRD analysis of the palladium component is consistent with either i) a higher dispersed palladium(II) phase or ii) contains a lower concentration of palladium(II)oxide. Hence, the XRD analysis shows that catalyst prepared by co-impregnation of palladium with zinc from the acetates results in the catalyst containing a $ZnO \Leftrightarrow Zn^0 \Leftrightarrow Pd^0 \Leftrightarrow PdO$ equilibria which are to the right. The XRD of Pd-Zn(0.5)/CeO$_2$/γ-Al$_2$O$_3$ is presented in figure 3. The XRD spectrum confirms that the sample, prepared from the nitrates, contains reflections at 41.2^0 and 44.1^0 assigned to the (111) and (020) planes of the PdZn alloy [16]. Hence we can conclude that the preparative method for the supported catalyst utilising an inorganic sol route, gives $ZnO \Leftrightarrow Zn^0 \Leftrightarrow (PdZn)^0 \Leftrightarrow Pd^0 \Leftrightarrow PdO$ supported on a solid solution of $Ce_{016}Zr_{0.84}O_2$.

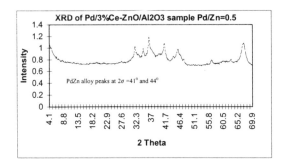

Figure 3 *XRD of Pd/Zn (0.5)-CeO₂/γ-Al₂O₃ sample confirming presence of PdZn alloy.*

2.2 Engine Test Data

Emissions data produced from a "close-coupled" position of the coated catalyst monolith are presented in Table 1.

Table 1. *Emissions data from Production vs $PdZn(0.5)-Ce_{0.16}Zr_{0.84}O_2/\gamma-Al_2O_3$ formulation fitted to an Astra 1.6l engine running on RON 98 fuel. Monolith details; material = NKG cordierite, volume = 1.7l; loading = 40g Pd/ft^3; close coupled: Ageing lean spike engine temperature=940^0C' t= 120 h. No ECU chip reprogramming for $PdZn(0.5)-Ce_{0.16}Zr_{0.84}O_2/\gamma-Al_2O_3$ system was performed.*

Catalyst	Test No.	NOx (g/km)	HC (g/km)	CO (g/km)	CO₂ (g/km)	NH₃ (g/km)
Production	1	0.0049	0.0296	1.0658	188.1	
	2	0.0044	0.0291	1.0635	188.6	
Aged Production	1	0.0107	0.0398	1.1323	193.67	
	2	0.0181	0.0369	0.9338	194.34	
Nanocat	1	0.3801	0.0510	1.832	187.07	0.0127
	2	0.3801	0.0522	1.751	184.58	0.0200
Aged Nanocat	1	0.6709	0.0652	2.196	188.63	0.0145
	2	0.6571	0.0671	2.1702	190.04	
EU 3 limit		0.15	0.2	2.3		
EU 4 limit		0.08	0.1	1.0		
Production Deactivation		3.06	1.3	0.97		
Nanocat Deactivation		1.60	1.28	1.07		

The eluent from the $PdZn(0.5)/Ce_{0.16}Zr_{0.84}/\gamma-Al_2O_3$ formulation achieves Euro 4 levels in hydrocarbons for the fresh and aged condition and Euro 3 level for carbon monoxide as fresh and aged condition. In the absence of rhodium from the catalyst formula, NOx conversion is relatively low, however two important results are worthy of mention at this point. For the case of thePdZn(0.5)-$Ce_{0.16}Zr_{0.84}O_2/\gamma-Al_2O_3$ formulation, the presence of ammonia (0.015 g km⁻¹, Table 1) is observed after "light-off". This is an important catalytic product for use in SCR catalysis (Selective Catalytic Reduction of NOₓ). The ageing process does not greatly reduce the formulation's ability to produce ammonia. Ageing analysis confirms that the deactivation ratio for PdZn(0.5)-Ce$_{0.16}$Zr$_{0.84}$O$_2$/γ-Al$_2$O$_3$ is 1.60 for nitrogen species, whereas the production catalyst gave a deactivation ratio of 3.06 under the same test conditions (Table 1).

Figure 4 shows the catalytically produced ammonia eluent concentration relative to that of the production formula. Ammonia is observed in the eluent stream from the production catalyst, reducing in concentration after "light-off", asymptotic with time-on-line. The PdZn(0.5)-Ce$_{0.16}$Zr$_{0.84}$/γ-Al$_2$O$_3$ formulation shows the opposite effect, in which the ammonia concentration increases after "light-off" and remains constant during the run. The presence of ammonia in the eluent is consistent with the surface chemistry of the respective production and PdZn(0.5)-Ce$_{0.16}$Zr$_{0.84}$O$_2$/γ-Al$_2$O$_3$ formulations operating through differing adsorbed states and reaction mechanisms. The supporting details of this surface chemistry will be published shortly.

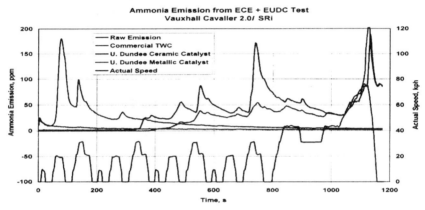

Figure 4: *NH$_3$ emissions from Production vs PdZn(0.5)-Ce$_{0.16}$Zr$_{0.84}$/γ-Al$_2$O$_3$ formulation*

The second important data set for the engine test provides confirmation that the PdZn(0.5)-Ce$_{0.16}$Zr$_{0.84}$O$_2$/γ-Al$_2$O$_3$ formulation runs cooler ,for the same inlet gas temperature, relative to that observed for the production catalyst (Figures 5a and 5b below).

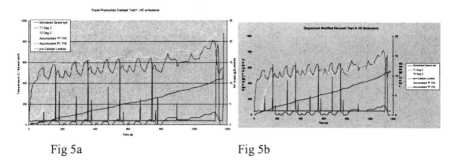

Fig 5a   Fig 5b

Figures 5a and 5b *Examples of the pre and post catalyst temperatures for the production vs PdZn(0.5)-Ce$_{0.16}$Zn$_{0.84}$/γ-Al$_2$O$_3$ formulations respectively. Monoliths positions close-coupled.*

Figure 5a shows that the inlet gas temperature to the production catalyst at *ca* 500^0C increased to around 800^0C over the period of the test. The catalyst performs the hydrocarbon conversion to Euro 4 performance. Unsurprisingly the production catalyst generates the expected exotherm such that the eluent gas maintains a temperature of *ca* 100^0C above the inlet gas temperature. For the PdZn(0.5)-$Ce_{0.16}Zr_{0.84}O_2$/γ-Al_2O_3 formulation (Fig 5b), the inlet gas temperature is the same as that for the production system, however, the eluent gas temperature is around 50^0C lower than the inlet gas temperature thus giving a decrease in the temperatures experienced by the catalyst of *ca* 150^0C relative to that of the production formulation. The lowered temperature experienced by the surface results in less thermal stressing of the catalyst as supported by the catalyst deactivation results (Table 1).

2.3 NO_x Removal: The Addition of a "Thrifted" Rhodium Catalyst Coat

The final stage in the development of the novel "thrifted" three-way-catalyst was to add a "thrifted" rhodium coating to that of the PdZn(0.5)/$Ce_{0.16}Zr_{0.84}O_2$/ γ-Al_2O_3 coating previously described in sections 2.1 and 2.2. The rhodium coating was prepared by the addition of 1g of rhodium(III) nitrate hydrate with 1.4l of the mother liquor of 16 mol% ceria stabilised zirconia. A cordierite monolith containing a loading of 40 g cu ft of palladium was coated to give a rhodium loading of 2.3 g cu ft of monolith. The Pd/Rh atom ratio being 20:1. Figure 6a and 6b give the NO_x and CO eluent analysis from a 1.6l VW engine running on RON 95 fuel. Figure 6a gives the NO_x and CO eluent concentrations without the rhodium loading applied as an overcoat, and figure 6b shows the eluent concentrations from the catalyst with the rhodium overcoat. The mean NO_x emissions at *ca* 120 ppm (Figure 6a). For the rhodium impregnated system (2.0 g / cu ft monolith; Figure 6b) the NOx emissions fall to *ca* 60 ppm.

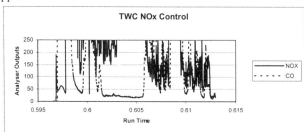

Figure 6a: *NO_x and CO eluent concentration from a 40g palladium loading of PdZn(0.5)/$Ce_{0.16}ZrO_{0.84}O_2$/γ-Al_2O_3 coating.*

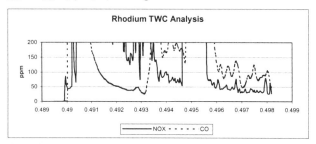

Figure 6b: *NO_x and CO eluent concentration from a 40 g palladium /cu ft loading and a 2.3 g rhodium /cu ft supported.*

The data sets from this initial engine work shows that the "thrifted" rhodium coating at this loading reduces the NO_x eluent from the palladium only monolith by *ca* 50%. Work is underway to optimise the rhodium coating and the data will be published elsewhere.

CONCLUSION

The preparation of solid solutions of ceria stabilised zirconia from the application of an inorganic sol approach has been demonstrated. The preparative route can produce the desired metastable tetragonal phase of stabilised zirconia required for three-way-catalysis in exhaust after treatment at greatly reduced temperatures. The sol route also facilitates the impregnation of palladium in the formation of the supported metal. The addition of zinc to the mother sol disperses the palladium crystallites to form a solid solution of palladium and zinc to give the respective alloy. The $PdZn(0.5)/Ce_{0.16}Zr_{0.84}O_2$ catalyst has been shown to perform the TWC reactions in vehicle exhaust clean-up at $\lambda = 1$. The reactions produce *in situ* ammonia and also operate through an apparent endothermic process resulting in reduction of the catalyst eluent gas temperature respective to the inlet temperature of the untreated gases. The addition of a 2g / cu ft monolith of rhodium as an "overcoat" layer to the $PdZn(0.5)$-$Ce_{0.16}Zr_{0.84}O_2/\gamma$-$Al_2O_3$ layer reduces the NO_x emissions by 50%.

References
1. J.G.McCarty, *Catal. Today,* 1995, **26 (3/4)**, 283
2. X.C.Guo and R.Madix, *Catal. Lett.,* 1996, **39**, 1
3. R.Burch and T. Watling, *Catal Lett.,* 1996, **37**, 21
4. B.J.Cooper, *Platinum Metal Rev.,* 1994, **38**, 2
5. G.C.Koltsakis and A. M. Stamelos, *Prog. Energy Combust. Sci.,* 1997, **23**, 1
6. P.Degobert "Automobiles and Pollution" Edn Techn., Paris, 1995, 1
7. J.Cairns and J. Thomson, "Supported Reagents and Catalysis in Chemistry", Eds. B.K.Hodnett, A.P.Kybett, J.H.Clark and K. Smith, Royal Soc. Of Chem., London (ISBN 0-85404-797-2), 1997, 72, J.A.Cairns, J.Thomson, H.Bradshaw, P.Goulding, I.McAlpine, P.Moles, SAE Paper 970468, 1997
8. A.E.R.Budd and M. Wyatt "Air Pollution by Nitrogen Oxides, Eds. T.Sneider and L. Grant, Elsevier, Amsterdam , 871
9. J.C.Sommers and K. Baron, *J. Catal.,* 1979, **57**, 380
10. R.Franklin, P.Goulding, J.Haviland, R.W.Joyner, I.McAlpine, P.Moles, C.Norman, T.Nowell, *Catal. Today,* 1991, **10**, 405
11. E.C.Subbarao "Science and Technology of Zirconia", *Adv. Ceram.,* 1981, **3**, 1
12. R.C.Garvie, *J. Phys. Chem.,* 1965, **69**, 1238
13. J.Cuif, G. Blanchard, O.Touret, M. Marczi and E. Quemere, *SAE Tech Paper Ser.No. 969106,* 1996
14. M.Pijolat, M.Prin, M. Soustelle, O.Touret and P. Nortier, *J.Chem Soc, Faraday Trans.,* 1995, **91**, 3941
15. G.Colon, M.Pijolat, F.Valdivieso, H.Vidal, J.Kaspar, E.Finocchio, M.Daturi, C.Binet, J.C.Lavalley, R.T.Baker and S.Bernal, *J.Chem Soc., Faraday Trans.,* 1998, **94**, 3717
16. N.Iwasa,S.Masuda,N.Ogawa and N. Takewaza, *Appl. Catal. A. General,* 1994, 125, 2385

G.A. Attard[1], D.J. Jenkins[1], O.A. Hazzazi[1], P.B. Wells[1], J.E. Gillies[2], K.G. Griffin[2] and P. Johnston[2]

[1]Department of Chemistry, Cardiff University, Cardiff, CF10 3TB, UK
[2]Johnson Matthey, Orchard Road, Royston, SG8 5HE, UK

1 INTRODUCTION

The enantioselective hydrogenation of pyruvate esters catalysed by cinchona-modified supported Pt has been intensively investigated from the standpoint of molecular mechanism.[1-5] Most models conceive of the active form of the modifier (e.g. cinchonidine) as adsorbed at the Pt surface with the quinoline ring system oriented parallel to a flat metal surface; D-tracer, NEXAFS and ATRIR studies have supported this view.[6-8] Cinchonidine molecules exhibit four low energy conformations and their equilibrium populations have been measured.[4,9] The populations of the conformations in the adsorbed state are not known, but are assumed to be similar to those in solution. For cinchonidine adsorbed in the open-3 conformation, adjacent surface sites exist at which pyruvate ester may undergo selective enantioface adsorption and, by subsequent hydrogenation, may give preferential formation of one enantiomer in the product. Such 1:1 reactant-modifier interactions require a substantial area of surface involving possibly as many as 25 metal atoms.[1,4] Enantioselective reaction might thus be expected to be structure-sensitive, and indeed enantiomeric excess (ee) tends to increase with increasing Pt particle size[10]. Whether catalyst particles contain sufficiently large terraces to accommodate the proposed 1:1 complexes is uncertain.

In terms of chiral performance, cinchonidine-modified Pt/alumina catalyses the hydrogenation of pyruvate esters giving an enantiomeric excess in the lactate product in the range 65 – 80% without optimisation[1], 70 – 90% with optimisation of solvent and other conditions, and as high as 98% by use of special techniques such as ultrasonication[11] or the use of size-specific colloids.[12] These treatments again suggest the importance of Pt particle size or shape. The present work was initiated especially to explore the relationship of enantioselectivity to Pt particle morphology.

Recently we have demonstrated that the cyclic voltammogram (CV) for Pt/graphite contains features which, by analogy with those for single crystal electrodes, can be assigned to redox processes at Pt steps and terraces[13]. These aspects of catalyst morphology are therefore accessible to experimental scrutiny. It is known from surface science that Bi atoms adsorb selectively at edge sites of Pt whereas S atoms adsorb selectively at terrace sites.[14] We have therefore studied the morphology and catalytic performance of Pt-Bi/graphite and Pt-S/graphite catalysts with a view to determining the effects on enantioselectivity and other reaction parameters (rates, polymer yields) of the occupation of Pt steps and terraces by catalytically inert atoms.

Kink sites are formed in surfaces where monatomic steps intersect. Such kinks are composed of three planes having atomic densities that increase in either a clockwise or an anticlockwise manner, and hence are chiral.[15] Pt single crystals have been cut so as to expose R- or S-surfaces, and molecular recognition between a such surfaces and D- or L-sugars during glucose oxidation has been demonstrated by cyclic voltammetry.[15,16] The question arises as to whether molecular recognition occurs between alkaloid modifier and R- or S-kink sites, and whether this contributes to enantioeselectivity in pyruvate hydrogenation. Attard has shown that cinchonidine does not discriminate in its adsorption between R- and S- Pt(531)[15]; tests with other chiral crystal surfaces are reported here.

2   EXPERIMENTAL

2.1   Materials and reaction conditions

5% Pt/graphite (Johnson Matthey type 287) had a surface area of 10 m^2 g^{-1} and a mean Pt particle size of 14 nm. Platinum single crystals exposing chiral surfaces, including Pt(976)^R and Pt(976)^S, were prepared by standard procedures.

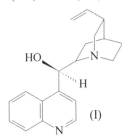

Ethyl pyruvate, MeCOCOOEt, cinchonidine (**I**), and quinuclidine, $HC(CH_2CH_2)_3N$ (Aldrich) were used as received. 10,11-Dihydrocinchonidine was prepared by hydrogenation of the vinyl group in cinchonidine in the usual way[17].

Standard reaction conditions. Catalysed reactions were conducted at 293 K in a Baskerville stainless steel reactor of volume 150 ml. 7.2 ml pyruvate ester (65 mmol) and alkaloid (0.17 mmol cinchonidine (50 mg) or 0.17 mmol quinuclidine (19 mg) or 0.17 mmol of each) were dissolved in 12.5 ml dichloromethane and added to the catalyst (250 mg) in the reactor. The reactor was operated at 30 bar in a constant pressure mode. Product analysis was by chiral gas chromatography to determine the enantiomeric excess (ee) and by GCMS to determine the masses of higher molecular weight products.

Enantiomeric excess, ee/%(R) = 100([R-] − [S-])/([R-] + [S-]).

2.2   Catalyst preparation and characterisation

Bismuth-containing catalysts. 2 g samples of Pt/graphite were immersed in a volume of 1.5mM bismuth nitrate solution for 3 h at room temperature. The catalyst was then filtered, washed in ultrapure water and dried in air at 353 K. The volumes of bismuth nitrate solution used were 2.5, 7.5, 10.0, 12.5 and 20.0 ml.

Sulphur-containing catalysts. Hydrogen and then nitrogen were passed, each for 0.5 h, through aqueous solutions of sodium sulphide in which was suspended 2 g Pt/graphite. Catalysts were filtered and dried as above.

Catalysts were characterised by cyclic voltammetry using a two compartment electrochemical cell. The working electrode consisted of a Pt-mesh basket containing 2.7 mg catalyst, the electrolyte was 0.5M sulphuric acid and the sweep rate 10 mV s^{-1}. CVs of the single crystals were obtained in 0.1M sulphuric acid at a sweep rate of 50 mV s^{-1}.

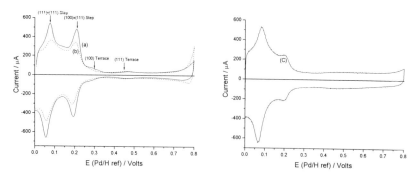

Figure 1   *CVs for Pt/graphite (curve (a)), Pt-Bi(θ_{Bi} = 0.26)/graphite (curve (b)) and Pt-S(θ_S = 0.29)/graphite (curve (c))*

CVs for the 5% Pt/graphite and the bismuthated and sulphided catalysts are shown in Figure 1.   That for Pt/graphite (curve (a)) has been reported previously and interpreted by reference to single crystal data;[13] the features at 0.06, 0.20, 0.28 and 0.47 V correspond to electrosorption sites associated with (111) x (111) step sites, (100) x (111) step sites, (100) terraces, and (111) terraces respectively.   The area of the voltammogram was reduced when Bi or S was adsorbed onto the catalyst. The coverage of Bi or S adatoms was calculated by use of the equation: $\theta_{Bi,S} = [(Q^o_H) - (Q^{Bi,S}_H)]/Q^o_H$ where (Q^o_H) and $(Q^{Bi,S}_H)$ are the hydrogen sorption charges before and after the adsorption of Bi or S onto the catalyst. The catalysts prepared contained θ_{Bi} = 0.09, 0.26, 0.35, 0.44, and 0.70, and θ_S = 0.19, 0.28 and 0.49. The CV for Pt-Bi(θ_{Bi} = 0.26)/graphite is shown in Figure 1 curve (b). Reduction in intensity occurred mostly in the features corresponding to the step sites, accompanied by a slight reduction in the 100-terrace feature; there was no reduction in the 111-terrace feature. Thus bismuth shows a high preference for adsorption at step sites and a low propensity for adsorption at terrace sites, in common with information from single crystal and polycrystalline electrode studies.[18-20]   The CV for Pt-S(θ_S = 0.29)/graphite is also shown in Figure 1. Adsorption occurred at both steps and terraces, but the feature for the (100)-terrace was absent, showing preferential adsorption on this face, and that for the (111)-terrace was greatly reduced. Thus, again in conformity with the literature[14], S adsorbed preferentially at terrace sites in that these were the first to be completely filled. The (111) x (111) step/kink sites remain unoccupied at all coverages.
        Quinuclidine had no effect on the CV of Pt/graphite, i.e. it was not adsorbed to a measureable extent.

3   RESULTS

3.1 Catalysis over supported Pt

Hydrogen uptake/time curves were of the expected form. Racemic reaction in the absence of alkaloid modifier was a slow self-poisoning reaction for which the initial rate was the maximum rate. Alkaloid-modified reactions were rapid and gave S-shaped uptake/time curves, i.e. they showed an initial acceleration, a period of reaction at the maximum rate, and deceleration as reactant was exhausted. Over Pt/graphite (Table 1) 0.17 mmol of cinchonidine or quinuclidine each enhanced the rate by a factor of ~30, the former reaction being enantioselective and the latter racemic. In the presence of a 1:1 mixture of cinchonidine and quinuclidine (0.17 mmol of each), the effect on the maximum rate was almost additive, and the enantiomeric excess was only slightly lower than for cinchonidine as the sole additive.

Table 1   *Characteristics of various racemic and enantioselective reactions over Pt/graphite under standard conditions*

alkaloid modifier	maximum rate /mmol $h^{-1} g^{-1}$	enantiomeric excess/%(R)	HMWP yield
none	25	0	6
CD	740	41	100
QN	685	0	84
CD + QN	1205	37	45

Cinchonidine-modified reactions over Pt-S/graphite and Pt-Bi/graphite showed similar S-shaped uptake/time curves. Table 2 shows that sulphur diminished the rate but raised the enantiomeric excess, whereas Bi enhanced the rate but diminished the enantiomeric excess.

Table 2   *Effect of sulphur and of bismuth on cinchidine-modified reaction under standard conditions*

bismuth loading, θ_S	sulphur loading, θ_{Bi}	maximum rate /mmol $h^{-1} g^{-1}$	enantiomeric excess/%(R)
none		910[a]	42[a]
0.19		740	51
0.28		285	51
0.49		290	54
	none	740[a]	46[a]
	0.09	930	41
	0.26	1350	39
	0.44	1635	32
	0.70	1950	23

[a] S- and Bi-modified catalysts prepared from different batches of catalyst

The effects of cinchonidine and quinuclidine, separately and in combination, on reactions over Pt-S/graphite and Pt-Bi/graphite are shown in Table 3. Over Pt-S/graphite, cinchonidine and quinuclidine each gave substantial rate enhancements; their effect on rate was almost additive and enantiomeric excess was again as high in the (CD + QN)-modified reaction as in the CD-modified reaction. Over Pt-Bi/graphite, cinchonidine gave a substantial rate enhancement, quinuclidine an even greater effect, and in combination the

Catalysis in Application

rate was so fast as probably to be diffusion controlled. The (CD + QN)-modified reaction gave an enantiomeric excess that was the mean of that afforded by the alkaloids separately, but the high reaction rate may render this value unreliable.

Table 3 *Characteristics of various racemic and enantioselective reactions over Pt-S/graphite and Pt-Bi/graphite under standard conditions*

elemental additive	alkaloid modifier	maximum rate /mmol h^{-1} g^{-1}	enantiomeric excess/%(R)	HMWP yield
S, $\theta = 19$	CD	470	51	
S, $\theta = 19$	QN	310	0	
S, $\theta = 19$	CD + QN	645	51	
Bi, $\theta = 0.35$	CD	1205	27	49
Bi, $\theta = 0.35$	QN	2200	0	36
Bi, $\theta = 0.35$	CD + QN	4600	15	33

Lactate formation was accompanied by the production of several higher molecular weight products, HMWPs, having molecular masses in the range 186 to 190 Da; their molecular identities have yet to be established. The total HMWP yields, defined as the sum of the peak areas in the GC traces and expressed relative to the value for the cinchonidine-modified reaction over Pt/graphite (= 100), are recorded in Tables 1 and 3. The following trends are evident. First, HMWP yield was low in racemic reaction involving no alkaloid. Second, yields were high and comparable in the presence of cinchonidine and of quinuclidine (whether reaction was enantioselective or racemic). Third, in the presence of both alkaloids the yield was reduced to about one-half under conditions where the rate is doubled which suggests that HMWP production is a zero order process independent of the pyruvate hydrogenation rate. HMWP yields over Pt-Bi/graphite were about half of those over Pt/graphite.

3.2.1 Adsorption of dihydrocinchonidine on Pt(976)^R and Pt(976)^S

Figure 2 shows the CVs of Pt(976)^R and Pt(976)^S and the effect of adsorption of dihydrocinchonidine. (The dihydro-derivative was used to preclude any complications in respect of adsorption via the vinyl moiety). The rate of adsorption was the same at the R- and S- surfaces and all features in the CV were affected equally, i.e. there was no molec-

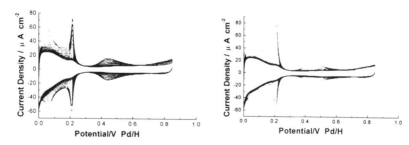

Figure 2 *CVs of Pt(976)^R (left) and Pt(976)^S (right) and the effect of adsorbing dihydrocinchonidine onto these single crystal surfaces*

ular recognition between the chiral adsorbate and the chiral surface. Similar negative results were obtained for dihydrocinchonidine adsorption on the R- and S- forms of Pt(321) and Pt(643) and for cinchinidine adsorption on the R- and S- forms of Pt(321), Pt(643) and Pt(11,7,1). This, with the published result for Pt(531)[16], brings to seven the number of surfaces to which this test of molecular recognition has been applied.

4 DISCUSSION

Cinchonidine-modified Pt/graphite gave values of the enantiomeric excess of 42 – 48%. Although lower than those typical for Pt/silica and Pt/alumina (70 – 90%)[1,2], they are typical for C-supported Pt. Orito and co-workers in their original work recorded 48% for Pt/C[21] and more recently values in the range 29 to 35% have been reported for Pt supported on a wide range of activated carbons[22]. The reason for the relatively poor performance of C-supported Pt is not understood.

4.1 Morphology and enantioselectivity

Bi adsorbed at steps in the polycrystalline Pt surface and lowered enantioselectivity. Thus cinchonidine adsorbed at or near a step was more effective as a modifier than cinchonidine adsorbed elsewhere on the terrace away from the step. On this basis S, which adsorbed preferentially on the terraces away from the steps, raised enantioselectivity because it blocked away-from-step sites of lower enantioselectivity leaving a larger proportion of reaction to occur at the near-step sites.

This interpretation has also to be consistent with the effects of Bi and of S on reaction rate. S is a well known catalyst poison, and its effect on rate may be both geometric a s i ndicated above, a nd e lectronic i n that i t may h ave p erturbed P t a toms at some distance from its point of adsorption[20]. The effect of Bi in enhancing rate is unexpected and is considered further below. The additional product represented by the enhanced rate showed a low enantiomeric excess and hence alkaloid *was* involved in its formation.

4.2 Morphology and reaction rate

In early studies it appeared that enantioselectivity and enhanced rate were directly related, and the first mechanisms were developed with a specific objective of simultaneously interpreting these reaction characteristics[4]. The rate acceleration is related to the activation of the α-keto group which may result either from the effects of H-bonding with the quinuclidine-N atom of the modifier[1,2,4] or as a result of the lowering of the energy of the π-orbitals of the α-keto group in the reactant/modifier diastereomeric complex[23]. In 1991 we reported that a variety of organic N-bases, including quinuclidine, cause substantial rate enhancement of racemic methyl pyruvate hydrogenation over Pt/sililca, which was attributed to H-bonding between adsorbed pyruvate ester and molecules of N-base in solution[24]. This type of rate enhancement is in evidence again here (Tables 1 and 3) but on this occasion t he y ields of higher m olecular weight products (HMWPs) have also been recorded and this information casts new light on the observations.

It is well known that pyruvate esters form high molecular weight products at the Pt surface. In surface science experiments these have been seen as highly ketonic polymers adsorbed to the Pt[25]. In enantioselective hydrogenations some HMWPs have been identified in reaction solutions[2,26]. The HMWPs from the present experiments have yet to be identified. However, something may be adduced from their yields. Racemic hydrogenation in the absence of alkaloid is slow, self-poisoned, and the HMWP yield in solution is low (Table 1) suggesting that the majority of the 'polymer' remains on the surface. Racemic hydrogenation in the presence of quinuclidine and enantioselective hydrogenation in the presence of cinchonidine is fast and the yield of HMWP is high. The rapid rate suggests that alkaloid inhibits or reduces the extent of pyruvate polymerisation, thereby increasing the number of sites for simple hydrogenation to lactate ester. The high yield may arise either because polymerisation is terminated earlier so that intermediate products desorb rather than undergo further propagation, or because polymerisation is inhibited altogether and natural keto-enol condensations are catalysed by the N-bases. When the alkaloid was quinuclidine, the experiment showed that the inherent activity for racemic ethyl pyruvate hydrogenation was very high. Finally, over Pt/graphite and Pt-S/graphite (but not over Pt-Bi/graphite where possible diffusion control occurred), when quinuclidine and cinchonidine were present together the enantiomeric excess was retained at or close to the value provided by cinchonidine alone even though the rate was much enhanced. Thus, all of the surface released by the polymer-inhibiting action of quinuclidine was available for cinchonidine and pyruvate adsorption, i.e. for enantioselective reaction. This result arises naturally out of the polymer-inhibition model proposed here but is not consistent with the H-bonding model advanced earlier[24]. According to the earlier model, the enantiomeric excess observed in the (CD + QN)-modified reaction should have been lower than that in the CD-modified reaction in proportion to the extent to which QN had further enhanced the rate.

On the basis that rate enhancement in these reactions is attributable in part to inhibition of pyruvate polymerisation by the alkaloid, the effect of Bi in *further enhancing* the rate (Table 3) indicates that the polymerisation may be preferentially initiated at sites adjacent to steps in the Pt surface. When these sites are blocked by bismuth, polymerisation occurs only at sites on terraces where it is initiated less efficiently. Polymerisation is a parasitical reaction in many catalytic systems and it would be of interest to know whether Bi could be used more widely in its inhibition.

4.3   Morphology and molecular recognition

The CV for dihydrocinchonidine adsorption on Pt(976)^R is identical to that for its adsorption on Pt(976)^S (Figure 2). The absence of an adsorption rate differentiation in this system, viewed alongside the presence of a reaction differentiation in the Pt/sugar system, suggests that molecular recognition in these chiral systems may require relatively weak interactions, and that perhaps the cinchona alkaloids are too strongly adsorbed to show delicate chiral-chiral interactions under these conditions.

5 CONCLUSIONS

This study has shown: (a) that sites at the polycrystalline Pt surface adjacent to steps give higher enantioselectivities than sites on terraces when both are available for cinchonidine and pyruvate ester adsorption; (b) that the rate enhancement in the presence of alkaloid is due at least in part to the ability of organic N-bases to inhibit pyruvate ester conversion to

HMWPs; (c) that sites adjacent to steps are particularly active for the initiation of HMWP formation, and (d) that there is currently no evidence for chiral-chiral recognition of kink sites when cinchona alkaloids adsorb at these surfaces .

Acknowledgements

DJJ was supported by Johnson Matthey and EPSRC, and OAH by the Government of Saudi Arabia and Umm Al-Qura University.

References

1   (a) P.B. Wells and R.P.K. Wells, *Chiral Catalyst Immobilisation and Recycling*, eds., D.E. De Vos, I.F.J. Vankelecom, P.A. Jacobs, Wiley-VCH, Weinheim, 2000, Chapter 6, pp. 123-154. (b) A. Baiker, loc. cit., Chapter 7, pp. 155-171.

2   M. von Arx, T. Mallat and A. Baiker, *Topics in Catal.*, 2002, **19**, 75.

3   H.U. Blaser, *J. Chem. Soc. Chem. Commun.* 2003, p. 293.

4   K.E. Simons, P.A. Meheux, S.P. Griffiths, I.M. Sutherland, P. Johnston, P.B. Wells, A.F. Carley, M.K. Rajumon, M.W. Roberts and A. Ibbotson, *Recueil Trav. Chim. Pays-Bas*, 1994, **113**, 465.

5   P.B. Wells and A.J. Wilkinson, *Topics in Catal.*, 1998, **5**, 39.

6   G. Bond and P.B. Wells, *J. Catal.*, 1994, **150**, 329.

7   T. Evans, A.P. Woodhead, A. Gutierrez-Sosa, G. Thornton, T.J. Hall, A.A. Davis, N.A. Young, P.B.. Wells, R.J. Oldman, O. Plashkevych, O. Vahtras, H. Agren and V. Carravetta, *Surf. Sci.*, 1999, **436**, L691.

8   D. Feri, T. Burgi and A. Baiker, *J. Catal.*, 2002, **210**, 160.

9   T. Burgi and A. Baiker, *J. Amer. Chem. Soc.*, 1998, **120**, 12920.

10 J.T. Wehrli, A. Baiker, D.M. Monti and H.U. Blaser, *J. Molec. Catal.*,1989, **49**, 195.

11 B. Torok, K. Balazsik, G. Szollosi, K. Felfoldi and M. Bartok, *Chirality*, 1999, **11**, 470.

12 X. Zuo, H. Liu and M. Liu, *Tetrahedron Letts.*,1998, **39**, 1941.

13 G.A. Attard, J.E. Gillies, C.A. Harris, D.J. Jenkins, P. Johnston, M.A. Price, D.J. Watson and P.B. Wells, *Appl. Catal. A: General*, 2001, **222**, 393.

14 E. Herrero, V. Climent and J.M. Feliu, *Electrochem. Commun.*, 2000, **2**, 636.

15 G.A. Attard, *J. Phys. Chem. B.*, 2001, **105**, 3158.

16 G.A. Attard, C.A. Harris, E. Herrero and J. Feliu, *Faraday Discuss.*, 2002, **121**, 253.

17 P.A. Meheux, A. Ibbotson and P.B. Wells, *J. Catalysis*, 1991, **128**, 387.

18 A. Steponavicius, L. Ciudaviciute, V. Karpoviciene and V. Kapocius, *Chemija*, 2001, **12**, 42.

19 J. Clavilier, J.M. Feliu and A. Aldaz, *J. Elecroanal. Chem.*, 1988, **243**, 419.

20 E. Lamy-Pitara and J. Barbier, *Appl. Catal. A*, 1997, **149**, 49.

21 Y. Orito, S. Imai and S. Niwa, *Nippon Kagaku Kaishi*, 1979, p.1118.

22 M.A. Fraga, M.. Mendes and E. Jordao, *J. Molec. Catal. A: Chemical*, 2002, **179**, 243.

23 A. Vargas, T. Burgi and A. Baiker, *New J. Chem.*, 2002, **26**, 807.

24 G. Bond, P.A. Meheux, A.Ibbotson, and P.B. Wells, *Catal. Today*, 1991, **10**, 371.

25 J.M. Bonello, R.M. Lambert, N. Kunzle and A. Baiker, *J. Amer. Chem. Soc.*, 2000, **122**, 9864.

26 J.A. Slipszenko, S.P. Griffiths, P. Johnston, K.E. Simons, W.A.H. Vermeer and P.B. Wells *J. Catal.*, 1998, **179**, 267.

THE EFFECT OF PREPARATION ON LANTHANUM AND LANTHANUM DOPED COBALTATES FOR APPLICATION IN THE WATER GAS SHIFT REACTION

M O' Connell[1], K.G. Nickel[1], J. Pasel[2] and R. Peters[2]

[1]Institut für Geowissenschaften, Arbeitsbereich Mineralogie,Wilhelmstr. 56 D-72074 Tübingen, Germany
[2] Institute for Materials and Processes in Energy Systems, IWV 3, Energy Process Engineering, Forschungszentrum Jülich, D-52425 Jülich, Germany

1   INTRODUCTION

Perovskite-type oxides, which have general formula of ABO_3, are an important class of catalytic materials and have been studied extensively as catalysts for complete oxidation of hydrocarbons, direct decomposition of NO, and nonselective reduction of NO by CO, H_2, and hydrocarbons [1]. In view of this, perovskite-type catalysts have been investigated in diversified areas such as the improvement in the preparation method, the promotion effects by partially substituting A or B cations [2], the adsorption phenomenon of oxygen [3] the characterization of the surface properties [4], etc. In addition, many basic studies have been carried out to elucidate relationships between the solid-state chemistry and the catalytic properties of perovskite-type oxides. So far, both unsubstituted (ABO_3) and substituted ($A_{1-x}A'_xBO_3$, $A_{1-x}A'_xB_{1-y}B'_yO_3$, etc.) systems have been investigated [1], almost all of which contain rare earth cations, especially La, at the A sites and 3d transition metal ions, especially Co and Mn, at the B sites.

In order to prepare these mixed oxides, the oxide-mixing method based on the solid state reaction between the component metal oxides is still utilized because of its lower manufacturing cost and simpler preparation process. However, the solid-state reaction method has some drawbacks, namely, a high reaction temperature to eliminate the unreacted starting oxides, a large particle size and a limited degree of chemical homogeneity in the final materials. . In order to overcome these inevitable disadvantages arising from the solid state reaction, chemical methods, such as sol–gel [5], the decomposition of precipitated precursors [6], polymeric gel [7], amorphous citrate [8] or coprecipitation [9] have been employed.

The amount of information available in the literature concerning the use of perovskite catalysts for the water gas shift reaction is very minimal. However it has been shown that some ABO_3 perovskites have exhibited significant activity for the reaction. $GdFeO_3$ is such an example whose activity is attributed to their p-type semi-conductivity due to a limited number of Fe^{4+} cations, existing in their crystal lattice and associated with the presence of some cation vacancies [10].

Here, we present preliminary characterisation results of perovskite catalysts developed with a view to application in PEM-based fuel cell applications, as developed in Forschungszentrum Juelich. CO is a poison to the Pt-based anode catalyst and thus deep removal of CO to the ppm level is necessary. On the other hand, the activity of existing commercial WGS catalysts is generally low, and as a result, the largest fraction of the reactor volume is occupied by the WGS part of the fuel processor for H_2 production. Development of more active catalysts is necessary for a more efficient WGS step in the fuel processing train. New ways of catalyst preparation could lead to more active or more selective catalysts for the WGS reaction.

2 EXPERIMENTAL

Three main preparation methods were employed for the manufacture of the perovskites ($LaCoO_3$) and the doped perovskites ($La_xCe_{1-x}CoO_3$, where x = 0.1, 0.2 and 0.3). These were (a) solvent evaporation, (b) co-precipitation and (c) citrate process. The first method involves simple thermal evaporation, of dissolved, hydrated precursor nitrates of the relevant components, in their respective molar amounts. Firstly the materials were intimately stirred for 30 minutes at room temperature and then transferred to an evaporating oven at 80°C where the samples remained overnight. The second method, co-precipitation, again involved using hydrated nitrate precursors, as before, but on this occasion, an appropriate mixture of nitrates were mixed together intimately and precipitated using tetramethyl ammonium hydroxide (TMAH). A violet precipitate was observed, which was isolated by filtration carefully and then extensively washed up to pH = 7. Subsequently, the precipitate was dried at 60°C and then aged at 800°C for 12 hours and in 12-hour intervals thereafter. Finally, perovskites were prepared via the citrate route. This synthesis is an easy way of obtaining high quality perovskite phases [11]. It involves no rigorous chemical procedure such as careful aging of gel or fine control of pH usually required for sol-gel methods. As before the relevant nitrate precursors were prepared. Citric acid was used to make the gel. The gel was obtained by keeping the solution at 100°C for several hours. Then, the gel was fired in air at 600-800°C for 12 h, at 100° intervals.

XRD analysis was then performed on all samples in the range from 10° to 80°, with a step size of 0.05° on a Philips PW 1050 model, using the normal θ/2θ geometry. This apparatus used high intensity monochromatic Cu Kα (λ = 1.54 Angstroms) with an anode current of 40 mA and a voltage of 40 mV. Again, all samples were ground thoroughly before the PXRD analysis. A phase analysis of the diffractograms obtained was then done, both qualitative and quantitative (where possible) using a commercial software package called Siroquant. Surface area measurements were taken on a Micromeritics Gemini 60 instrument. Typical sample weights were 0.3 g in all cases. After outgassing the sample, a mixture of nitrogen and helium was applied to the sample, which was immersed in a liquid nitrogen bath. X-ray photoelectron spectroscopy was performed using a Vacuum Science Workshop HA 150 electron energy analyser with a Vacuum Science Workshop Mg/Al X-ray source. Binding energies in XPS were calibrated using known reference data for the relevant materials. All samples were pressed and sintered prior to analysis.

3   RESULTS

3.1 PXRD Analysis

Previous work has shown the influence of the A/B metal ratio on the nature of the prepared perovskite [12]. Here we show results, which demonstrate the marked effect the preparation method has on the final perovskite. When $LaCoO_3$ was prepared via the evaporation route, only pure perovskite bulk structures, as R-3C trigonal, were observed in diffractograms measured after calcinations at 800°C for 12 hours, as can be seen in Figure 1. Stoichiometric $LaCoO_3$ has been quoted as having a = 5.37Å [13] compared to a = 5.49Å here, as calculated from Rietveld analysis, suggesting significant expansion of the perovskite lattice. This would imply considerable defect chemistry in these materials. When the same perovskite was prepared via co-precipitation, markedly different results were seen. PXRD studies show that perovskite (as R-3C trigonal), Co_3O_4 (spinel) and $La_2O_3/La(OH)_3$ (hexagonal) are present. Since there is a large amount of lanthana and lanthanum hydroxide visible, the data would suggest that the perovskite formed is cobalt rich.

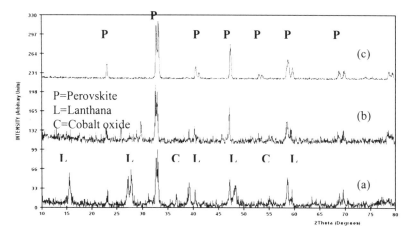

Figure 1: *PXRD"s of $LaCoO_3$ prepared, respectively via (a) precipitated. (b) evaporated and (c) citrate processes, after calcinations at 800°C for 12 hours.*

Again significant expansion of the perovskite lattice (a = 5.44Å) was observed. $LaCoO_3$ prepared via the citrate route, as can be inferred from the sharpness of the peaks, exhibit a large increase in the degree of crystallinity when compared with those prepared via both the co-precipitation and the thermal evaporation route. Similar results were also seen at lower temperatures, namely 600°C and 700°C, again an observation not seen for the other two preparation routes. However this time the lattice expansion was less significant (a = 5.41Å).

When $La_{1-x}Ce_xCoO_3$ materials were then prepared via the same three routes, where x = 0.1, 0.2 and 0.3. After 12 hours of calcination, in air, only perovskite structures were seen. Table 1 summarises the XRD data collected for this system.

Sample	Preparation Route	Phase(s) present	Surface Area/m^2 g^{-1}
La$_9$Ce$_1$CoO$_3$	Precipitation	Perovskite, ceria	6.40
La$_9$Ce$_1$CoO$_3$	Evaporation	Perovskite, ceria	6.66
La$_9$Ce$_1$CoO$_3$	Citrate	Perovskite (97.5%), ceria (2.5%)	7.54
La$_8$Ce$_2$CoO$_3$	Precipitation	Perovskite (91%), ceria (9%)	4.83
La$_8$Ce$_2$CoO$_3$	Evaporation	Perovskite (90.1%), ceria (9.9%)	5.1
La$_8$Ce$_2$CoO$_3$	Citrate	Perovskite, ceria	7.01
La$_7$Ce$_3$CoO$_3$	Precipitation	Perovskite (91%), ceria (9%)	3.51
La$_7$Ce$_3$CoO$_3$	Evaporation	Perovskite (79.9%),ceria (20.1%)	3.77

Table 1: *A summary of XRD data for the Ce doped perovskite system, after aging at 800°C for 12 hours. Phase amounts, as calculated by Siroquant are given. However these can only be approximate values since data collection time, step sizes etc. were too small for very accurate calculations. Also included are BET surface areas for the doped perovskites.*

As can be seen on inspecting the phase percentages, significant amounts of CeO$_2$ as a separate phase (as FM3M fluorite phase) are seen, especially as the molar ratio of ceria is increased. This is not surprising considering the differences in ionic radius between Ce^{4+} and La^{3+}. Effectively, Ce^{4+} substitution into the perovskite structure does not occur, or is minor. Therefore formation of cation vacancies becomes significant and a part of the trivalent Co ion in the B site must change to the tetravalent state. It is believed that the presence of Co^{4+} ions enhances catalytic activity but as this state is unstable, one cannot form large amounts of this. However, it should also be noted that at least some Ce^{3+} must substitute for La^{3+}. On inspection of the quantitative phase amounts for the doped perovskites in Table 1, the amounts were smaller than expected on assuming a quantitative segregation, knowing that only about 3 at% or x = 0.03 of cerium substitution can be accommodated by the perovskite phase [13]. The peaks of the segregated phases were also broader than those of corresponding oxides prepared separately, such as in mechanical mixtures, indicating a high level of dispersion of these segregated phases (or a highly disordered structure). Secondly, the presence of ceria as a minor phase may prove significant as a WGS reaction model has been proposed by Bunluesin et al [14], whereby CO adsorbed on the metal is oxidized by ceria, which in turn is oxidized by water.

Also included in Table 1 are BET surface areas for the doped perovskites. A maximum in the surface area occurs, in both preparation methods, when the dopant level is 10% i.e.

La$_{.9}$Ce$_{.1}$CoO$_3$. It is believed that when Ce is the dopant, that the most active catalysts are those with 10% doping [13]. Since all such prepared perovskites have a defective structure, one reason for an increase in activity is this increased surface area, relative to the other materials. What should also be noted here is that perovskite formation occurs at a lower temperature for perovskites prepared via the citrate route, with a corresponding resultant increase in surface area. These were 23.53 m^2/g and 22.29 m^2/g for 10% and 20% doped; respectively calcination was performed at 600°C for 12 hours. Again, as before, 10% doped perovskites, gave the largest values. However, the most marked difference is that the surface area is 3 times larger for 600° compared to 800°. Of course, this is not surprising but where it does become relevant, is that at the low temperature catalytic reactions envisaged for these perovskites, the lower the calcination temperature for perovskite formation, the better.

3.2 XPS Analysis

Data collected from undoped lanthanum–cobalt perovskites are discussed elsewhere [15]. Here data was collected from the cerium doped lanthanum–cobalt perovskites. The O1s spectra depicted in Figure 4 shows two main photo lines at 529.1 eV and 531.2–531.7 eV.

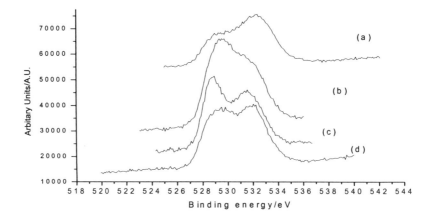

Figure 2 *O1s XP spectra for (a) La$_{.8}$Ce$_{.2}$CoO$_3$, citrate, (b) La$_{.9}$Ce$_{.1}$CoO$_3$, evaporated, (c) La$_{.9}$Ce$_{.1}$CoO$_3$, citrate, and (d) La$_{.9}$Ce$_{.1}$CoO$_3$, co-precipitated*

According to Tejuca et al. [16] the low binding energy peak must be assigned to lattice oxygen, and the high binding energy one to adsorbed oxygen species normally designated as **α**-oxygen. This peak is associated with cobalt presence and the release of oxygen from weakly adsorbed species. It should also be noted that the peak ascribed to bulk oxygen decreases with cerium content (spectrum a), indicating that on increasing cerium content the amount of α-oxygen decreases. There is also a slight decrease in the amount of surface oxygen present for the evaporated sample as compared to those prepared via the citrate and co-precipitated routes, a decrease, which can be ascribed to a lower level of molecular mixing seen in this preparation.

The La 3d spectra of La oxide, illustrated in Figure 3, consist of two doublets, which are broadly similar (in shape) for all 4 perovskites. The energy loss peaks appearing on the high-energy side of the $3d_{5/2}$ (BE = 850.6 eV and 854.48 eV) and $3d_{3/2}$ peaks (BE = 833.7 eV and 837.5 eV) are called satellite peaks [17]. These peaks are believed to result from core-hole screening by nearly degenerate ligand O 2p and empty La 4f orbit. The general explanation for high binding energy satellite peaks of lanthanum oxide is that there is charge transfer from O 2p to empty 4f of La leading to the $3d^9 4f^1$ final state. The spin-orbit splitting of La 3d is 16.8 eV. Again, the data indicates that lanthanum ions are present in the trivalent form for all the samples. It seems that La is mainly at the surface of the catalysts and it generally decreases with a higher degree of Ce substitution.

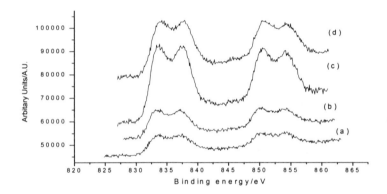

Figure 3 *La3d XP spectra for (a) $La_{.8}Ce_{.2}CoO_3$, citrate, (b) $La_{.9}Ce_{.1}CoO_3$, evaporated (c) $La_{.9}Ce_{.1}CoO_3$, citrate, and (d) $La_{.9}Ce_{.1}CoO_3$, co-precipitated*

Ce 3d spectra for $La_{.9}Ce_{.1}CoO_3$, (citrate route) were also performed and gave rise to multiple states. These multiple states arise from different Ce 4f level occupancies in the final state [18]. Pure ceria exhibits a complex 3d peak envelope. Much work has been carried out to assign these peaks. Allowing for charging effects, these peaks were assigned, with reference to the C1s peak, as follows, 882.3 eV, 888.1 eV, 897.8 eV, 900.9 eV, 907.0 eV and 916.1 eV. The Ce 3d spectrum can be fully described by six peaks consisting of three spin-orbit doublets [19]. The assignments of these peaks are somewhat controversial but the consensus of the experimental [19] and theoretical literature [18] that the peaks at 916 and 897 eV result from Ce^{4+} ionisations from 3/2 and 5/2 states and satellites, peaks at 907 and 888eV are believed to be associated with Ce^{3+} ionisations from 3/2 and 5/2 states and satellites and the remaining peaks are associated with so called shake down features. No peaks of noticeable intensity were observed in the 105-115 eV region, indicating that no Ce 4d peaks were present.

Co2p spectra are shown in Figure 4. Co2p region showed a position (BE Co2p$_{3/2}$=780.1 eV) and a band shape typical of Co$_3$O$_4$ [20]. There is also an increased shoulder at around 778 eV, where the peak for Co^0 is expected. There is a possibility, due to the shape of the shake-up satellite structure, that divalent Co is present in small quantities as the satellite structure is typical of CoO, which is very sensitive to defects [20].

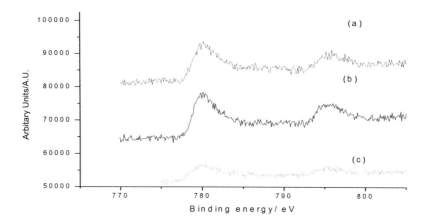

Figure 4 *Co2p XP spectra for (a) La$_{.8}$Ce$_{.2}$CoO$_3$, citrate, (b) La$_{.9}$Ce$_{.1}$CoO$_3$, citrate, and (c) La$_{.9}$Ce$_{.1}$CoO$_3$, co-precipitated*

It is believed that catalytic activity increases with increasing binding energy difference between the Co2p3/2 signal and the O1s lattice signal and increases in proportion to the ionicity of the Co-O bond [13]. Here, in order of increasing BE is, La$_{.8}$Ce$_{.2}$CoO$_3$, citrate (249.78 eV) < La$_{.9}$Ce$_{.1}$CoO$_3$, evaporated (251.55 eV) < La$_{.9}$Ce$_{.1}$CoO$_3$, precipitated (252.86 eV) < La$_{.9}$Ce$_{.1}$CoO$_3$, citrate (253.1 eV). Also, the values of spin-orbit splitting of Co 2p peaks can provide clues to predicting catalytic activity for these oxides. Both La$_{.9}$Ce$_{.1}$CoO$_3$, precipitated and La$_{.9}$Ce$_{.1}$CoO$_3$, citrate have values (15.3 eV and 15.5 eV respectively) for the splitting, similar to that of Co$_2$O$_3$, 15.2 eV. In contrast, similar values for La$_{.9}$Ce$_{.1}$CoO$_3$, evaporated and La$_{.8}$Ce$_{.2}$CoO$_3$, citrate (15.7 eV and 16.2 eV, respectively), are close to that of CoO, 16 eV. Therefore, it is possible that reduction from Co^{3+} to Co^{2+} can occur, resulting in a structural change and a predicted decrease in catalytic activity as the reduced cobalt ions cannot be reoxidised to the active state at ambient temperature.

4 CONCLUSIONS

A series of perovskites were prepared but care must be taken in preparing these perovskites as it has been shown that composition and structure of these perovskites are highly dependent on the preparation route and the presence/concentration of dopants is critical. If all the perovskites studied here have a defective structure, then it is believed that the Co based perovskites with minor phases such as ceria, lanthana and cobalt oxide offer the most promise for catalytic application. PXRD analysis shows the presence of ceria as a minor phase in ceria doped perovskites and is believed to be important for catalytic application. Secondly, XPS analysis has shown that the nature of the oxygen species on the perovskite surface is quite complex and variable but in brief, it can be ascribed to a lattice and a bulk component. La appears to be present in trivalent form and there is evidence to suggest that there is a small concentration of divalent cobalt but in the main is trivalent. However the nature of the Co species appears to be important for catalytic applications.

5 ACKNOWLEDGEMENTS

We gratefully acknowledge financial support of the Tuebingen group from Forschungszentrum Jülich GmbH. In particular we would like to thank Prof. D. Stolten for his help and guidance.

REFERENCES

[1] T Seiyama, *Catal. Rev.-Sci. Eng.*, **34** 4(281)1992

[2] H Tanaka and M Misono, *Current Opinion in Solid State and Mat. Sci.*, **5**, 5(2001)38

[3] R. Leanza, I. Rossetti, L. Fabbrini, C. Oliva and L. Forni, *App. Cat B: Env.*, **28**, 1(2000)55

[4] Y. Zhang-Steenwinkel, J. Beckers and A. Bliek, *App. Cat. A: Gen.*, **235**, 1-2(2002)79

[5] H. P. Beck, W. Eiser and R. Haberkorn, *J. Euro. Cer. Soc.*, **21**, 6(2001)687

[6] S Nakayama, M Okazaki, Y Aung and M Sakamoto, *Solid State Ionics*, 2003, in press

[7] M Kakihana, M Arima, M Yoshimura, N Ikeda and Y Sugitani, *J.Alloys and Cpds*, **283**, 1-2(1999)102

[8] M Popa and M Kakihana, *Solid State Ionics*, **151**, 1-4(2002)251

[9] A. K. Norman and M. A. Morris, *J. Mat. Process. Tech.*, **92-93**(1999)91

[10] J. Tsagaroyannis, K-J. Haralambous, Z. Loizos, G. Petroutsos, N. Spyrellis, *Mat. Lett.* **28**(1996) 393

[11] J Kirchnerova, M Alifanti and B Delmon, *App. Cat. A: Gen.*, **231**, 1-2(2002)65

[12] M O'Connell et al, in press

[13] K. Tabata, I. Matsumoto, S. Kohiki and M. Misono. *J. Mater. Sci.* **22** (1987) 4031.

[14] T. Bunluesin, R. J. Gorte and G. W. Graham, *App. Cat. B: Env.*, **15**, 1-2(1998)107

[15] M O'Connell, K.G. Nickel et al, submitted

[16] L.G. Tejuca, J.L.G. Fierro and J.M.D. Tascón. *Adv. Catal.* **36** (1989), p. 237

[17] H. Taguchi, S. Yamada, M. Nagao, Y. Ichikawa, K. Tabata, *Mat. Res. Bull.*, **37**(2002) 69

[18] A. Kotani, T. Jo and J.C. Parlebas. *Adv. Phys.* **37** (1988), p. 37

[19] A. Pfau and K.D. Schierbaum. *Surf. Sci.* **321** (1994), p. 71

[20] V.M. Jimenez, J.P. Espinos and A.R. Gonzalez-Elipe *Surf. Interface Anal.* **26** (1998), p. 62

THE INFLUENCE OF CATALYST GEOMETRY AND TOPOLOGY ON THE KINETICS OF HYDROCARBON HYDROGENATION REACTIONS

A.S. McLeod

Department of Chemical Engineering and Chemical Technology, Imperial College, London, SW7 2AZ ,UK.

1 INTRODUCTION

As a structure insensitive reaction with a reasonably well understood mechanism, the hydrogenation of ethene is commonly employed as a reference reaction for assessing the catalytic behaviour of supported metals. On the basis of the original kinetic study of ethene hydrogenation it was proposed that the reaction proceeded via a Horiuti-Polanyi mechanism in which dissociated hydrogen atoms sequentially hydrogenate ethene via an ethyl intermediate to ethane. Subsequent surface science studies strongly support this hypothesis. Recent work has demonstrated that the active intermediate is σ−bonded ethene, which is hydrogenated on the metallic surface vacancies located within a predominant carbonaceous overlayer composed of inactive partially dehydrogenated species[1,2].

This model is not, however, entirely consistent with the observed reaction kinetics, which exhibit a far richer behaviour than the standard Horiuti-Polanyi model would imply[3,4]. In several hydrocarbon hydrogenation reactions, including that of ethene, it has been shown that there exists a clear kinetic discontinuity separating two regions of distinct kinetic behaviour: one associated with high levels of hydrocarbon coverage and one associated with lower hydrocarbon coverage[3,4]. Kinetic behaviour of this kind has also been observed for the hydrogenation of propyne[5] and acetylene[6]. These experimental data are incompatible with the Langmuir-Hinshelwood kinetic model derived from the Horiuti-Polanyi mechanism and, instead, suggest that either the proposed reaction mechanism is incorrect or the assumptions underlying the implied kinetic model are invalid. Given the mass of supporting experimental data, we suggest that it is the limitations of the Langmuir model that account for this discrepancy.

In recent work, we have proposed a Monte Carlo model for the hydrogenation of hydrocarbons based on the Horiuti-Polyani which relaxes the mathematical constraints imposed by the Langmuir adsorption model[7-9]. We demonstrated that the discontinuity is due to the formation of a hydrocarbon overlayer which inhibits the adsorption of hydrogen, effectively stopping the reaction which is then rate limited by the slow desorption of ethene. In the current paper we extend this work to develop a model of ethene hydrogenation on a highly dispersed catalyst. We focus on the role of particle size, hydrogen spillover and adsorbate diffusion on the hydrogenation kinetics. Temperature programmed desorption spectra of silica supported dispersed Pt/silica catalysts and additional kinetic data are presented. On the basis of these data we argue that: (i) the dynamics of hydrogen diffusion on silica can be modelled by assuming the exchange of hydrogen between adjacent hydroxyl groups and (ii) the predominant differences in kinetic behaviour between single crystal catalysts and supported catalysts are due to the finite metal particle sizes and not hydrogen spillover onto or diffusion on the support. We demonstrate that on small metal particles, an irreversibly adsorbed hydrocarbon overlayer can readily form. This leads to a reduction in the ethene/hydrogen ratio at which the transition between the two kinetic regimes occurs. This result may explain the findings of previous studies of ethene hydrogenation by supported metals, which fail to consistently predict either the existence or the exact location of the transition[3,4].

2 EXPERIMENTAL

2.1 Materials

A series of 2.5 wt% Pt/silica catalysts were prepared either by the impregnation of the silica (Grace, type 254) in an aqueous medium (hexachloroplatinic acid) or by anchoring of an organic precursor (Pt(acac)$_2$)). Following contact with the precursor solution the catalysts were calcined at various temperatures between 30°C and 400°C before reduction in a stream of 20% hydrogen in helium. After reduction, the metal particle sizes were obtained by X-ray diffraction and independently verified by electron microscopy. The quoted values for the average particle diameters were obtained using a Phillips powder diffractometer (Cu K$_\alpha$ source, λ=0.154nm). A summary of the catalyst preparation conditions and properties is provided in table 1.

Table 1 *Summary of catalyst preparation conditions and particle sizes. Catalysts Pt-A and Pt-B were prepared by aqueous impregnation of hexachloroplatinic acid and Pt-C and Pt-D by anchoring of Pt(acac)$_2$*

Catalyst	Calcination Temp.(°C)	Particle size (nm)	Dispersion
Pt-A	30	50	0.20
Pt-B	200	63	0.16
Pt-C	200	163	0.061
Pt-D	400	187	0.053

The catalyst samples have been previously characterised by ^{2}H and ^{29}Si NMR spectroscopy[9]. Neither the ^{2}H NMR study of adsorbed deuterated benzene dynamics nor the ^{29}Si study of the surface silica structure indicated significant differences in the surface chemistry of the samples. Any differences in the catalytic activity of the two materials are thus ascribed to the variations in particle size and not to the catalyst preparation routes.

2.2 Kinetic Study of Ethene Hydrogenation

Reaction data for the hydrogenation of ethene were obtained using a narrow bore quartz tube reactor operating under differential conditions with the conversion of ethene kept below 5%. The reactor was surrounded by both a heating element and cooling jacket to maintain a constant temperature. All reaction rate measurements were conducted at atmospheric pressure using helium as a carrier. Gas composition measurement was by a chromatography column (Puraspec) connected to a thermal conductivity detector.

Kinetic data for the hydrogenation of ethene by the series of catalysts are presented in Figure 1. The presence of the discontinuity appears to depend on the structure of the catalyst, and consequently must either be associated with the preparation route or with particle size. The discontinuity between two kinetic regimes is observed for samples Pt-A and Pt-B, the two samples with the smallest particle sizes. The independent data of Jackson *et al.*, obtained for the hydrogenation of ethene by a range of supported metal catalysts, also show a correlation between the presence of a sharp discontinuity and small metal particle size[3]. These data, taken together with the results of the NMR characterisation of the materials[9] would tend to support the conclusion that it is metal particle size, not support interactions that are responsible for the presence of the kinetic discontinuity.

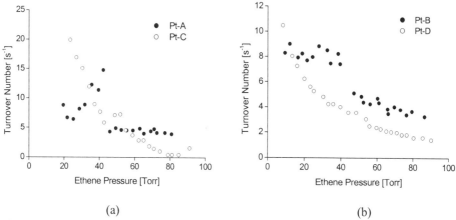

(a)   (b)

Figure 1   *Rate of ethene hydrogenation catalyst by the series of highly dispersed silica supported Pt. Data were obtained at 283K at atmospheric pressure under a hydrogen partial pressure of 100 torr*

2.3   Temperature Programmed Desorption

In the absence of significant electronic interactions with the support, a justifiable assumption for Pt/silica catalysts, a further potential influence of the catalyst support on the reaction kinetics is hydrogen spillover and diffusion on silica. In order to develop a model for the spillover and diffusion of hydrogen on silica a temperature programmed desorption (TPD) study of deuterium desorption from silica was conducted. The TPD experiments were conducted using a narrow bore silica reactor in a small furnace. The catalysts were first reduced *in-situ* in deuterium at the reduction temperature for 1 hour and then cooled to ambient temperature, again in deuterium. The reactor was then evacuated and heated at $20^{\circ}C$/min. Analysis of the desorbed gases was mass-spectrometer.

The TPD spectra obtained for the desorption of $^{2}H_{2}$ and of $^{2}H_{2}O$ from the catalyst surfaces (Pt-B is shown, the others do not differ significantly) are revealing. Figure 2(a) is the spectra of deuterium desorption obtained from the catalyst after reduction of the sample at $300^{\circ}C$ and $400^{\circ}C$. Three desorption features are evident: a low temperature peak at approx $100^{\circ}C$ corresponding to the desorption of desorption of hydrogen from metallic Pt[10] and two features at approx 300°C and 600°C. The two peaks at higher temperature coincide with the desorption of water formed during dehydroxylation of the silica surface by condensation of vicinal and geminal silanol groups. Figure 2(b) shows the spectra for desorption of ^2H_2O from the catalyst surface. There is a clear correlation between the desorption rate of $^2H_2$ and of ^2H_2O. The isotopic scrambling is evidence of the rapid spillover and diffusion of adsorbed deuterium.

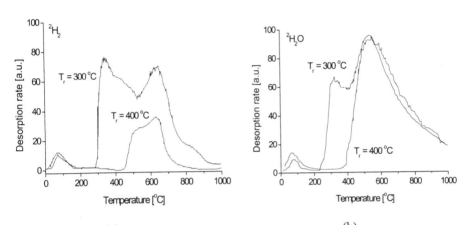

(a)   (b)

Figure 2   *Temperature programmed desorption spectra of (a) $^2H_2$ and (b) ^2H_2O from catalyst Pt-B. Data are shown for reduction in deuterium at 300°C and 400°C*

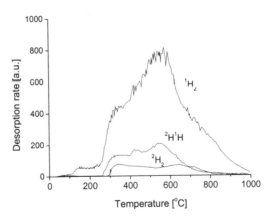

Figure 3   *Desorption spectra of the various hydrogen isotopes from catalyst Pt-A reduced at 300°C in deuterium*

Recalling that the reduction of the sample is carried out in deuterium we would expect little $^1H_2$ to be evolved, but the majority of hydrogen evolved from the surface is $^1H_2$ (Figure 3). This result suggests that the higher temperature $^2H_2$ desorption peaks are due to the decomposition of water liberated from the support, as this is the only possible source of the

[1]H shown in Figure 3. A similar conclusion was reached by Alexeev *et al.* following a study of hydrogen desorption from Pt supported on γ-alumina[11].

In view of the TPD isotope data, which indicate the free diffusion of spillover hydrogen at temperatures lower than those required to dehydoxylate the surface, we propose to model the migration of hydrogen on the silica surface as the exchange of dissociated hydrogen atoms with the surface hydroxyl groups. Martin and Duprez used isotopic exchange to estimate the activation energy of hydrogen diffusion mediated by surface hydroxyl groups on silica to be 30 kJ mol[-1]. This value will subsequently be used as the activation energy for hydrogen jumping between adjacent hydroxyl groups[12].

3   A MODEL FOR ETHENE HYDROGENATION ON HIGHLY DISPERSED METAL CATALYSTS

A Monte Carlo model for the hydrogenation of hydrocarbons has been described in detail previously[7-9]. Here we extend the model by conducting simulations on small metal particles and by including a model for hydrogen spillover and diffusion. The kinetic Monte Carlo model simulates the ethene hydrogenation reaction by a series of transformations of a regular square lattice, here composed of 256^2 discrete adsorption sites. To account for the effect of particle size the lattices were generated from Voronoi tessellations of the plane[13]. This partitions the lattice into active regions (metal particles) and inactive regions (support). The simulation can therefore account for the influence of the constrained geometry on which the reaction is occurring. Each step of the reaction mechanism corresponds to a specific transformation of this lattice. The transformations are intuitive, i.e. adsorption of hydrogen occupies two vacant locations of the lattice with dissociated hydrogen atoms.

The transformations of the lattice occur with a frequency that is proportional to the number of events of each type (as dictated by the configuration of the lattice) and by the event rate constant. The various processes that may occur on the metal particles are: hydrogen adsorption and desorption, ethene adsorption and desorption, hydrogenation of ethene to ethane via an ethyl intermediate and the diffusion of hydrogen. The only event that may occur on the inactive (support) regions of the surface is the diffusion of hydrogen. Hydrogen transport by hydrocarbon mediated hydrogen exchange is neglected, an assumption which will be justified later. The only inputs required to the model are the rate constants for each step, for which we use the data of Rekoske *et al.*[3] supplemented by the data of Martin and Duprez[12] for the diffusion of hydrogen on silica (E_a=30 kJmol[-1]). Adsorbed hydrocarbons are assumed to be immobile.

All possible diffusion and reaction events are grouped into two categories (i) those which do not occur sufficiently rapidly for equilibrium to be reached, and so must be explicitly simulated, and (ii) those which occur at such a rapid rate such that they may be assumed to be fully equilibrated after each time step. In practice, the only event of type (ii) is the diffusion of hydrogen on the metal particles. The simulation begins by checking the lattice for all possible events of type (i) and assigning each type (i) event to one of a number of lists. The rate of each possible event, r_j, is calculated and assigned to a list, l_n , such that $n = \mathrm{int}\left(\log_2\left(r_j\right)\right)$. This bounds the rates of all events on the same list by $2^n < r_j < 2^{n+1}$ which ensures that the rates of any two members of a given list do not differ by more that a factor of 2.

The probability of selecting a particular list is proportional to the rate associated with a given list and the number of events on that list, Σ_l. Once a list is selected, an event is selected from the list, at random, and the lattice updated. The distribution of waiting times between subsequent events, δt, follows a Poisson distribution and is given by

$$\delta t = -\frac{\ln \rho}{\sum\limits_{k=1}^{l} \bar{r}_k \Sigma_k} \qquad (1)$$

where ρ is a uniformly distributed random number between 0 and 1 and $\bar{r}_l$ the mean rate of the events on each list. Once an event is the lattice updated and the total simulation time, t, increased by δt. The overall reaction rate, expressed as a turnover number, is obtained by counting the number of ethane molecules produced, n_e, and dividing by the total simulation time multiplied by the number of lattice sites, N, to give

$$R = \frac{n_e}{Nt} \qquad (1)$$

After the execution of a single Monte Carlo event, events of type (ii) are executed. Hydrogen atoms adsorbed on the metal crystallites are allowed to diffuse by hopping between adjacent Pt adsorption sites. The number of these diffusion events is given by the ratio of the rate of hydrogen diffusion to the rate of the next most rapid event on the lists. Typically this value is typically around 10^4, i.e. 10^4 hydrogen diffusion events occur per lattice site between each Monte Carlo step. The next event of type (i) is then executed and the process repeated until the rate of hydrogenation reaches a steady state.

4   RESULTS

The results obtained for the simulation of ethene hydrogenation as a function of particle size are presented in Figure 4. In Figure 4(a) the rate of reaction is shown to be a strong function of particle size below 35-40nm. This is not the typical structure insensitive behaviour expected for ethene hydrogenation. The simulation data are shown compared to the predictions of a simple ensemble model, which assumes the reaction is limited by the adsorption of a dimer on the metal particles, this is the ideal behaviour expected for a structure insensitive reaction. The reduction in activity clearly cannot be attributed to the simple inhibition of reactant adsorption. It is proposed that the rapid decline in activity is associated with the formation of an immobile hydrocarbon overlayer onto which no further hydrogen can adsorb[7]. The diffusion limited growth of this layer, and eventual blocking of the particle will be more rapid on small particles than on large particles[14], the probability of the hydrocarbon displacing the hydrogen being statistically greater.

This conclusion is supported by the data presented in Figure 4(b), where the rate of reaction is shown for a series of perfect square particles of various side lengths. Smaller particles reduce the ethene pressure required to form the hydrocarbon overlayer which blocks further hydrogen adsorption and greatly reduces the reaction rate. The chemical structure of the deposited hydrocarbon overlayer is of no consequence, the only important consideration is the geometric structure of the overlayer. Once there is enough hydrocarbon on the surface to block all adjacent pairs of vacant sites (i.e. potential hydrogen adsorption sites) the reaction becomes limited, not by the surface reaction, but by the much lower rate of ethene desorption. There is then a discontinuous jump to the lower reaction rate.

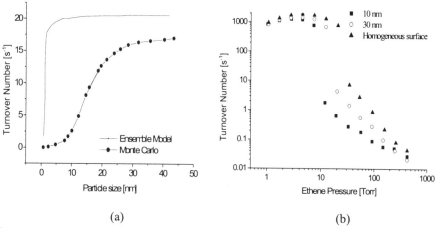

(a)   (b)

Figure 4   *Monte carlo and ensemble model predictions of the dependence of the rate of ethene hydrogenation on particle size. Data are presented for T=280K, hydrogen pressure = 100 Torr, ethene pressure = 10 Torr*

In the light of these results, it was anticipated that hydrogen supply from the support would have a significant influence on the reaction kinetics. Spillover hydrogen would provide an alternative source of hydrogen, diffusing from the larger particles, to react on the small particles where further adsorption of hydrogen is inhibited by the hydrocarbon overlayer. The influence of surface diffusion is, however, found to be only marginal (Figure 5). Whether or not hydrogen is permitted to spillover onto the support or to diffuse on the metal particles the results obtained by Monte Carlo simulation, with and without diffusion, or by a Lattice Gas model[8] are in close agreement.

The assumption that hydrogen can diffuse only on the Pt surface, and not via hydrocarbon overlayer mediated hydrogen exchange[15], has not yet been justified. Relaxing this constraint is, however, incompatible with the observed kinetic data. Permitting uninhibited diffusion of hydrogen atoms by hydrogen exchange would imply either that the mean-field assumption held or that the diffusion of surface hydrogen was the rate limiting step in the reaction mechanism. The first conclusion is incompatible with the presence of the kinetic discontinuity and the second is not supported by deuterium tracing experiments[4].

For comparison, the solution of the mean-field (Langmuir) micro-kinetic model described by Rekoske *at al.* is also shown in Figure 5. In the absence of mass or heat transfer limitations there can be no discontinuity in the mean-field model, yet such a discontinuity is observed. The mean-field model predicts the kinetics of the reaction reasonably well in the high activity regime, where the reactant distribution is random. The predicted order in ethene (-0.17) is in good agreement with some previously published experimental data[2]. The mean-field model, however, clearly fails to predict either the transition or the reaction rate in the low activity regime, where the absence of adjacent vacant adsorption sites in the hydrocarbon overlayer prevents the further adsorption of hydrogen.

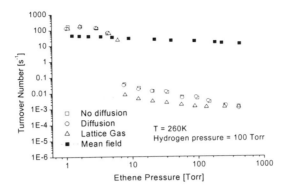

Figure 5 *Influence of the surface diffusion model on the ethene hydrogenation rate*

5 CONCLUSIONS

A model for the hydrogenation of ethene on supported metal catalysts that accounts for finite metal particle sizes, hydrogen spillover and diffusion on the catalyst support is proposed. The model is able to reproduce the observed kinetic discontinuity by explicitly describing the distribution of isolated hydrogen adsorption sites on the catalyst surface. The location of the discontinuity is associated with the metal particle size, small particles inducing the transition at lower hydrocarbon pressures. This is due to the ease of forming an inhibiting hydrocarbon overlayer in a confined geometry. Hydrogen spillover and diffusion on the support are not found to significantly influence the reaction kinetics.

References

1   P.S. Cremer and G.A. Somorjai, JH. Chem. Soc. Faraday Trans, 1995, **91**, 3671
2   J.E. Rekoske, R.D. Cortright, S.A. Goddard, S.B. Sharma and J.A. Dumesic, *J. Phys. Chem.*, 1992, **96**, 1880.
3   S.D. Jackson, G.D. McLellen, L. Conyers, M.T.B. Koegan, S. Simpson, P.B. Wells, D.A. Whan and R. Whyman, *J. Catal.*, 1996, **127**, 10.
4   D. Vassilakis, N. Barbouth and J. Oudar, J. Chim. Phys., 1991, **88**, 209.
5   D.R. Kennedy, B. Cullen, D.Lennon, G. Webb, P.R. Dennison and S.D. Jackson, *Stud. Surf. Sci. Catal., 1999,* **122**, 125
6   R.B. Moyes, D.W. Walker, P.B. Wells, D.A. Whan and E.A. Irvine, *Appl. Catal.*, 1989, **55**, L5.
7   A.S. McLeod and L.F. Gladden, *J. Chem. Phys.*, 1999, **110**, 4000.
8   A.S. McLeod and L.F. Gladden, *Stud. Surf. Sci. Catal.*, 1999, **122**, 167.
9   A.S. McLeod, K. Y. Cheah and L.F. Gladden, *Stud. Surf. Sci. Catal.,*1998, **118**, 1.
10 J.T. Miller, B.L. Meyers, F.S. Modica, G.S. Lane, M. Vaarkamp and D. Koningsberger, *J. Catal.*, 1993, **143**, 395.
11 O. Alexeev, D.-W. Kim, W. G.W. Graham, M. Shelef and B.C. Gates, *J. Catal.*, 1999, **185**, 170.
12 D. Martin and D. Duprez, *J. Phys. Chem. B*, 1997, **101**, 4428.
13 A.S. McLeod and L.F. Gladden, *J. Catal.*, 1998, **173**, 43.
14 K. Fichthorn, E. Gulari and Y. Barshard, *Phys. Rev. Lett.*, 1989, **63**, 1257.
15 T.V.W. Janssens, D. Stone, J.C. Hemminger and F. Zaera, 1998, **177**, 284.

ADSORPTION/DESORPTION BASED CHARACTERISATION OF HYDROGENATION CATALYSTS

J.M. Kanervo, R.I. Slioor and A.O.I. Krause

Helsinki University of Technology, Laboratory of Industrial Chemistry, P.O. Box 6100
FIN-02015 HUT, Finland

1 INTRODUCTION

Chemisorption of hydrogen is widely utilised in various experimental techniques to characterise hydrogenation catalysts. Data from conventional static chemisorption, temperature programmed desorption (TPD) and pulse chemisorption together constitute an extensive collection of qualitative and quantitative information. The kinetic analysis of a heterogeneously catalysed reaction can be supported by separate studies of adsorption/ desorption kinetics of reactants and products. In this sense detailed knowledge of hydrogen chemisorption provides essential information for the kinetics of catalytic hydrogenation reactions.

Static chemisorption gives information on the adsorption capacities and the adsorption equilibrium behaviour. TPD measures either desorption alone or combined desorption and re-adsorption. Re-adsorption during TPD may even maintain the adsorption equilibrium between surface and gas phase. In this case TPD provides temperature-dependent information on the adsorption equilibrium and may be used to determine the equilibrium parameters.[1]

TPD experiments can be carried out either in ultra high vacuum or at ambient pressure under flow condition. Thus they can bridge the material and pressure gaps between surface science and heterogeneous catalysis. Despite the popularity of TPD as a catalyst characterisation method its application in kinetic analysis for porous samples is often being discouraged. There are indeed important methodical considerations such as the selection of the reactor model and the intrinsic kinetic model and the evaluation of mass transfer limitations.[2-7] Furthermore, experimental data with sufficient information content should be collected in a carefully selected and standardised manner, since the pretreatment and the adsorption step prior to TPD significantly influence the TPD patterns.

We have collected experimental data regarding the adsorption and desorption of hydrogen on commercial hydrogenation catalysts, especially concentrating on developing methods for kinetic analysis of temperature programmed reactions. Previously we have carried out kinetic analysis of H_2-TPD data for a commercial 17 wt-% Ni/Al_2O_3 catalyst and introduced a model formulation describing quasiequilibrium desorption from multiple adsorption states under the assumption of differential reactor operation.[8]

In the present paper we report kinetic analyses of H_2-TPD for two commercial hydrogenation catalysts. The quasiequilibrium adsorption is still assumed in the analysis, but the reactor is now modelled as a plug flow reactor. Preliminary indications on the microkinetic relevance of hydrogen adsorption to the toluene hydrogenation are presented.

2 EXPERIMENTAL

Two supported nickel catalysts were investigated: one with a nickel content of 17 wt-% (Cat 1) and one with a nickel content of 55 wt-% Ni catalyst (Cat 2).

H_2-TPD experiments were carried out using AMI-100 catalyst characterisation system (Altamira Instruments). 50 mg (Cat 1) or 20 mg (Cat 2) of the catalyst was loaded in a U-tube. In all the experiments the following procedure was used. At first, the catalyst sample was dried under argon flow at 423 K for 30 minutes. Then the sample was reduced under hydrogen flow for 3 h at 673 K (Cat 1) or 573 K (Cat 2). After reduction it was kept at the reduction temperature under argon flow for 2 h and then cooled to the adsorption temperature T_0 of 303, 323 or 343 K under argon (standard procedure) or hydrogen flow (procedure 2). The adsorption was done under hydrogen flow for 20 minutes at the starting temperature of the TPD. The sample was flushed with argon flow at the adsorption temperature for 80 minutes. Subsequently the temperature was raised linearly to 673 K (Cat 1) or to 573 K (Cat 2) under argon flow and held constant for 70 minutes. Five TPD runs were performed consecutively. Between two TPD runs the sample was cooled to the adsorption temperature under argon (standard procedure) or hydrogen flow (procedure 2) and then kept under hydrogen flow for 20 minutes. The nominal heating rate was 10 K/min in the first, third and fifth TPD runs, 5 K/min in the second and 15 K/min in the fourth run. Preliminary experiments[8] had shown that the spectra from the second to the fifth runs were reproducible, and therefore the spectrum from the first TPD was always discarded. In all the steps the argon or hydrogen flow rate was 30 cm^2/min. The amount of desorbed hydrogen was measured by a thermal conductivity detector.

Static volumetric hydrogen chemisorption experiments were carried out with Coulter OMNISORP 100CX equipment. The chemisorption temperatures were 303, 333 and 363 K. The catalysts were pretreated in a similar manner as in the TPD experiments, except that the heating and cooling were done under vacuum. After the measurement of the total hydrogen adsorption the sample was degassed to remove weakly adsorbed hydrogen, and the isotherm for the reversible adsorption was measured.

Temperature programmed hydrogenation of toluene was studied with the AMI-100 equipped with a quadrupole mass spectrometer (Balzers MSC 200 Thermo-Cube). A gas mixture of hydrogen, argon and toluene was fed into the reactor (a U-tube) in which the catalyst sample was loaded, and the temperature of the sample was raised linearly from 373 to 473 K. The exiting gas stream was analysed by the mass spectrometer utilising Quadstar™ 421 software. The conversion was calculated from the toluene molar flows.

3   RESULTS AND DISCUSSION

The static chemisorption measurements carried out at three temperatures for both catalysts showed that the total hydrogen uptakes were almost independent of the temperature. However, the amounts of irreversibly adsorbed hydrogen displayed different temperature dependence. For Cat 1 the amount first slightly increased up to a temperature of 333 K and thereafter decreased. For Cat 2 the irreversibly adsorbed amount decreased with temperature.

The TPD experiments were first carried out according to the standard procedure for both catalysts (= cooling in argon and isothermal adsorption of hydrogen at the starting temperature of TPD). For Cat 2 the amounts of hydrogen desorbed during TPD decreased with increasing adsorption temperature (303, 323, 353 K). The adsorbed amounts thus followed the expected equilibrium trend, which was the same as for the irreversibly adsorbed hydrogen in the static chemisorption. When the standard procedure was applied for Cat 1, the total desorbed amounts first increased and then decreased as a function of adsorption temperature indicating that the surface did not reach equilibrium with hydrogen at 303 K. Consequently, the adsorption procedure 2 (= cooling in hydrogen followed by isothermal adsorption of hydrogen at the starting temperature of TPD) was applied for Cat 1. This adsorption procedure resulted in decreasing amounts of desorbed hydrogen with increasing adsorption temperature, meaning that the adsorption procedure influenced both the qualitative characteristics of TPD patterns and the total amounts of desorbed hydrogen for Cat 1, as previously found.[8]

The desorbed amounts obtained with the standard procedure were equal to the irreversible hydrogen uptake for both catalysts (Table 1). This is reasonable because the adsorption in static chemisorption and the adsorption step prior to TPD in the standard procedure are both performed isothermally. The desorbed amounts using procedure 2 (Cat 1) increased, approaching the total hydrogen uptake in static chemisorption.

Table 1   *Adsorbed amounts of hydrogen for the studied catalysts.*

	Metal content, wt-%	Hydrogen uptake, total, 303 K mmol H_2/ g_{Ni}	Hydrogen uptake, irreversible, 303 K mmol H_2/ g_{Ni}	Desorbed amount TPD T_0 = 303 K mmol H_2/ g_{Ni}
Cat 1	17	1.55	1.18	1.18 (1.63)*
Cat 2	55	1.30	0.87	0.86

* This value is related to the TPD procedure 2.

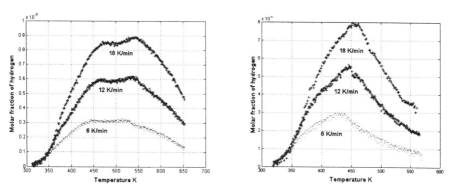

Figure 1   *TPD patterns of a) Cat 1 (procedure 2) and b) Cat 2 (standard procedure)*

The TPD data of Cat 1 and Cat 2 (Figure 1) were analysed kinetically to obtain more information on the adsorption behaviour. The kinetic analysis was based on the following assumptions: 1) intraparticle diffusion limitations were negligible in the experiments, 2) readsorption was significant during TPD, 3) the reactor dynamics could be described as a plug flow reactor model, and 4) the initial coverage represented equilibrium under the initial conditions of TPD. The assumptions were verified either experimentally or by the use of appropriate criteria.[8] Surface heterogeneity was accounted for by introducing a sufficient number of adsorption states into the model. The readsorption was assumed to maintain the quasiequilibrium adsorption during TPD and thus the coverage θ_i of each adsorption state i was expressed as:

$$\theta_i = \frac{(K_i x)^{1/2}}{1 + (K_i x)^{1/2}} \tag{1}$$

with the temperature-dependent adsorption equilibrium constant:

$$K_i = \frac{A_i}{\sqrt{T}} \exp\left(\frac{-\Delta H_i}{RT}\right) \tag{2}$$

where x is the molar fraction of hydrogen in the gas phase, A is the pre-exponential factor and ΔH is the adsorption enthalpy. The adsorption model was embedded into the plug flow reactor model:

$$\frac{\partial x}{\partial t} = -\frac{\partial (u\,x)}{\partial z} - \left(\sum_i v_{mi} \frac{\partial \theta}{\partial t}\right) / (pV_r / RT) \tag{3}$$

Nonlinear regression analysis of the H_2-TPD was implemented in MATLAB® 6 and parameter estimation was carried out to determine A_{refi}, ΔH_i and the adsorption capacity v_{mi} for the required number of states. (Temperature mean-centring for K_i was applied in the estimation)

The TPD of both Cat 1 and Cat 2 could be described by the quasiequilibrium model with two adsorption states. The TPD curves of Cat 2 were best described by two significant adsorption states whereas those of Cat 1 were best described by one major and one minor adsorption state. The parameters of the minor adsorption state were poorly identified and the pre-exponential factor was therefore fixed to match the maximum ΔS. The estimated parameters with their 95 % confidence intervals are reported in Tables 2 and 3. The enthalpy estimates for state 1 were nearly the same for Cat 1 and Cat 2, whereas those for state 2 differed. Figure 2 presents the model solution and the TPD data of Cat 2. Compared to the previous results obtained by applying the differential reactor model[8] the fit and the number of required states remained the same despite the reactor model.

The estimates of the adsorption enthalpies appeared to be of realistic magnitude. Comparison to published values is not straightforward, because enthalpy estimates have been determined from different experiments and using different methods than ours. A distinctive feature of our results is that the two adsorption states appear to release hydrogen almost simultaneously during TPD. The enthalpies of states 1 (-99 and -95 kJ/mol) are close to the isosteric heats of adsorption derived from adsorption isotherms for nickel single crystals (-90, -96 kJ/mol)[9]. The enthalpies for 14 wt-% Ni/Al$_2$O$_3$ reported by Weatherbee and Bartholomew[10] were determined from the temperatures of two distinct desorption maxima in TPD (-70, -125 kJ/mol as assigned to maxima at the lower and the higher temperatures, respectively). Lee et al.[11] determined adsorption enthalpies for ~50 wt% Ni/SiO_2 using desorption rates at constant coverages (-55... -89 kJ/mol). The calorimetrically determined total heat of H_2 adsorption on 23.3 wt-% Ni/SiO_2 (108...50 kJ/mol) was reported by Prinsloo and Gravelle.[12] Our results for the two states remain between all these estimates.

The sum of estimated adsorption capacities agrees with the total desorbed amounts. Pre-exponential factors are more difficult to evaluate. The calculated pre-exponential factors of the two states were of a very different order of magnitude (~5.8e-8 and 3.8 for Cat 2). However, comparing these to the transition state theory estimates of pre-exponential factors[13], they are both found acceptable, and they are attributed to immobile and mobile adsorption, respectively.

Table 2   *Results of kinetic modelling with a two-state model for Cat 1*
T_0=323 and 343 K and β =6, 12, and 17 K/min.

State index (i)	Parameter estimates		
	$A_{i,ref}$ ($K^{1/2}$)	ΔH_i (kJ/mol)	v_{mi} (mmol H_2/ g_{Ni})
1	$(1.7 \pm 0.6)e4$ *	-95 ± 7	0.18 ± 0.03
2	$(5.5 \pm 0.3)e5$	-58.7 ± 0.4	1.53 ± 0.03

$K_i = A_{i,ref} * T^{-1/2} * \exp(\Delta H_i/R(1/T_{ref}-1/T))$, where T_{ref}= 475 K. * fixed value.

Table 3   *Results of kinetic modelling with a two-state model for Cat 2*
T_0=303, 323 and 343 K and β =6, 12, and 17 K/min.

State index (i)	Parameter estimates		
	$A_{i,ref}$ ($K^{1/2}$)	ΔH_i (kJ/mol)	v_{mi} (mmol H_2/ g_{Ni})
1	$(3.5 \pm 0.4)e4$	-99 ± 5	0.35 ± 0.04
2	$(2.9 \pm 0.5)e5$	-41 ± 2	0.68 ± 0.04

$K_i = A_{i,ref} * T^{-1/2} * \exp(\Delta H_i/R(1/T_{ref}-1/T))$, where T_{ref}= 440 K

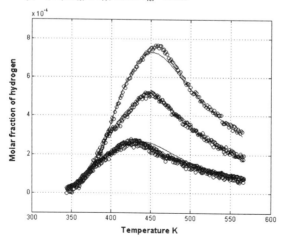

Figure 2   *TPD data and the model solution for Cat 2.*

The collected TPD data could be adequately described by the quasiequilibrium adsorption with two adsorption states with physically acceptable parameters. This is one suitable approach to account for the observed surface energetic heterogeneity, but is not conclusive evidence for the adsorption/desorption mechanism.

One objective of the kinetic studies of H_2-TPD was to gain more understanding of catalytic hydrogenation of toluene. Therefore, the conversion of toluene was measured in temperature programmed hydrogenation on Cat 1 and on Cat 2 (Figure 3). There is a clear maximum in the hydrogenation rate around 408 K under the conditions applied for both catalysts. Above 408 K the conversion falls drastically as a function of temperature and drops below 10% at 460-475 K. The decline in catalytic activity as a function of temperature has been previously explained by the adsorption equilibrium characteristics of hydrogen on a catalyst similar to Cat 1.[14]

The behaviour of the hydrogenation activity as a function of temperature and especially the coinciding maxima for different nickel catalysts are subjects of interest. As the TPD results (Figure 1) reveal, significant amounts of hydrogen desorb above 500 K. Thus total adsorption behaviour of hydrogen does not fully explain the observed maximum in the catalytic activity.

The obtained adsorption parameters were used to predict adsorption behaviour of hydrogen under hydrogenation reaction conditions (~10 % H_2). Figure 4 illustrates the coverage of the adsorption states as a function of temperature. The coverage of the adsorption state 1 (for both catalysts) dropped more in the temperature range of 375-520 K than the coverage of the adsorption state 2. The behaviour of state 1 consequently correlates better with the hydrogenation activity. If hydrogen adsorbed to state 1 is responsible for the hydrogenation activity, Cat 2 should be more active than Cat 1 based on the estimated adsorption capacities. Figure 3 supports this. However, no definite conclusion on the origin of the catalytic activity can be drawn.

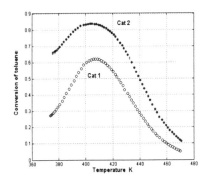

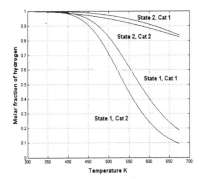

Figure 3   *Temperature programmed hydrogenation of toluene.*

Figure 4   *Coverage of states as a function of temperature in 10 % H_2.*

4 CONCLUSIONS

In this paper we presented results of kinetic analyses for H_2-TPD of two commercial nickel catalysts. TPD patterns could be described by quasiequilibrium desorption from two kinds of adsorption states with physically acceptable adsorption parameters. It was shown that TPD offers, in addition to conventional characterisation, rich kinetic information, which can be treated according to general methods of transient kinetics. The adsorption/desorption behaviour of hydrogen is essential for operation and development of hydrogenation catalysts. Indeed a correlation between the adsorption behaviour and the hydrogenation activity was found.

Acknowledgements

The authors thank Ms. K. M. Reinikainen for carrying out a part of the experiments and Ms. K. M. Vilonen for providing the hydrogenation data.

References

1   J.L. Falconer and J.A. Schwarz, Temperature-programmed desorption and reaction: applications to supported catalysts, *Catal. Rev. Sci. Eng.*, 1983, **25**, 141.
2   R-K. Herz, J.B. Kiela and S.P. Marin, Adsorption effects during temperature programmed desorption of carbon monoxide from supported platinum, *J. Catal.*, 1982, **73**, 66.
3   J.S. Rieck and A.T. Bell, Influence of adsorption and mass transfer effects on temperature-programmed desorption from porous catalysts, *J. Catal.*, 1984, **85**, 143.
4   R.J. Gorte, Design parameters for temperature programmed desorption from porous catalysts, *J. Catal.*, 1982, **75**, 164.
5   R.A. Demmin and R.J. Gorte, Design parameters for temperature-programmed desorption from a packed bed, *J. Catal.*, 1984, **90**, 32.
6   E. Tronconi and P. Forzatti, Experimental criteria for diffusional limitations during temperature-programmed desorption from porous catalysts, *J. Catal.*, 1985, **93**, 197.
7   Y.-J. Huang, J. Xue and J.A. Schwarz, Experimental procedures for the analysis of intraparticle diffusion during temperature-programmed desorption from porous catalysts in a flow system, *J. Catal.*, 1988, **109**, 396.
8   J.M. Kanervo, K.M. Reinikainen and A.O.I. Krause, Kinetic analysis of temperature programmed desorption, manuscript.
9   K. Christman, O. Schober, G. Ertl and M. Neumann, Adsorption of hydrogen on nickel single crystal surfaces, *J. Chem. Phys.*, 1974, **60**, 4528.
10   G.D. Weatherbee and C.H. Bartholomew, Effects of support on hydrogen adsorption/ desorption kinetics of nickel, *J. Catal.*, 1984, **87**, 55.
11   P.I. Lee and J.A. Schwarz, Adsorption-desorption kinetics of H₂ from supported nickel catalysts, *J. Catal.*, 1982, **73**, 272.
12   J.J. Prinsloo and P.C. Gravelle, Volumetric and calorimetric study of the adsorption of hydrogen, at 296 K, on supported nickel and nickel-copper catalysts containing preadsorbed carbon monoxide, *J. Chem. Soc., Faraday Trans. I*, 1980, **76**, 512.
13   J.A. Dumesic, D.F. Rudd, L.M. Aparicio, J.E. Rekoske and A.A. Trevino, The microkinetics of heterogeneous catalysis, ACS Professional Reference Book, Washington D. C., 1993.
14   S. Smeds, T. Salmi, L.P. Lindfors and O. Krause, Chemisorption and TPD studies of hydrogen on Ni/Al₂O₃, *Appl. Catal. A: General*, 1996, **144**, 177.

ETHYL ETHANOATE SYNTHESIS BY ETHANOL DEHYDROGENATION

S .W. Colley[1], M. W. M. Tuck[2]

[1]Davy Process Technology, Princeton Drive, Stockton-on-Tees TS17 6PY, UK. E-mail:
steve.colley@davyprotech.com
[2]Davy Process Technology, 20 Eastbourne Terrace, London W2 6LE, UK
E-mail: mike.tuck@davyprotech.com

1 INTRODUCTION

Ethyl ethanoate is an industrially important bulk chemical[1] used primarily as a solvent in the paints, coatings and inks industry. Commercial synthesis of ethyl ethanoate was until recently limited to esterification of acetic acid, the Tischenko reaction of ethanal and addition of ethene to acetic acid. Davy Process Technology has commercialised a process based on the dehydrogenation of ethanol[2] to synthesise ethyl ethanoate without the need for acetic acid or ethanal. The process consists of 3 linked unit operations – ethanol dehydrogenation to produce an impure ethyl ethanoate stream, selective hydrogenation of this steam to remove a troublesome by-product, and finally a novel distillation scheme that separates pure (99.95%+) ethyl ethanoate from a mixture of ethyl ethanoate, ethanol, higher alcohols and esters. The development of the dehydrogenation and selective hydrogenation reactions, and some aspects of the operation of the selective hydrogenation reactor at a commercial scale are discussed.

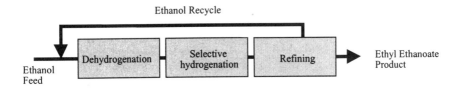

Figure 1   *DPT ethanol to ethyl ethanoate unit operations*

2 THE ETHYL ETHANOATE FROM ETHANOL PROCESS

2.1 Ethanol Dehydrogenation

The ethyl ethanoate process relies on three main process steps, the first of which is the dehydrogenation of ethanol to form a crude ethyl ethanoate product. Development of the dehydrogenation stage was carried out at DPT laboratories in Stockton-on-Tees during 1996 and 1997. The initial concept of ethanol dehydrogenation to provide a route to ethanoic acid had been performed at the same laboratories in 1984. The earlier development had been shelved at the concept stage as it was not judged to be commercially viable. Subsequent development of catalysts and separation technology, and interest from an ethanol producer, reignited interest in the dehydrogenation technology for ethyl ethanoate production.

Ethanol dehydrogenation to ethyl ethanoate is carried out in the vapour phase over heterogeneous copper based catalysts such as copper chromite[3]. Scheme 1 shows the main reactions; overall, two moles of ethanol react to form one mole of ethyl ethanoate and 2 moles of hydrogen. Side reactions, mainly involving ethanal, form C_3 to C_5 ketones and alcohols. The reaction is general – all copper-containing heterogeneous catalyst precursors tested synthesise ethyl ethanoate to some degree, with selectivity to the desired product ranging from 40% to 96% depending on the detailed physical and chemical form of the catalyst. Those catalyst precursors that contain large numbers of acidic sites (Raney copper, unmodified copper-chromite) tend to exhibit low selectivity to ethyl ethanoate. Modified copper-chromite catalyst precursors, and perhaps surprisingly copper oxide (after activation in hydrogen), exhibit high selectivity to ethyl ethanoate (90-96%) at acceptable conversion of ethanol (30-40%).

$$CH_3CH_2OH_{(g)} \rightarrow CH_3CH_2O_{(ad)} + H_{(ad)} \tag{1}$$
$$CH_3CH_2O_{(ad)} \rightarrow CH_3CHO_{(ad)} + H_{(ad)} \tag{2}$$
$$CH_3CH_2O + CH_3CHO \rightarrow CH_3CH(O)OCH_2CH_{3(ad)} \tag{3}$$
$$CH_3CH(OH)OCH_2CH_{3(ad)} \rightarrow CH_3COOCH_2CH_{3(ad)} + 2H_{(ad)} \tag{4}$$
$$CH_3COOCH_2CH_{3(ad)} \rightarrow CH_3COOCH_2CH_{3(g)} \tag{5}$$
$$4H_{(ad)} \rightarrow 2H_{2(g)} \tag{6}$$

$$2\ CH_3CH_2OH \rightarrow CH_3COOCH_2CH_3 + 2H_2 \tag{7}$$

Scheme 1

Catalyst precursors used for ethyl ethanoate development were selected from those commercially available. All catalyst precursors used were in the form of 3mmx3mm cylindrical pellets. A charge of 100cm^3 of each catalyst precursor was loaded into an electrically heated fixed bed reactor, ID 18mm, and activated in a stream of 5% hydrogen in nitrogen at 200-220°C. Following activation, reactor pressure was raised to the desired value using hydrogen and ethanol vapour passed over the activated catalyst at a rate equivalent to a LHSV of 0.5 to 2.0 (50 to 200 cm^{3}h^{-1}). The organic products from dehydrogenation were condensed and collected for analysis by GC and GC-MS; hydrogen produced during dehydrogenation was measured by a wet gas meter.

Dehydrogenation reactions are typically carried out at low pressure (sub atmospheric to 450kPa) and high temperature, 240-280°C. During the initial development, we found that low pressure and high temperature gave relatively low selectivity to ethyl ethanoate, and favoured the production of ethanal (an intermediate), ethanal condensation products (aldehydes, ketones and higher alcohols) and ethers. Figures 2 and 3 illustrate the effect of pressure and operating temperature on selectivity and ethanol conversion over a copper oxide based catalyst.

Figure 2 shows the expected relationship of decreasing selectivity to ethyl ethanoate and increasing conversion of ethanol with increasing temperature. All of the tests shown in figure 2 were carried out at a pressure of 680kPa and a volumetric flow of $100cm^3h^{-1}$ ethanol. Above about 240°C, selectivity to ethyl ethanoate falls to below 90 mol% - the limit for a practical industrial process.

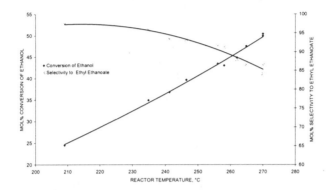

Figure 2   *Dependence of ethanol conversion and selectivity to ethyl ethanoate at 680kPa with temperature*

The effect of pressure on the dehydrogenation of ethanol is more surprising. Figure 3 plots the effect of pressure on conversion of ethanol and selectivity to ethyl ethanoate. Both selectivity and conversion reach a peak at about 1300-1400 kPa.

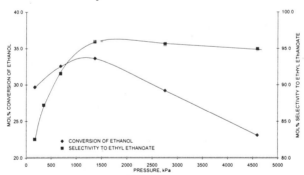

Figure 3   *Dependence of selectivity to ethyl ethanoate and conversion of ethanol with pressure*

An explanation for this behaviour can be deduced from Scheme 1. Ethanol decomposes at the surface of the catalyst, forming ethoxy then acyl species. This reaction is an equilibrium, the concentration of ethoxy and acyl species dependent on temperature and operating pressure. For ethyl ethanoate formation, an adsorbed acyl species reacts with an adsorbed ethoxy species to form an intermediate hemiacetal; subsequent dehydrogenation forms ethyl ethanoate. By-products, such as ketones, aldehydes and alcohols are formed by the reaction of two acyl species (see Scheme 2). Another class of by-products, ethers, are formed by the reaction of two ethoxy species. The relative concentrations of acyl and ethoxy species on the catalyst surface appear to be at an optimum for ethyl ethanoate formation at circa 1300-1400 kPa.

$$CH_3CH_2OH_{(g)} \rightarrow CH_3CH_2O_{(ad)} + H_{(ad)} \tag{1}$$
$$CH_3CH_2O_{(ad)} \rightarrow CH_3CHO_{(ad)} + H_{(ad)} \tag{2}$$
$$CH_3CHO_{(ad)} + CH_3CHO_{(ad)} \rightarrow CH_3CH(O)CH_3CHO_{(ad)} \rightarrow CH_3CH:CH_3CHO_{(ad)} \tag{8}$$
$$CH_3CH_2O_{(ad)} + CH_3CH_2O_{(ad)} \rightarrow CH_3CH_2OCH_2CH_{3(ad)} \tag{9}$$

Scheme 2

Table 1 *Typical dehydrogenation product*

Component	Wt%
Ethanol	58.44
Ethyl Ethanoate	39.49
Carbonyls	1.23
Alcohols	0.76

Commercial operation of the dehydrogenation reactor was fixed by this experimental work at 1380 kPa and 220-245°C. Even though the reactions taking place over the dehydrogenation reactor are relatively complex, commercial experience with the full scale reactor has been straight forward. The reactor and catalyst performed at almost exactly the selectivity and conversion predicted from experimental work. This contrasts with the experience of the selective hydrogenation reactor which is detailed in Section 2.2.

2.2 Selective Hydrogenation

A selective hydrogenation step was introduced to the ethyl ethanoate process to hydrogenate butanone to 2-butanol. Butanone has a similar boiling point to ethyl ethanoate, and cannot be separated by simple distillation with a commercially viable distillation column. However, 2-butanol has a higher boiling point and can be separated easily. The selective hydrogenation step also converts other ketones and aldehydes present in the dehydrogenation product to their respective alcohols, simplifying the overall distillation scheme. Selective hydrogenation of aldehydes and ketones in the presence of alcohols and esters is not widely practised. A review of commercial and academic[4] literature suggested supported ruthenium as a likely candidate; nickel was also tested for the reaction.

Experimental work on the selective hydrogenation was performed in a trickle bed reactor, using a supported ruthenium catalyst. A charge of $100cm^3$ of the catalyst precursor was loaded into the reactor and activated under a stream of hydrogen. Liquid dehydrogenation product was fed to the top of the reactor, with an excess of hydrogen at 1000-5000kPa and 60-120°C. The product from hydrogenation was collected and analysed by GC and GC-MS. Hydrogen exiting the reactor was collected and analysed by GC and GC-MS.

The extent of reaction of ketones was found to be highly dependent on reactor temperature. Figure 4 illustrates the effect of temperature on butanone concentration of the product of reaction. The testwork was performed over a supported ruthenium catalyst at

4100kPa and a liquid feed rate of 100cm^{3}h^{-1}. The liquid feed used had a similar composition to that shown in Table1.

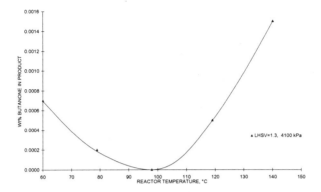

Figure 4   *Butanone concentrations in selective hydrogenation reactor product*

The detailed features of the experimental results suggest that at low temperature the reaction is kinetically controlled, while at higher temperatures the reaction appears to be thermodynamically controlled. The equilibrium between butanone/hydrogen and butanol is temperature dependent and it would seem reasonable that as the reaction temperature increased the position of equilibrium would shift to favour butanone. Given that the specification for butanone in ethyl ethanoate is typically 50ppm, the maximum butanone content of the reactor product was fixed at 17ppm, a stiff target.

The commercial selective hydrogenation reactor was constructed such that the catalyst was divided into a series of independent beds, with liquid and gas redistribution trays between. At start-up, the reactor performed well, hydrogenating butanone to <10ppm at the design temperature. However, an apparent slow decrease in catalyst activity over several months forced the operators to increase the reactor temperature to the maximum design limit. Detailed analysis of the selective hydrogenation product indicated that a previously minor side reaction, hydrogenolysis of ethanol to methane, had become significant. Samples of catalyst from the commercial production run were tested at Davy Process Technology's laboratories and were found to behave similarly to fresh catalyst in the laboratory equipment. Measured rates of both the desired reaction, butanone hydrogenation, and the side reaction were found to be indistinguishable from fresh catalyst. It seemed that changes in the chemical or physical nature of the catalyst were not causing the lack of activity in the commercial reactor.

Attention turned to liquid and gas distribution between the catalyst beds. The precise arrangement of the distribution trays is known to have a marked effect on liquid distribution and hence catalyst effectiveness. A close examination of the distributor trays suggested that leakage between sections of the trays may have been significant, leading to uneven distribution of liquid onto the catalyst below.

The distribution trays were constructed in 3 parts to allow them to be fitted and removed from the reactor during maintenance or catalyst removal. The design of the trays were such that at the joints the trays fitted together tightly to eliminate or reduce leakage through the seams. Due to a minor difference between the gauge of metal specified in the tray design and that used for construction, small gaps were introduced at the circumference of the distribution trays, resulting in the liquid distribution pattern to the catalyst being disrupted. Figure 5 shows the predicted distribution pattern at 150mm into the catalyst bed based on the original distributor design, and figure 6 that obtained by experiment on the as-installed distributor.

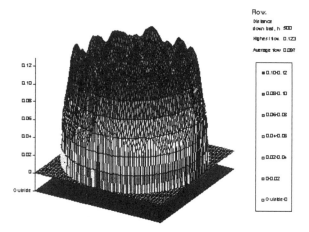

Figure 5   *Predicted liquid flow pattern through selective hydrogenation reactor*

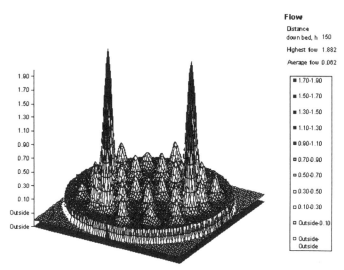

Figure 6   *Measured liquid flow through selective hydrogenation reactor*

A kinetic model derived from information gathered in the laboratory was applied to the model that predicted the liquid flow through the reactor. In true trickle bed flow, gas and liquid are evenly distributed in the reactor, and the ratio of hydrogen to liquid is uniform. However, when the actual flow patterns from the distributors were modelled, the prediction was that there was reduced liquid flow and increased hydrogen flow at the reactor wall. This deviation from the design flow pattern altered the stoichiometry of hydrogen to liquid;

at the circumference of the reactor the ratio of hydrogen to liquid was high, and at the centre, low. The effect of this variation from design was twofold; at the centre of the reactor, catalyst was starved of hydrogen, reducing its effectiveness. This hydrogen starvation was the major causal factor in the initial underperformance. At the reactor wall, the high hydrogen to liquid ratio promoted ethanol hydrogenolysis turning it from a minor to significant side reaction. The hydrogenolysis reaction is highly temperature sensitive, and as the reactor temperature was raised the rate of methane formation increased. The reaction consumed hydrogen and produced inert methane; as the reactor temperature was raised the partial pressure of hydrogen fell, causing a further loss in catalyst activity. A cycle of incremental reactor temperature increases, increased methane formation, and subsequent fall in catalyst effectiveness was set up.

The solution to the problems experienced in the commercial hydrogenator was simple - liquid distributors were redesigned and refabricated, and the operating procedure changed to ensure that liquid flow through the reactor was maintained at design flows at all times. Testwork carried out on the new distributors in the laboratory confirmed that flow distribution conformed closely to design, and that the distributors were suitable for operation. Stable operation for over 1 year has been achieved since the new distributors were fitted to the commercial reactor and the plant restarted.

3 CONCLUSIONS

The laboratory development of an ethyl ethanoate from ethanol process has been successfully transferred to the industrial scale, at a scale factor in excess of 70,000. The commercial reactors, when operating as designed, performed identically to the laboratory models. Problems with the hydrodynamics of the hydrogenator caused a loss of catalyst activity and selectivity. The problem of poor hydrodynamics was in part diagnosed by an awareness of the effects of maldistribution on the process chemistry.

The importance of good distribution of both gas and liquid in hydrogenation reactors is recognised as an important consideration in plant design. What is perhaps not fully appreciated is that poor distribution will not only reduce the effectiveness of the catalyst, but may also promote competing reactions that in well distributed reactors and the laboratory, are insignificant. Laboratory work to determine the effect of a large change in the stoichiometry within hydrogenation reactions is strongly recommended. While this will not prevent maldistribution, it allows the chemical engineer to design safeguards where necessary, and alerts the operator to potential changes in behaviour of a catalyst and their causes.

References

1   WO 0020375 (2000) C.R. Fawcett, M.W.M. Tuck, C. Rathmell, S.W. Colley;
    WO 0020374 (2000) N. Harris, C. Rathmell, S.W. Colley;
    WO 0020373 (2000) C.R. Fawcett, M.W.M. Tuck, D.J. Watson, C.M. Sharif, S.W. Colley, M.A. Wood
2   W. Johnston, T. Esker, *Chemical Economics Handbook*, Research Report, 1999, 610.7000 A
3   K. Takeshita, *J. Sci. Hiroshima Univ. Ser. A,* 1990, **54 (1)**, p. 99-123
4   P. N. Rylander, *Catalytic Hydrogenation over Platinum Metals*, Academic Press, New York, 1967, Chapter 15, p. 258

OBSERVING HETEROGENEOUS CATALYSTS AT WORK: *IN-SITU* FUNCTIONAL ANALYSIS OF CATALYSTS USED IN SELECTIVE OXIDATION

Robert Schlögl, Fritz-Haber Institut der MPG, Faradayweg 4-6, 14195 Berlin
acsek@fhi-berlin.mpg.de

ABSTRACT

Many catalyst systems used in selective oxidation of small organic molecules are bulk oxides with complex compositions and excessive chemical complexity. Concepts of "lattice oxygen" availability, "site isolation" and "phase co-operation" have driven the technical development in the past. The functional analysis of such systems is, however, still in its infancy as the application of model systems is barely possible and few techniques are available giving access to the details of the real structure (defects) that govern the operation of these systems. The application of a combination of *in-situ* techniques allows insight into the structural dynamics of these systems. Even in the simple example case of MoO_3 the definition of the matrix material hosting the active sites is different from *a priori* assumptions derived from structural analysis of the as-synthesized catalysts. It is further possible to address the electronic structure of a working catalyst. Using copper/oxygen as example, transient oxygen species were identified that are directly responsible for selective oxidation. Such species were characterised before by surface science experiments.

1   DEFINITION OF "*IN-SITU*" ANALYSIS

In-situ catalyst characterisation is frequently required and was systematically expressed first in reference one[1]. The definition there was that catalysts should be characterised in their working state. This definition was often relaxed to an experiment conducted in the presence of a gas phase similar to the educt gas phase of an operating system. The conflicting requirements for chemical kinetics on one side and geometric requirements of the *in-situ* cell together with compromises in gas pressure, composition and reaction temperature on the other side led to flexible definitions of "working state". In many experiments no proof of the working state is given, it is simply *a priori* assumed that the provision of some reaction environment brings a catalytic material into the same active state where it is used in conventional reactor experiments.

In-situ studies on so-called model systems are usually forgiving for not maintaining realistic testing conditions. The sensitivity of a catalyst to changes in its reaction environment is closely related to its efficiency. Highly effective catalysts are materials far from their equilibrium state. They contain much "chemical energy" and transform easily into less reactive or differently active materials. This is one origin of the often-quoted "material gap" between model and real-world catalysts. Model systems prepared with the emphasis on a rigorous structural definition are *per definitionem* equilibrium materials with the exception of the inevitable surface defect. Hence, their bulk is often non-reactive and stable to modifications of the reaction environment or the structural response is slow within the time window of *in-situ* observation.

True *in-situ* experiments on practical systems must be conducted in such a way that the catalytic performance is measured simultaneously with the spectroscopic or structural property of the experiment. Reliable *in-situ* experiments are performed at multiple steady states and a quantitative correlation between catalytic function (activity, selectivity) and spectroscopic/structural property is established. The preferred way of doing this should be a modulation of the reaction conditions coupled with observation of the temporal evolution of the spectroscopic signal. Only then, it is proven that a direct and physically meaningful correlation exists between structure/spectral property and catalytic function.

2   WHY *IN-SITU* ANALYSIS?

Many negative experiences with the relation of characterisation results to functional performance have their origin in the fact that a "non-relevant" property of the catalyst has been analysed with no physical relation to the function. The typical example is the application of X-ray powder diffraction phase analysis used to correlate structure with catalytic performance. In general, there exists a relation between bulk structure and surface properties. The details of this relation are, however, difficult to unravel, as reconstructions that minimise the free energy tend to modify grossly the bulk structure in the surface-near region. In Figure 1(A) the structure determination of iron oxide is exemplified[2]. It should be noted that the distances between the top atomic layers deviate in both directions from the respective bulk values by a factor of 2. It is the great virtue of surface science studies of catalytic model systems to provide such data. They give evidence that the surface chemistry and so the disposition of active sites cannot be deduced from cuts through the bulk phase and thus render it difficult to correlate the information about the average solid phase with the structure of the active phase or even with that of the active sites.

Surface reconstructions are controlled by free energy minimisation allowing the conclusion that there exists no unique surface structure (such as under static LEED analysis conditions) but rather a dependence of the surface structures on testing conditions[3]. This is illustrated in Figure 1(B,C) where structural and catalytic results of the application of the well-ordered initial model film in the dehydrogenation of ethylbenzene (EB) to styrene are shown. The conversion was measured in a special microreactor[4] that allowed testing single crystal samples of 1 cm² surface area under technical reaction conditions. Significant conversion occurs only after a long induction period. After interrupting the active state, the structure of the film was re-analysed. The LEED analysis revealed massive disorder of the phase. The induction period was required to produce the catalytically relevant disorder. Unfortunately, there is no way of ascertaining that the surface structure of the activated (under reaction condition transformed) model film is identical with that shown in the top panel of Figure 1. The experiment indicates the limitation of model surface studies with rigorously defined geometric structures. Such experiments[5] describe catalytic processes in boundary conditions of minimal activity and fail to capture essential parts of structural (non-periodic) complexity[6] of active phases. Consequently, it is important to conduct structure-sensitive *in-situ* experiments that can look at defective and disordered ("dirty") objects.

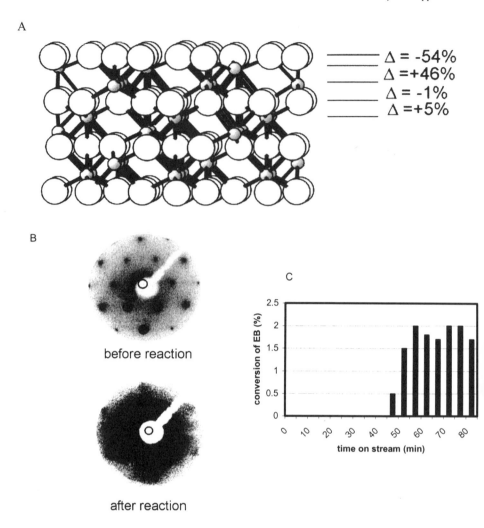

Figure 1 *A: Model of the surface crystallography of iron oxide (Fe$_2$O$_3$) as determined from LEED analysis of a thin film grown at 1mbar oxygen pressure. The numbers indicate the positional changes of the atomic layers in percent with respect to the position in the bulk structure. B: LEED image of a film before and after use as catalyst of dehydrogenation of ethylbenzene (EB) to styrene; reaction temperature 873 K, reactant pressure 1bar, composition steam to EB 10:1, LHSV 0.5 h^{-1}. The unit cell reflections for (001) Fe$_2$O$_3$ are indicated by circles. C: Evolution of the conversion to styrene as function of time on stream under the conditions given in (B)*

Analytical tools may probe the relevant surface property of the material but the test conditions are so vastly different from the reaction conditions that a non-relevant state of the system is probed. In Figure 2, this is illustrated for the NEXAFS spectrum of a (001) single crystal surface of MoO$_3$ prepared by standard UHV conditions and annealed in situ at 10 mbar oxygen pressure and at a temperature of 673 K. This system[7] is used frequently as model for the complex partial oxidation catalysts employed in technical selective oxidation reactions[8]. The average bonding state of oxygen is different in the two cases despite the spectra indicating no substantial loss of oxygen atoms from the surface under UHV conditions (height of the edge jump). After reoxidation, the local electronic structure is different as the Mo d-states are less involved in the metal-oxygen interaction (increase in white line). It is obvious that two different materials are studied under the two different partial pressures of oxygen. As for most studies, the *in-situ* preparation is not accessible, those studied deal with oxygen-deficient and partly reduced oxide surfaces although they are considered to be intact MoO_3 surfaces. This discrepancy has serious consequences when, for example, bulk-sensitive experiments, catalytic data on nominally MoO_3 samples, or theoretical results are compared with such "faulted" data.

These findings have their exact analogy in the bulk chemistry. High quality vapour-phase grown crystals of MoO_3 were defected by thermal annealing in inert atmosphere at 823 K[9]. The electron diffraction data in Figure 2 reveal that despite of the retention of the average phase, defects are present that manifest themselves in a violation of the symmetry extinction rules and hence in extra spots that do not indicate a change in phase. Such subtle effects are "invisible" in powder X-ray diffraction and hence inaccessible to in situ observation by diffraction but may be of major consequences to the electronic structure of "MoO_3". The results of Figure 2 indicate that there are correlations between the real bulk structure and the electronic state of the surface. It is evidently necessary for functional description of active catalysts to describe the real structure of the system in detail.

3   CORRELATION OF SOLID-STATE REACTIVITY AND CATALYTIC PERFORMANCE

The statement about a connection between bulk reactivity and catalytic performance needs confirmation by an *in-situ* experiment. Pure MoO_3 was used as a model substance for selective oxidation of C3 compounds approximated by propene oxidation to acrolein. Well-ordered single crystalline MoO_3 is inactive in this reaction[9]. Defective varieties were prepared by *in-situ* decomposition of ammonium heptamolybdate (AHM) or $H_3[PMo_{12}O_{40}]$ (HPA).

The bulk reactivity of these oxides was determined by *in-situ* observation of the re oxidation of MoO_2 to MoO_3. The reverse of the expected partial reduction was chosen as the dioxide is structurally well defined and allows a quantitative determination of the fractional degree of conversion (α). The data were obtained by time-resolved NEXAFS at the Mo K-edge[10]. The raw data were subject to principal component analysis and correlated with the spectral weights of MoO_2 and MoO_3 respectively. No stable intermediates occur under these conditions. Figure 3 shows results for a series of isothermal experiments. The phase change between MoO_2 and MoO_3 occurs instantaneously with the changing chemical potential of oxygen as seen by comparing the gas pulse profile with the normalised conversion curves. Their shape at various temperatures is qualitatively different exhibiting a diffusion-controlled kinetics for low temperatures and a nucleation-controlled kinetics at higher temperatures. The inset in Figure 3 shows the curves on a re-normalised timescale highlighting the qualitative difference in line shape. Without *in-situ* time resolved NEXAFS it would have been impossible to monitor the change in reaction kinetics in such a fast process being completed in about one minute for a free standing solid pellet.

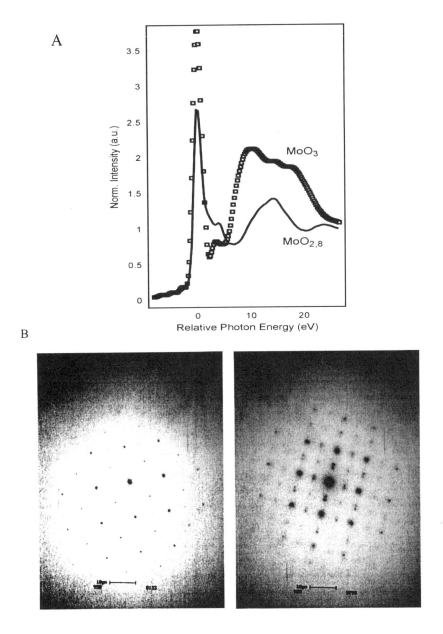

Figure 2 *A: XAS at the oxygen K-edge (530 eV) of a MoO₃ (0001) single crystal. The crystal was investigated after thermal cleaning and after reoxidation at 10 mbar oxygen at 873 K. B: Electron diffraction data of vapour-phase grown single crystals of MoO₃ in their as-prepared state (left) and after heating to 673 K in He (right)*

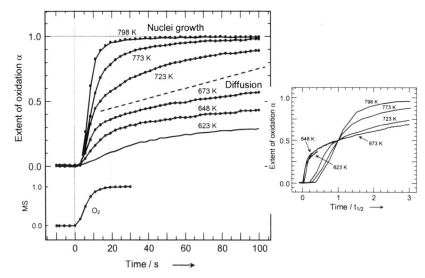

Figure 3 *Series of isothermal conversion experiments of MoO₂ into MoO₃ observed by in-situ time resolved EXAFS. The oxygen gas profile is given at the bottom. The inset shows the same data in a re-normalised time scale*

The velocity of the bulk transformation highlights that reaction-isolation studies as conventionally done with catalytic materials are inadequate if reactive forms of functional solids are under study.

With this knowledge it is possible to perform an *in-situ* conversion experiment of propene oxidation and apply the usual non-stationary conversion experiment to unravel the effects of the bulk structure on the monitored catalytic performance. In Figure 4 (A) the transients for partial oxidation to acrolein are shown as measured for the defective MoO_3 ex AHM produced in the *in-situ* EXAFS cell. Interrupting the oxygen flow causes no irreversible damage (reduction) of the catalyst as can be seen from the conversion levels before and after the oxygen-free time. No selective oxidation occurs from the lattice oxygen reservoir below a critical lower temperature. Above that temperature there is 100% selective (see Figure 4 (B)) conversion of propene to acrolein using the lattice oxygen of the oxide. It is highly significant that this critical temperature is identical with the temperature at which the redox kinetics of the bulk oxide changes from diffusion to nucleation control (see data in Figure 3). The notion[11] that a *diffusion* process is the physical explanation of the Mars-van-Krevelen (MvK) kinetics is not valid in all cases. The experiment shows that the oxygen available at the surface for selective oxidation is created in a bulk process[12] controlled by the growth of MoO_2 nuclei and not by a transport process through any type of defects. This finding depends heavily on the defects state of the oxide and may be different if better-defined starting materials are used[13] that will undergo the Magnelli defect series (crystallographic shear structures)[14] before they convert into the dioxide.

The correlation of the lattice oxygen consumption with the average electronic structure as sensed by the shift of the Mo K-edge NEXAFS is illustrated in Figure 4 (B). The loss of oxygen is related to a chemical reduction of the bulk oxide that is, however, much larger than to be expected from the conversion of the propene. This indicates that catalysis with lattice oxygen and thermally induced decomposition are at least partly independent. The control experiment of degradation in He is partly inconclusive as the extent of reaction is much smaller. This is no surprise as the removal of oxygen from the bulk is kinetically hindered through the gas-solid interface. A rapid adaptation of the oxygen partial pressure between solid oxide and gas phase can occur if precisely at this barrier a chemical process like propene oxidation removes the inhibition. The re-oxidation in the bulk upon admission of gas phase oxygen is instantaneous with this sample. This casts doubt on the notion that the rate-determining step of the MvK mechanism is the re-oxidation[11] of the catalyst i.e. of the surface-near region of the oxide. A detailed inspection of the traces reveals that indeed two kinetically different processes of re-oxidation are operative. The simple assignment to bulk and surface processes is inadequate as the transient for total oxidation of propene follows this double feature exactly indicating the both reactions have effects on the surface electronic structure. It can further be concluded from the comparison of the transients for total and partial oxidation which are not identical in shape that the two propene reactions are parallel and not consecutive processes, i.e. that at least part of the total oxidation occurs directly from propene in analogy to observations on propane oxidation[15] and not through combustion of acrolein. This is corroborated by the time profile of the transients after the halt of oxygen admission where the traces for the two reactions cross over and hence cannot be consecutive in nature. If better crystalline samples are used then the direct correlation between bulk redox chemistry and surface reaction is lost as the bulk processes are kinetically inhibited[15,16] (by diffusion or nucleation as shown in Figure 3). This is not necessarily felt by the catalytic probe reactions at the surface. The situation in Figure 4 is thus a fortunate case in which a highly defective bulk reacts so fast to changes in the gas-phase chemical potential that a correlation occurs between bulk and surface reactions that become the overall rate-determining steps. The findings stress that it does matter what model material is used and that the application of structurally well-defined materials may not be the optimum choice for every experimental strategy. It is pointed out that only the structure determination by EXAFS done "on the go" with the experiments in Figure 4 (data not shown, confirmation of MoO_3 as phase amongst many other possibilities[16] with defects in the polyhedra network) credits the term "model study" for this experiment.

With respect to mechanistic studies it must be concluded that the method of instationary changes of partial pressure of reactants is inadequate as severe consequences for the bulk chemistry will arise and the assumption of an independence of bulk and surface chemistry is not adequate. Conducting the experiments in the presence of propene and oxygen leads to steady state bulk structures containing mixtures of MoO_3 and Magnelli phases. The way to identify the role of each of these phases is the steady state isotope transient method that does not affect the abundance and dynamics of the defect structure given by the overall chemical potential of the gas phase, the thermodynamic stability of the solid phase(s) and by the solid-state dynamics leading to equilibrium.

4   A SURFACE-SENSITIVE *IN-SITU* EXPERIMENT

The limitations of using intrinsically bulk-sensitive methods to investigate surface catalytic processes became apparent in the previous example. Surface-sensitivity is achieved if not photons (except in special cases) but electrons are used as the carriers for the spectral information. There is a fundamental physical limitation of the application of electron probes under *in-situ* conditions as the mean free path of electrons of relevant energies in gases is very limited (a dimension of millimetres at millibar pressure). Complementing the developments in non-linear photon-in photon-out techniques, an international co-operation between the present group and the University of Berkeley set out to design the very useful XPS technique into a "high-pressure" *in-situ* version. The new idea was to use besides the conventional[17] set-up a high-pressure (10 mbar) flow reactor chamber with an aperture for letting the photoelectrons produced from a highly focused synchrotron radiation source escape. This was combined with an electron optics that collects all photoelectrons through the multiple differential pumping stages.

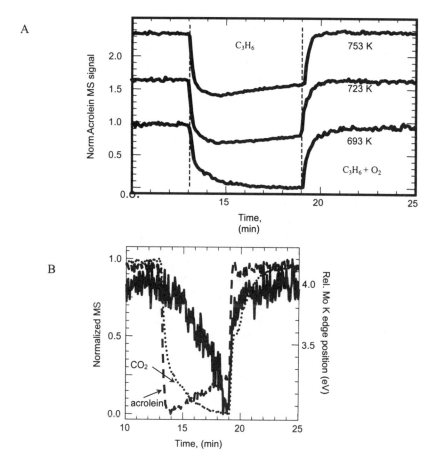

Figure 4 *A: Yield in acrolein during three isothermal in situ time resolved EXAFS experiments with interruption of the supply of oxygen from relative time 13 to 19 min.*
B: Correlation of the average Mo valence as given by the Mo K-edge position relative to that of Mo metal (left ordinate) with the catalytic activity during an instationary experiment as conducted in (A) at 733 K. The sample was defected MoO₃ ex thermal degradation of AHM. The time-resolved mode of DXAFS allowed the determination of the Mo valence every 2s and so to resolve the kinetics of the oxygen removal

A test case is the investigation of selective oxidation of methanol to formaldehyde over metallic copper. This process occurring as dehydrogenation and/or oxydehydrogenation was studied extensively by *in-situ* NEXAFS[18] and earlier by surface science[19]. These studies brought about the relevance of unique oxygen species termed as "sub oxide"[20] for the selective catalytic reaction. Conventional copper oxides were identified to catalyse the total oxidation and bare metal was found to be inactive. A target for *in-situ* XPS was thus the search for a sub oxide, the confirmation of its relevance for selective oxidation and the determination of surface electronic structure of the copper (metallic or oxidic) during reaction.

Figure 5 summarises relevant results in the oxygen 1s region. The richly structured spectrum occurs due to the presence of gas phase XPS lines besides the relevant surface information. Table 1 summarises the peak assignment. The role of the three oxygen species residing at the surface for the catalytic process was clarified using different gas compositions. Table 2 reports the relevant performance data. The following findings were corroborated by changing the reaction temperature and the total pressure as additional variables all giving rise to a unique structure-function relation. Traces of potassium and silicon as impurities on copper foil form surface oxides without detectable effects on the surface reactivity. The copper oxide Cu_2O, is relevant for the catalytic action. It dominates the spectrum when an excess of oxygen over methanol is applied. From a comparison of the spectral weights in Figure 5 with the data in Table 2 it appears that the oxide catalyses the total oxidation. Relating the amount of oxide to the 10% conversion to CO_2 occurring from it indicates that it is an inefficient catalyst. A plausible morphological picture of this state of the system is a patchy surface with thick islands of copper oxides between the other species. The surface is free of copper oxide when excessive methanol is present. Then the sub oxide species occurs in abundance parallel to the oxygen content of the gas phase. This species is a very effective catalyst and induces the selective oxidation. The sub oxide peak is only present when the system is catalytically active; when the gas pressure or any component is removed from the cell its abundance falls to zero within seconds after interrupting the reaction. This highlights the relevance of *in-situ* spectroscopy bringing about the existence of highly metastable species during catalysis under "realistic" conditions. It is obvious that the present results fully corroborate the NEXAFS results and that additional spectroscopic information (chemical shifts and surface abundances) were obtained. The data are, without any extrapolation, directly related to the functional performance as catalytic data and XPS both probe the surface of the catalyst.

BE (eV)	Assignment	Surface	Gas
540.0	Oxygen		XX
539.4			
536.6	CO_2		XX
535.2	H_2O		XX
534.6	CH_2O		XX
534.2	CH_3OH		XX
531.5	Hetero oxide	XX	
530.6	Cu_2O	XX	
529.9	Sub oxide	XX	

Table 1 *Assignment of the peaks in the oxygen 1s spectrum displayed in Figure 5 and location of the species*

These validated results can be compared to investigations from the surface science community and to draw some conclusions about their relevance for practical catalysis. The species assigned here as sub oxide is by its spectroscopic signature identical to the Oδ- investigated extensively by the group of Roberts[17,19]. Ample evidence was found in these studies for the participation of this species in the conversion of methanol and other molecules. From the fact that a positive correlation is found between abundance of oxygen and extent of selective oxidation it can be concluded that the generation of this species may be rate determining as otherwise the surface abundance of a fast-reacting species should be low at high conversions. The metastable character of negatively charged oxygen atoms or "hot oxygen" at surfaces was evidenced in model experiments with STM allowing the observation of surface dynamics and even desorption of such species[21] at conditions much milder than used here (but under UHV pressure). These observations scale well with the sensitivity of the surface abundance of the sub oxidic species to changes in reaction conditions.

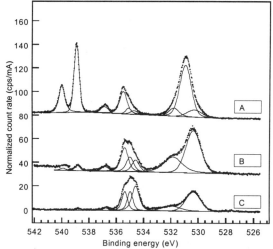

Figure 5 *In-situ XPS data in oxygen 1s region of methanol oxidation over polycrystalline copper. Total pressure 0.6 mbar, reaction temperature 673 K, flow-through mode. Gas composition as methanol : oxygen molar ratio: A=1:2, B=3:1,C=6:1*

State	Feed ratio CH_3OH : O_2	Conversion CH_3OH (%)	Yield CH_2O (%)	Yield CO_2 (%)
A	1:2	68	46	22
B	3:1	58	45	13
C	6:1	38	35	3

Table 2 *Conversion data of the Cu sample under the conditions used for the data in Figure 5*

The *in-situ* XPS experiment allows validation of the notion that the predominant electronic structure of the active copper catalyst is indeed the metallic state and not an oxide. This is the tacit assumption in all model experiments from surface science[22], carried out such that no oxide scale formation can occur. Figure 6 reports a temperature scan experiment. An initially metallic surface is oxidised first in the feed at intermediate temperatures. At 573 K, the activation is sufficient to remove the oxide scale and to arrive at a metallic surface despite the presence of oxygen in the gas phase (see trace B in Figure 5). This change associated with the occurrence of a Fermi edge coincides with the evolution of a steady state catalytic activity that is absent during the state of oxide coverage. The fact that lowering the temperature to 523 K did not bring back the oxide scale where it was present during the initial heating of the sample indicated that the surface has undergone restructuring making it less susceptible to deep oxidation under catalytic conditions. The SEM image of Figure 6 gives a typical impression of copper metal restructured under methanol oxidation conditions at atmospheric pressure. The complex pattern exhibits predominant facets of [111] and as minority [112] orientation as could be deduced from EBSD analysis of the facets. It is apparent that the active catalyst is a modified version of copper metal[23] and not oxide. Copper oxide is present in inactive states of the system and its formation is affected by morphological changes of the copper that occur during catalytic action much as in the silver system[24] used for the same reaction.

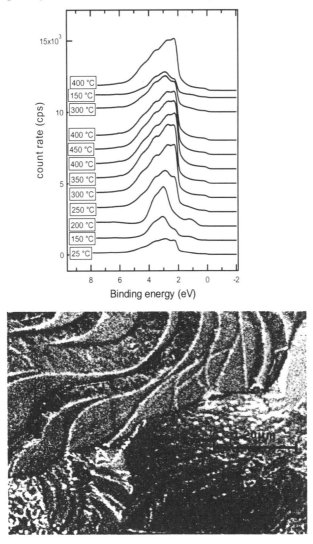

Figure 6 *In-situ valance band data of a copper foil under methanol oxidation conditions (mixture B in Figure 5) during a temperature scan. The temperature was increased linearly with 5 K min, the acquisition of each spectrum took less than 2 min. The line indicates the occurrence of the Fermi edge. The SEM image stems from a 100 μm Cu sphere held for 24 h under methanol oxidation conditions of stoichiometric feed and 573 K reaction temperature. The flat terraces are of [111] orientation, the rough parts expose facets of [112] orientation. An oxide scale from isolation out of the reactor covers the sample*

5 CONCLUSIONS

In-situ experiments are now available probing the bulk and surface reactivity of "real" catalysts. They yield detailed information on geometric and electronic structure of working systems. It is of minor relevance that a pressure gap still exists between a few millibars and atmospheric pressure. Significant conversions are achieved during *in-situ* experimentation under a few millibar pressures, at reaction temperature, and in the correct feed gas composition. Thus, all processes affecting the structure of the surface and of the sub surface region are operative. As the geometry of the gas-solid interaction is far from that in a practical catalytic reactor it cannot be expected that the kinetic parameters match between macrokinetic experiments and the *in-situ* experiments. Reservation is required for this reason when interpreting structure-function correlations in quantitative ways. The experiments prove, however, very clearly how dynamically bulk and surface of catalytically active systems behave. The findings underline that surface science model systems are rigorous boundary condition states of a system and the corresponding results should not be extrapolated quantitatively to practical operation. The present findings indicate, however, that the qualitative conclusions drawn from model studies are valid in pictures of the complex practical situation. There should be a co-operation between surface science giving insight into the mode of operation of catalysts and a combined *in-situ* analytical-preparative effort providing the relevant systems. In such a way, the gaps between model and real catalysts can be bridged effectively and there is no room left for speculations about possible missing fundamental insight into the mode of operation of heterogeneous catalysts.

Acknowledgements

The work presented here would be impossible without the engaged efforts of the co-workers of the group. Results from W. Ranke, Ch. Kuhrs, G. Ketteler, T. Ressler, J. Wienold, F. Girgsdies, R. Jentoft, H. Bluhm, A. Knop-Gericke, M. Haevecker and E. Kleimenov were used for illustration. The author is indebted to them and to their respective co-workers and to all contributing to the design and building of the necessary equipment. The fruitful co-operation with M. Salmeron and D.F. Ogletre was mentioned. Support is acknowledged from the synchrotron radiation facilities BESSY II, HASYLAB and ESRF, from the Deutsche Forschungsgemeinschaft (SPP 1051), from the Fonds der Chemischen Industrie and from the ATHENA co-operation.

References

1   "Characterisation of Catalysts" ed. J.M. Thomas and R.M. Lambert, Wiley, (1980)
2   C. Kuhrs, W. Arita, W. Weiss, W. Ranke, R. Schlögl, *Topics in Catalysis*, **14**, (2001), 111
3   K. Reuter, M. Scheffler, *Phys. Rev. Lett.*, 9004:6103, (2003), Jan. 31
4   W. Weiss, M. Ritter, D. Zscherpel, M. Swoboda, R. Schlögl, *J. Vac. Sci. Technol.*, **A16**, (1998), 21
5   Y. Joseph, M. Wühn, A. Niklewski, W. Ranke, W. Weiss, C. Wöll, R. Schlögl, *PCCP*, **2**, (2000), 5314
6   W. Weiss, R. Schlögl, *Topics in Catalysis*, **13**, (2000), 75
7   R. Tokarz, K. Hermann, M. Witko, A. Blume, G. Mestl, R. Schlögl, *Surf. Sci.*, **489**, (2001), 107

8   R. Schlögl, A. Knop, M. Hävecker, U: Wild, D. Frickel, T. Ressler, R. Jentoft, J. Wienold, A. Blume, G. Mestl, O. Timpe, Y. Uchida, *Topics in Catalysis*, **15**, (2001), 219

9   Y. Uchida, G. Mestl, O. Ovsitzer, J. Jäger, A. Blume, R. Schlögl, *J. Mol. Catal.*, **A, 187**, (2002), 247

10   T. Ressler, O. Timpe, T. Neisius, J. Find, G. Mestl, M. Dieterle, R. Schlögl, *J. Catal.*, **191**, (2000), 75

11   A. Bielanki, J. Haber „Oxygen in Catalysis", M. Dekker, (1991), p.121

12   R.K. Grasselli, *Catal. Today*, **49**, (1999), 141

13   T. Ressler, J. Wienold, R.E. Jentoft, F. Girgsdies, *Eur. J. Inorg. Chem*, **2** (2003), 301

14   D. Wang, D.S. Su, R. Schlögl, *Crystal Res. Technol.*, **38**, (2003), 153

15   K. Chen, E. Iglesia, A.T. Bell., *J. Chem. Phys.*, **B 105**, (2001), 646

16   J. Guo, P. Zavalij, M.S. Whittingham, *J. Solid State Chem*, **117**, (1995), 323

17   A.F. Carley, A.W. Owens, M.K. Rajumon, M.W. Roberts, S.D. Jackson, *Catal. Lett.*, **37**, (1996), 79

18   A. Knop-Gericke, M. Haevecker, T. Schedel-Niedrig, R. Schlögl, *Topics in Catalysis*, **15**, (2001), 27

19   A.F. Carley, A. Chambers, A. Davis, P.R. Mariotti, R. Kurian, M.W. Roberts, *Faraday Disc.* **105** (1996), 225

20   T. Schedel-Niedrig, T. Neisius, I. Boettger, E. Kitzelmann, G. Weinberg, D. Demuth, R. Schlögl, *PCCP*, **2**, (2000), 2407

21   T. Greber, R. Grobecker, A. Morgante, A. Boettcher, G. Ertl, *Phys. Rev. Lett.*, **70**, (1993), 1331

22   M. Bowker,*Top. Catal.*, **3**, (1996), 461

23   A.F. Carley, P.R. Davies, G.G. Mariotti, S. Read, *Surf. Sci.*, **364**, (1996), L525

24   A. J. Nagy, G. Mestl, R. Schlögl, *J. Catal.*, **188**, (1999), 58

REACTIONS OF 1,2-DICHLOROETHENE ON CU(110): *CIS* VERSUS *TRANS* ISOMER.

S. Haq[1], S. C. Laroze[1], C. Mitchell[2], N.Winterton[2†*], and R. Raval*[1]

[1] Leverhulme Centre for Innovative Catalysis and Surface Science Research Centre, Department of Chemistry, University of Liverpool, Liverpool, L69 7ZD, UK.
[2] ICI Chlor Chemicals, Runcorn Technical Centre, The Heath, Runcorn, Cheshire, WA7 4QE, UK.

† current address: Leverhulme Centre for Innovative Catalysis, University of Liverpool
*corresponding authors: N.Winterton@liverpool.ac.uk; Raval@liv.ac.uk

ABSTRACT

The adsorption and reactions of *cis*-1,2-dichloroethene on Cu(110) have been investigated using Reflection Absorption Infrared Spectroscopy, Temperature Programmed Desorption and Molecular Beam Adsorption Reaction Spectroscopy, and the behaviour compared with that of the *trans* isomer. No isomerisation between the *cis* and *trans* isomers was seen. The data reveal that, although the related positions of the Cl atoms force different adsorption geometries for the two isomers, the behaviour of both compounds follows a very similar general reaction pathway which is critically temperature-dependent and can be described in terms of three main regimes: Regime I occurring at low temperatures in which intact molecular adsorption occurs; Regime II, extending over the temperature range 155 - 280 K, in which desorption/dechlorination events are triggered; and, finally, Regime III, over 280 - 450 K, is governed, for both isomers, by the behaviour of the acetylene intermediate at the surface which, trimerises to form benzene and desorbs above 350 K.

1   INTRODUCTION

Chlorinated organic compounds are industrially important both in their own right and as feedstocks or intermediates,[1] though the poor selectivity of some processes leads to unwanted waste by-products. Constraints on existing production technology, changes in demand and the need to minimise waste all require new processes to be developed both to effect more selective transformations and to use by-products as feedstocks. For all these reasons, hydrodechlorination, the conversion C-Cl + H_2 → C-H + HCl, represents a reaction of increasing industrial and technological importance. Group VIII(8-10) metals, particularly palladium, platinum and copper, appear to be the most attractive hydrodechlorination catalysts in terms of activity, selectivity and stability.[2,3] Fundamental studies[4-19] on well defined single crystal surfaces and UHV conditions using surface spectroscopic techniques are aiding our understanding of the interactions of chlorinated compounds with the surfaces of these metals, enabling insights to be gained into mechanistic changes associated with the metal, surface structure, temperature, coverage and the molecular structure of the chlorinated compound.

Dichloroethenes (DCEs) are good candidates for delineating the behaviour of simple unsaturated chlorinated species with metal surfaces and, in particular, they provide a means for comparing and contrasting the reactivity patterns of different isomers. Previous work has studied the adsorption and reaction of chloroethenes on Pt(111),[15] Pt(100),[15] Pd(110),[11] Cu(100)[16] and Cu(110)[9,19]. In this paper, the adsorption, reactive transformation and product evolution of *cis*-1,2-dichloroethene on Cu(110) are reported and compared particularly with related studies on the *trans* isomer.[19] This work represents part of an on-going effort to characterise and understand the interactions of a series of chlorinated ethenes with well-defined metal surfaces, including Cu(110),[9,10,19] Pd(110)[11] and CuPd(110).[12-14]

2   EXPERIMENTAL

The experimental details of the characterisation methods and equipment used to carry out Molecular Beam Adsorption Reaction Spectroscopy (MBARS), Reflection Absorption Infrared Spectroscopy (RAIRS), Low Energy Electron Diffraction (LEED), Auger Electron Spectroscopy (AES) and Temperature Programmed Desorption (TPD), have been described in detail previously.[19] The experiments were performed in a ultra-high vacuum (UHV) chamber and the Cu(110) crystal was cleaned by sputtering (500 eV, 8 μA), annealed at ca. 873K and characterised using LEED and AES. TPD experiments were carried out using a heating rate of 2 Ks^{-1}. For RAIRS measurements the UHV chamber is interfaced with a FTIR spectrometer (Mattson-Galaxy) using KBr optics. A liquid nitrogen cooled mercury-cadmium-telluride (MCT) detector was used, accessing the spectral range of 4000-650 cm^{-1}. Spectra were acquired at a resolution of 4 cm^{-1} and by co-addition of 256 scans. The liquid *cis*-1,2-dichloroethene sample (Supelco, 99% purity) was subjected to freeze-pump-thaw cycles prior to admission into the chamber and the purity of each sample was confirmed using mass spectrometry. The exposures are quoted in Langmuirs (L) where 1L = 1x10^{-6} torr.s.

3   RESULTS AND DISCUSSION

Our previous work on *trans*-dichloroethene on Cu(110)[9,19] and Pd(110)[11] clearly demonstrates that the reaction of this molecule occupies a complex phase space, with five different reaction regimes identified for Pd(110) and three on Cu(110). Here, we show that the behaviour of the *cis* isomer is broadly similar to the *trans*-isomer on Cu(110). In contrast, our studies on 1,1-dichloroethene suggest[21] that its behaviour on Cu(110) is substantially different from that of *cis*- and *trans*-dichloroethene.

The adsorption of *cis*-1,2-dichloroethene on Cu(110) was followed as a function of exposure over the temperature range of 85 K to 450 K. RAIRS was used to obtain chemical and molecular information on the adlayer while TPD and MBARS were utilised to detect the evolution of products. From these data, three main reaction regimes can be identified, as shown in Figure 1 which displays the TPD trace of molecular desorptions following 14 L exposure of *cis*-1,2-dichloroethene to Cu(110) at 83 K. Regime I, occurring between 85 and

155 K characterises molecular adsorption in the monolayer and the multilayer. In Regime II, the retained monolayer undergoes desorption/dechlorination processes; and, finally, in Regime III (280 - 450 K) and the reaction of the acetylene intermediate dominates the system behaviour. Each of these reaction regimes is described in more detail below.

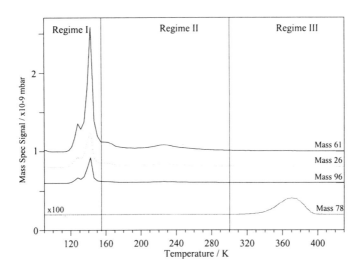

Figure 1 *TPD spectra after adsorption of 14 L cis-1,2-dichloroethene on Cu{110} at 85 K. The main reactivity regimes shown by this system are marked.*

3.1 Regime I: Non dissociative Molecular Adsorption; 85 K – 155 K.

The RAIR spectra obtained after 0.75 – 8.75L exposure of *cis*-1,2-dichloroethene at 85 K (Figure 2a) display a continuous increase in the intensity of all IR bands, with the highest exposure corresponding to an adsorbed multilayer. The infrared bands observed for the multilayer are similar to those reported for *cis*-1,2- dichloroethene in the gas and liquid phase, [22-25] with bands at 715, 848, 1292, 1589 and 3081 cm^{-1} due to fundamental modes of vibrations (Table 1). The molecular orientation adopted in the multilayer may be deduced by using the dipole selection rule as applied to a metal surface.[20] In particular we point out the almost equal intensities of the B$_1$ asymmetric ν_{as}C-Cl stretch and the A$_1$ symmetric ν_sC-Cl stretch at 848 and 715 cm^{-1}, respectively, suggests that the molecular plane and the C=C bond are, on average, highly tilted away from the surface plane. This behaviour is different from that observed for *trans*-DCE which shows a strong preference for adopting an orientation in which the molecular plane and the C=C bond are held parallel to the surface. Clearly, the relative positioning of the Cl atoms leads to a very significant difference in packing. However this packing difference does not affect the desorption temperature of the *cis*-1,2,DCE multilayer which occurs at 142 K, Figure 1, and is identical to *trans*-DCE within experimental error.

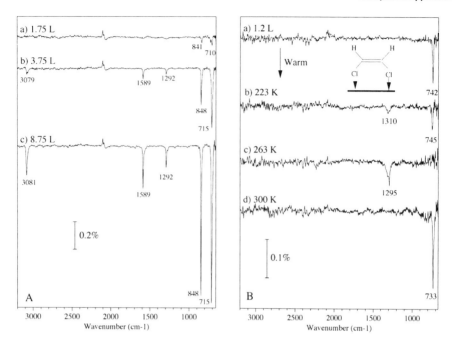

Figure 2 *RAIR spectra following a) increasing exposure of cis-1,2-dichloroethene on Cu{110} at 85 K, b) adsorption of 1.2 L at 177 K and the effect of heating.*

Table 1 *Assignment of vibrational frequencies observed in this work and those listed in reference 23.*

Assignment in the liquid /gas phase	Normal mode description	Liquid /gas phase frequency	Observed on Cu{110}multilayer
v_1 (A_1)	v_s (C-H)	3077 (IR, R) vs	
v_2 (A_1)	v (C=C)	1590 (IR, R) s	1589
v_3 (A_1)	δ (C-H)	1183 (IR, R) s	1179
v_4 (A_1)	v_s (C-Cl)	714 (IR, R) s	715
v_5 (A_1)	δ (C=C)	173 (IR, R) s	
v_6 (A_2)	τ (HClC=CClH)	876 (R)	
v_7 (A_2)	γ (C-Cl/H)	406 (R)	
v_8 (B_1)	v_{as} (C-H)	3072 (IR) s	3081
v_9 (B_1)	δ (C-H)	1294 (IR) s	1292
v_{10} (B_1)	v_{as} (C-Cl)	848 (IR) vs	848
v_{11} (B_1)	δ (C-Cl)	571 (IR) s	
v_{12} (B_2)	γ (C-H)	697 (IR) vs	

3.2 Regime II: Desorption Limited Dechlorination Events in the Monolayer; 155 K - 280K.

Upon heating to 177K, desorption of the weakly-held multilayer occurs, leaving a more strongly adsorbed monolayer at 177 K. Figure 2b shows that the RAIR spectrum obtained for this adlayer is quite different from that of the multilayer, displaying a single strong absorbance at 742 cm^{-1}. We attribute this band to the A$_1$ symmetric v$_s$C-Cl stretch and suggest that the upshift in frequency from the gas phase value arises as a result of direct interaction between the Cl atoms on the molecule and the Cu atoms in the surface. This behaviour is consistent with density functional theory (DFT) calculations carried out on trichloroethene on CuPd(110)[4] where the surface interaction is dominated by Cl-Cu bonding and leads to a strong perturbation of the C-Cl bond. It is proposed that in order to maximise this interaction, the *cis*-1,2-DCE orients itself with the C=C bond parallel but the molecular plane normal to the surface allowing both Cl atoms to interact directly. Such an adsorption geometry would also explain why the A$_1$ symmetric v$_s$C-Cl stretch is observed but not the B$_1$ asymmetric v$_{as}$C-Cl stretch. This appears to be an activated process as this adsorption symmetry is not observed at 85 K.

Upon increasing temperature to 223 and 263K, changes are observed in the RAIRS spectra, with the molecular band at 745 cm^{-1} decreasing rapidly so that it is no longer visible at the latter temperature, accompanied by growth of a new band at ~ 1295 cm^{-1}. On the basis of our previous studies of *trans*-DCE on Cu(110)[19] and using the fingerprint analysis of Sheppard,[26,27] we assign this band to the v(C≡C) stretch of a di-σ/di-π acetylene intermediate arising from dechlorination of the molecule. We also note that the vibrational literature reported for C$_2$H$_2$ adsorbed on Cu single crystal surface all show a significant 1292 cm^{-1} peak.[26-31] From the TPD data, Figure 1, it can be seen that the changes in the RAIRS spectra coincide with a broad molecular desorption peak in this temperature range. This suggests that the observed dechlorination is desorption-limited, *i.e.* vacant surface sites to be released to accommodate the products of a surface reaction. No desorption of chlorine is seen in this temperature range, suggesting that it is still attached to the copper surface.

Regime II thus comprises of the overall following process:

x C$_2$H$_2$Cl$_2$(a) → n C$_2$H$_2$Cl$_2$(g) + m C$_2$H$_2$Cl$_2$(a) + (x-n-m) C$_2$H$_2$ (a) + 2(x-n-m) Cl(a)

3.3 Regime III: Reactions of Molecular Intermediates; 280 K – 450 K.

Regime III is dominated by the reactions of the surface-bound molecular intermediates and was investigated using structural spectroscopic studies, TPD and thermal molecular beam experiments.

3.3.1   Structural Studies of the Adsorbed Layer

RAIRS data show that at 300 K, a new species is created at the surface, producing a single dominant band at 733 cm^{-1}, Figure 2b, closely resembling the out-of-plane γ(C-H) bending mode observed for benzene on Cu(110).[33] This is the major dipole-allowed mode of a flat-lying benzene species, also seen at the same position in related experiments using *trans*-DCE on Cu(110), suggesting that a trimerisation reaction of the adsorbed acetylene has occurred. The frequency of the out-of-plane γC-H mode is upshifted compared to the gas-phase molecule (671 cm^{-1}) or to adsorbed benzene on clean Cu(110)[33] where it occurs at 685 cm^{-1}. This increased frequency has been directly attributed to a stronger interaction of benzene[26,27,36] with a Cl-covered surface where the electrostatic field of the halogen increases the local work-function, inducing an increase in charge donation from benzene to Cu. TPD data, Figure 1, directly support this by showing that benzene is evolved at a temperature of 370 K when mass 78 (C$_6$H$_6$) and the known benzene fragments at amu 50,

51 and 52 desorb simultaneously although they are not shown. In contrast, benzene adsorption is limited to <300 K on clean Cu(110)[33] and Cu(111).[34] The increased chemisorption strength of benzene means that the evolution of product from the cyclotrimerisation process is desorption-limited on the Cl-contaminated Cu(110) surface, whereas it is reaction-limited on clean Cu(110).[28,33,35] Other possible desorption products, such as hydrogen, hydrogen chloride or carbon-containing compounds, are not seen.

The mechanism of trimerisation of acetylene on extended surfaces is thought to be similar to that observed in homogeneous catalysis using transition-metal cluster compounds.[32,37] A stepwise mechanism, shown below is thought to occur:

$$2C_2H_2(a) \rightarrow C_4H_4(a) \qquad \text{STEP 1}$$
$$C_4H_4(a) + C_2H_2(a) \rightarrow C_6H_6(a) \qquad \text{STEP 2}$$

We were unable to observe the metallopentacycle species in the IR spectra but the data in our studies of *cis*- and *trans*-DCE are in agreement with the general mechanism shown above, with Step 1 appearing to be the rate determining step for benzene formation.

3.3.2   *Molecular Beam Studies*

The monitoring of reactive adsorption as a function of temperature using molecular beam studies gives valuable information about sticking coefficients and general adsorption kinetics. Figure 4 illustrates the apparatus and procedure used and the typical uptake profile obtained for adsorption of *cis*-1,2-dichloroethene at 296 K. The beam enters the analysis chamber on opening flag 1 and is totally reflected from the inert flag 2 in front of the crystal. When this second flag is removed the beam strikes the crystal and the signal decreases as adsorption proceeds. Once the surface is saturated the signal recovers back to the level observed from reflection off flag 2.

Sticking probability curves obtained from such MBARS data are shown for both the *trans* and the *cis* isomer collected at 296 K. Within experimental error and variation of beam flux between the two experiments, it can be concluded that both isomers show very similar behaviour, namely that the initial sticking probability, S_0, is high at ~ 0.75, and the shape of the S *versus* time curve is indicative of precursor adsorption kinetics.

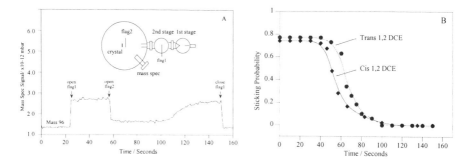

Figure 3 *A)Schematic of molecular beam apparatus and procedure used. B) The change in sticking probability with time for cis- and trans-1,2-dichloroethene.*

4   CONCLUSIONS

The adsorption and reaction behaviour of *cis*-1,2-dichloroethene on Cu(110), studied over the temperature range 85 - 450 K, shows that the reaction of this molecule follows a reaction pathway that is broadly similar to that exhibited by *trans*-1,2-dichloroethene on Cu(110). In both cases, the reaction phase space can be divided into three regimes where different events occur: Regime I (85 - 155K) representing molecular adsorption in the monolayer and the multilayer; Regime II (155 - 280 K) in which desorption and dechlorination occur to leave acetylene and coadsorbed Cl atoms at the surface; and, Regime III (280 - 450 K) where acetylene trimerises to benzene, which subsequently desorbs. Overall, copper is found to be extremely effective at cleaving the C-Cl bond (observed as low as 220 K in our work), while the C-H bond remains intact. In this respect, the behaviour of Cu differs significantly from group VIII metals such as Pd (110)[11] where a large part of the reaction phase diagram is dominated by HCl or H_2 production, testifying to the ease with which C-H bond-breaking occurs.

Acknowledgements

We are grateful to EPSRC for equipment grants and a postdoctoral fellowship for SH. We thank ICI Chemicals and Polymers Ltd and the University of Liverpool for a research studentship for SCL.

References

1   *'Kirk-Othmer Encyclopaedia of Chemical Technology'*, 4th Edition, 2000, *'Chlorocarbons and Chlorohydrocarbons'*, Vols 5 and 6, Wiley-Interscience, New York, USA.
2   S.C. Fung and J.H. Sinfelt, *J. Catal.*, 1987, **103**, 220.
3   L.S. Vadlamannati, V.I. Kovalchuk and J.L. d'Itri, *Catal. Letts.*, 1999, **58**, 173.
4   L.A.M.M. Barbosa, D. Loffreda and P. Sautet, *Langmuir*, 2002, **18**, 2625.
5   L.A.M.M. Barbosa and P. Sautet, *J. Catal.*, 2002, **207**, 127.
6   F.H. Ribiero, C.A. Gerken, G. Rupprechter, G.A. Somorjai, C.S. Kellner, G.W. Coulston, L.E. Manzer and L. Abrams, *J. Catal.*, 1998, **176**, 352.
7   C.W. Chan and A.J. Gellman, *Catal. Letts.*, 1998, **53**, 139.
8   V.H. Grassian and G.C. Pimentel, *J. Chem. Phys.*, 1988, **88**, 4478.
9   Y. Jugnet, N.S. Prakash, J-C. Bertolini, S.C. Laroze and R. Raval, *Catal. Letts.*, 1998, **56**, 17.
10   S.C. Laroze, S. Haq, R. Raval, Y. Jugnet and J-C. Bertolini, *Surf. Sci.*, 1999, **433-435**, 193.
11   L.H. Bloxham, S. Haq, C. Mitchell and R. Raval, *Surf. Sci.*, 2001, **489**, 1.
12   C.J. Baddeley, L.H. Bloxham, S.C. Laroze, R. Raval, T.C.Q. Noakes and P. Bailey, *Surf. Sci.*, 1999, **433**, 827.
13   T.C.Q. Noakes, P. Bailey, S. Laroze, L.H. Bloxham, R. Raval and C.J. Baddeley, *Surface and Interface Analysis*, 2000, **30**, 81.
14   C.J. Baddeley, L.H. Bloxham, S.C. Laroze, R. Raval, T.C.Q. Noakes and P. Bailey, *J. Phys. Chem.*, 2001, **105**, 2766.
15   A. Cassuto, M.B. Hugenschmidt, Ph. Parent, C. Laffon and H.G. Tourillon, *Surf. Sci.*, 1994, **310**, 390.

16   M.X. Yang, P.W. Kash, D.H. Sun, G.W. Flynn, B.E. Bent, M.T. Holbrook, S.R. Bare, D.A. Fischer and J.L. Gland, *Surf. Sci.*, 1997, **380**, 151; M.X. Yang, J. Eng, Jr., P.W. Kash, G.W. Flynn, B.E. Bent, M.T. Holbrook, S.R. Bare, J.L. Gland and D.A. Fischer, *J. Phys. Chem.*, 1996, **100**, 12431.

17   J.L. Lin and B.E. Bent, *J. Phys. Chem.*, 1992, **96**, 8529.

18   K.T. Park, K. Klier, C.B. Wang and W.X. Zhang, *J. Phys. Chem. B*, 1997, **101**, 5420.

19   S. Haq, S.C. Laroze, C. Mitchell, N. Winterton and R. Raval, *Surf. Sci.*, accepted for publication.

20   R.G. Greenler, D.R. Snider, D. Witt and R.S. Sorbello, *Surf. Sci.*, 1982, **118**, 415.

21   S. Haq, S.C. Laroze, C. Mitchell, N. Winterton and R. Raval, Surf. Sci., unpublished work.

22   G. Herzberg, *Infrared and Raman Spectra of Polyatomic Molecules 9D*, Van Nostrand Co. Inc., New York, 1945, p330

23   H.J. Bernstein and D.A. Ramsay, *J. Chem. Phys.*, 1949, **17**, 556.

24   H.J. Bernstein and D.E. Pullin, *Can. J. Chem.*, 1952, **30**, 963.

25   M.J. Hopper, J. Overend, M.N. Ramos, A.B. M. S. Bassi, and R.E. Bruns, *J. Chem. Phys.*, 1983, **79**, 19.

26   N. Sheppard and C. de la Cruz, *Adv. In Catal.*, 1996, **42**, 181.

27   N. Sheppard, *Ann. Rev. Phys. Chem.*, 1988, **39**, 589.

28   N.R. Avery, *J. Am. Chem. Soc.*, 1985, **107**, 6711.

29   Ts.S. Marinova, P.K. Stefanov, *Surf. Sci.*, 1987, **191**, 66.

30   B.J. Bandy, M.A. Chesters, M.E. Pemble, G.S. McDougall and N. Sheppard, *Surf. Sci.*, 1984, **139**, 87.

31   M.AChesters and E.M. McCash, *J. Electron Spectrosc. Rel. Phenom.*, 1987, **44**, 99.

32   G. Frenking, in '*Modern Coordination Chemistry: The Legacy of Joseph Chatt*', Eds. J. Leigh and N. Winterton, Royal Society of Chemistry, Cambridge, 2002, p111.

33   S. Haq and D.A. King, *J. Phys. Chem.*, 1996, **100**, 16957.

34   M. Fujisawa, T. Sekitani, Y. Morikawa and N. Nishijima, *J. Phys. Chem.*, 1991, **95**, 7415.

35   J.R. Lomas, C.J. Baddeley, M.S. Tikhov and R.M. Lambert, *Langmuir*, 1995, **11**, 3048.

36   J.R. Lomas and G. Pacchioni, *Surf. Sci.*, 1996, **365**, 297.

37   R.M. Ormerod and R.M. Lambert in *Surface Reactions; Springer Series in Surface Science*, Ed. R.J. Madix, Springer Verlag, Berlin, 1994, Vol. 34.

ALDOL CONDENSATION OF ALDEHYDES AND KETONES OVER SOLID BASE CATALYSTS

G. J. Kelly[1] and S. D. Jackson[2]

[1]Johnson Matthey, PO Box 88, Haverton Hill Road, Billingham, Cleveland, TS23 1XN
[2] Department of Chemistry, Joseph Black Building, The University of Glasgow, Glasgow, G12 8QQ, UK

1 INTRODUCTION

Industrially, condensation reactions are of great importance in the production of a number of key compounds. These include 2-ethyl hexanol, methyl isobutyl ketone (MIBK) and Guerbet alcohols. Over 1.5 million tonnes of these chemicals are produced world-wide every year using homogeneous bases such as NaOH and Ca(OH)$_2$ [1]. It has been estimated for these compounds that 30% of the selling price is product purification, recovery and waste treatment[2]. For every 10 tonnes of product formed the current homogeneous catalysts generate about 1 tonne of spent catalyst[2]. Although solid base catalysts have a number of advantages over conventional homogeneous (NaOH, KOH) systems[3,4,5,6], according to a recent review of industrial acid and base catalysis[7] of the 127 processes identified only 10 were solid base catalysed. Solid base catalysis is therefore an area of chemistry that offers an excellent opportunity for exploitation if suitable catalysts and processes can be identified and developed. The mechanistic understanding of the processes taking place on solid base catalysts is, however, very poor. This paper probes the mechanism of gas phase aldol condensation reactions over solid base catalysts using a range of aldehydes and ketones.

1.1 Solid Base Selection

A suitable solid base must have the appropriate base strength for the reaction under investigation. If the initial reaction step is the removal of a proton from a reactant of the form R$_1$-CH$_2$-R$_2$ then the acidity of the proton to be removed depends on the identity of the R$_1$ and R$_2$ groups[8] (Table 1). The solid base selected should have sufficient base strength to carry out the reaction but should not have excessive base strength as this may lead to rapid catalyst deactivation or to side-product formation. For aldehyde and ketone condensation reactions therefore with a pK_a of 19.7 – 20 a strong base is required but not a superbase material.[9] Caustic can be used to carry out reactions with reactants with the removable proton having a pK_a of up to around 20.

Table 1   *Base Strength required to remove proton from R_1-CH_2-R_2 reactant molecule*

R_1	R_2	pK_a	Base Required	
-CH_3	-CH_3	42		
-CH_3	$CH=CH_2$	35.5	SUPER	
-C_6H_5	H	35	BASE	
-C_6H_5	-C_6H_5	33		
-CH_3	-CN	25		
-CH_3	-COOR	24.5	STRONG	
-CH_3	-$COCH_3$	20	BASE	NaOH
-CH_3	-COH	19.7		
-COOR	-COOR	11.5		
-CN	-CN	11.2	MEDIUM	
-CH_3	-NO_2	10.6	BASE	
-COR	-COR	9		
-NO_2	-COOR	5.8		
-COH	-COH	5	MILD	
-NO_2	-NO_2	3.6	BASE	

The strength of surface base sites on solids can be measured by the use of Hammett indicators[10] and expressed in terms of the acidity function (H_) proposed by Paul and Long.[11] There are few available solid base materials in the strong solid base area (H_ 20 to 25) that would be suitable for the aldol condensation of aldehydes or ketones (Fig. 1).

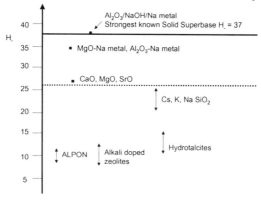

Figure 1   *Strength of solid base catalysts (H_ scale)*

1.2 Reaction Scheme

A series of condensation reactions were performed in the gas phase over fixed-bed solid base catalysts. The organic feeds tested were acetone, butanone, *n*- and *iso*-butanal. The solid base catalysts were alkali doped silica materials. A reaction scheme common to all of the organic feeds is shown in Fig. 2. The first step in the reaction is the base catalysed formation of the 'aldol intermediate'. It should be noted that from an asymmetric ketone (i.e. butanone) two possible aldol intermediates can be formed depending on which side of the carbonyl group reacts. Under our reaction conditions rapid dehydration of the aldol intermediate normally occurs resulting in the formation of an α,β – unsaturated carbonyl compound.

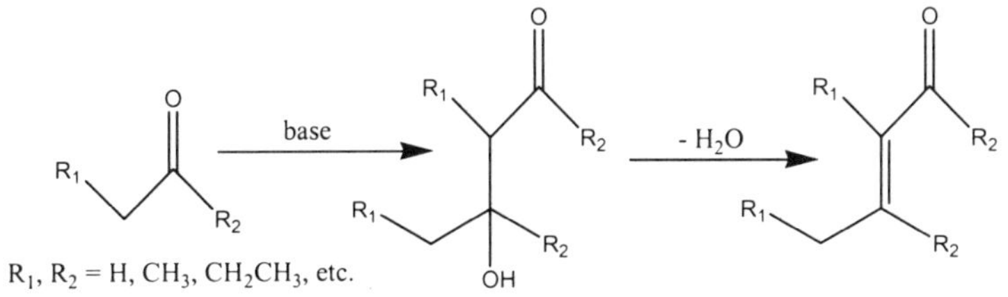

R$_1$, R$_2$ = H, CH$_3$, CH$_2$CH$_3$, etc.

Figure 2 *Reaction of aldehydes and ketones over solid base catalysts*

2 EXPERIMENTAL

2.1 Catalyst Preparation

A 4 wt% Na/SiO$_2$ catalyst and a 4 wt% K/SiO$_2$ catalyst were prepared by an impregnation method using nitrate salts. Fuji Q10 silica spheres (2-3 mm) were used which have a surface area of 359 m^2 g^{-1} and a pore volume 1.01 cm^3 g^{-1}. Both catalysts were dried overnight at 100 °C after impregnation and calcined at 450 °C for 3 hours.

2.2 Catalyst Testing

Reaction testing was carried out in a stainless steel fixed bed microreactor system that allowed testing at a range of pressures and organic and carrier flow rates[12,13]. The catalysts were crushed to a particle size of 0.6 to 1mm. H$_2$ or N$_2$ was used as a carrier gas. The organic flow was controlled using a HPLC pump, the typical liquid feed rate was 0.05 cm^3 min^{-1}. Reaction products were analysed off-line using a combination of GC and GC-MS. A larger scale test unit was used for the reaction of *iso*-butanal and mixtures of *iso*-butanal and *n*-butanal. This stainless steel fixed bed microreactor allowed the testing of the full size catalyst pellets (2-3 mm). A 10 cm^3 bed of catalyst was loaded into the reactor tube. In this unit N$_2$ was used as a carrier (250 cm^3 min^{-1}) and the liquid flow rate was 0.33 cm^3 min^{-1}.

3 RESULTS

The reaction of acetone was carried out over a 3 ml bed of the 4 wt% K/SiO$_2$ catalyst. Acetone was fed to the reactor at a feed rate of 0.05 ml min^{-1} in a H$_2$ carrier flow of 62 ml min^{-1} and at a reaction temperature of 300 °C. The results are shown in Table 2. The highest selectivity to the dehydrated aldol product, mesityl oxide (CH$_3$)$_2$CCHCOCH$_3$, was 82% at 10 bar pressure.

Table 2   *Reaction of acetone over a 4 wt% K/SiO₂ catalyst*

Run Time (hrs)	Pressure (bar)	Conversion (%)	Selectivity to mesityl oxide (%)	Selectivity to isophorone (%)
3	6	17.0	67.7	10.2
19	6	15.5	72.2	10.2
42	8	15.6	76.8	12.4
49	10	14.4	81.6	11.8
68	10	15.0	82.0	11.8

The main side product was a cyclic trimeric aldol product isophorone (see Fig. 3), which was formed with a selectivity ~ 12% throughout the test.

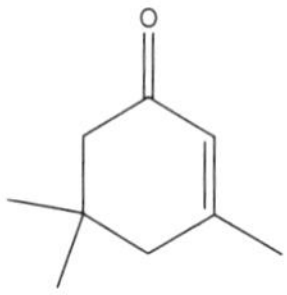

Figure 3   *Structure of isophorone*

The reaction of butanone was carried out over a 4 ml bed of a 4 wt% Na/SiO₂ catalyst. Butanone was fed to the reactor at a flow rate of 0.05 ml min^{-1} in a N₂ carrier flow of 50 ml min^{-1}, a pressure of 6 bar and at temperatures between 325 and 400 °C. The temperatures were tested in the order 375-400-325-350 and the catalyst was held for 3-4 hrs at each temperature before sampling. The key intermediates in the reaction of butanone are shown in Fig. 4. The dehydration of the terminal intermediate can produce three different unsaturated ketones depending on which proton is lost. The major product formed is the α,β – unsaturated product 5-methyl-4-heptene-3-one formed by the loss of proton c in the terminal intermediate.

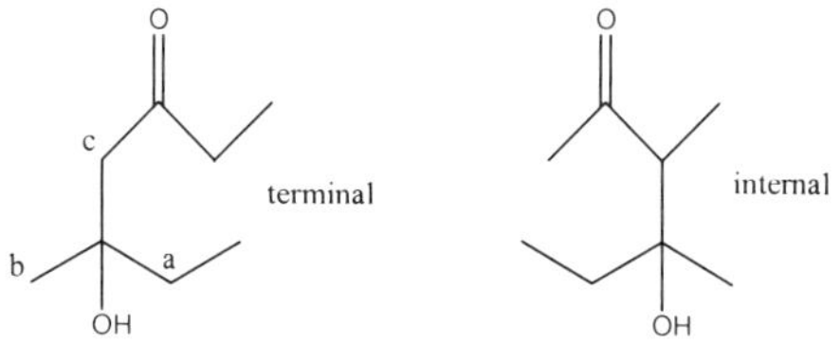

Figure 4   *Terminal and internal aldol products from reaction of butanone*

The other products from the loss of protons a and b on the terminal intermediate are also formed in lesser amounts (Table 3). Only very low levels of internal aldol products are formed. Various trimeric products were formed at all temperatures tested.

Table 3 *Reaction of butanone over 4 wt% Na/SiO₂ catalyst*

Temp (°C)	Conversion (%)	Selectivity (%)				
		terminal aldol products			internal aldol products	trimers
		a	b	c		
325	1.3	25.4	2.1	39.9	5.3	18.3
350	4.1	33.9	3.1	51.2	2.2	9.1
375	6.5	35.1	3.5	50.8	1.5	9.1
400	7.8	30.2	3.5	52.3	1.5	12.5

The aldol condensation reaction of *n*-butanal was carried out over 3 ml of the 4 wt% Na/SiO₂ at 7 barg, 350 °C, a *n*-butanal liquid feed of 0.05 ml min⁻¹ and a H₂ flow rate of 42 ml min⁻¹. This reaction achieved considerably higher conversions than the reaction of butanone over the same catalyst and a high selectivity to a single product (Table 4).

Table 4 *Reaction of n-butanal over 4 wt% Na/SiO₂ catalyst*

Time (hrs)	Conversion (%)	Selectivity to 2-ethyl hex-2-enal (%)
2	45	71.2
4	58	84.6
6	50	88.6
22	38	98.2

A larger scale test unit was used for the reaction of *iso*-butanal and a 1:1 molar mixture of *iso*-butanal and *n*-butanal which allowed the testing of the full size pellets (2-3 mm) of the 4 wt% Na/SiO₂ catalyst . The unit was ran at 4 bar. For *iso*-butanal although there was ~ 10% conversion throughout the 20 hour run there was no production of aldol products (Table 5). In fact there was no products detected at all during the run. The high carbon laydown found on the discharged catalyst (96.6 wt%) and the carbon found elsewhere in the discharged reactor tube explain the loss of mass balance.

Table 5 *Reaction of iso-butanal over a 4 wt% Na/SiO₂ catalyst*

Time (hrs)	Temp (°C)	Conversion of *iso*-butanal (%)	% Selectivity to aldol products
5	200	11.3	0
10	200	8.0	0
15	250	8.8	0
20	300	12.4	0

In contrast the reaction of a 1:1 molar mixture of *iso*-butanal and *n*-butanal ran successfully for 20 hours with no gross carbon laydown on the discharged catalyst (3.1 wt%). The products formed were 2-ethyl hexenal as expected from the self-condensation reaction of n-butanal and 2-ethyl-4-methyl-pent-2-enal, the dehydrated product from the crossed aldol reaction of *n* and *iso*-butanal (Table 6). The overall selectivity of this reaction to aldol products is lower than the selectivties achieved with crushed 4 wt %

Na/SiO₂ for the reaction of *n*-butanal (Table 4). This loss in selectivity is likely to be due to the diffusion properties of the full pellets.

Table 6   *Reaction of a 1:1 iso : n-butanal feed over a 4 wt% Na/SiO₂ catalyst*

Time (hr)	Temp (°C)	Conversion iso-butanal (%)	Conversion n-butanal (%)	% Selectivity to aldol products	% Selectivity of aldol products	
					2-ethyl hexenal	crossed aldol
5	200	10.9	28.0	87.2	85.8	14.2
10	200	8.3	15.6	77.9	85.6	14.4
15	250	16.4	26.9	68.7	79.3	20.7
20	300	12.9	30.4	48.5	71.0	29.0

4   DISCUSSION

According to Table 1 there is a small difference in the pK_a of the hydrogen on the carbon adjacent to the carbonyl group between aldehydes ($pK_a = 19.7$) and ketones ($pK_a = 20$). This will partly account for the much higher conversion of *n*-butanal over the 4 wt% Na/SiO₂ catalyst in comparison with acetone and butanone. Over the solid base catalysts only moderate selectivities are achieved with acetone and butanone. This is due to the formation of trimers and, in the case of butanone, a range of dehydrated products. If the dehydrated aldol product is to be hydrogenated further to a saturated alcohol the formation of a series of dehydrated products is less of an issue as they would all hydrogenate to the same alcohol. The formation of trimers is, however, more problematic. Trimer formation is not an issue with the reaction of *n*-butanal over the solid base catalyst. As 2-ethyl hexen-2-al is formed with high selectivity there is no –CH group present adjacent to the carbonyl group to allow further condensation to occur. *iso*-butanal does not react over the solid base catalyst but interestingly it can be incorporated as part of a crossed aldol product. This suggests that for a self-condensation reaction over a solid base catalyst the carbon adjacent to the carbonyl group on the reactant aldehyde has to have two protons.
A proposed mechanism for the self-condensation reaction of *n*-butanal over a solid base catalyst is shown in Fig. 5. This mechanism is consistent with the inability of *iso*-butane to carry out a self-condensation as *iso*-butane is unable to form a carbanion species on the catalyst surface in this way. Although the reaction of *n*-butanone has high selectivity towards the formation of terminal aldol products (>85%), the formation of trimers limits the overall selectivity. Higher selectivties should be achieved if the reactant ketone was branched on the internal carbon adjacent to the carbonyl group (i.e. 3-methyl butanone). According to the mechanism shown in Fig. 5 this would prevent the formation of trimers.

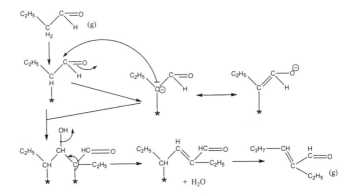

Figure 5 *Proposed Mechanism for the self-condensation of n-butanal*

References

1 K. Weisssermel and H.-J. Arpe, "Industrial Organic Chemistry", 3rd edition, Wiley, 1997.
2 J. J. Spivey and M. R. Gogate, Research Triangle Institute, US EPA Grant, Pollution Prevention in Industrial Condensation Reactions, 1996.
3 K. Tanabe, "Solid Acids and Bases", Kodansha, Tokyo and Academic Press, New York London , 1970.
4 H. Hattori, "Heterogeneous Catalysis and Fine Chemicals III", Elsevier, 1993
5 F King , G J Kelly, Catalysis Today, **73**, 75-82 (2002).
6 G J Kelly, F King and M Kett, Green Chemistry, **4**, 392-399 (2002).
7 K. Tanabe and W. F. Hoelderich, Applied Catalysis A: General **181**, 399-434 (1999).
8 A. J. Gordon, The Chemists Companion, Wiley, New York, 1972
9 K .Tanabe, Catalysis by Acids and Bases, Elsevier Science Publishers B.V., Amsterdam, 1985
10 L.P. Hammett, Physical Organic Chemistry, McGraw-Hill, New York, 1940
11 M. A. Paul and F. A. Long, Chem. Rev. 1957, **57**, 1
12 G. J. Kelly, patent WO 00/31011, June 2000
13 G. J. Kelly, patent WO01/87812, Nov 2001

FRIEDEL–CRAFTS ACYLATION AND FRIES REARRANGEMENT CATALYSED BY HETEROPOLY ACIDS

I.V. Kozhevnikov,* J. Kaur and E.F. Kozhevnikova

The Leverhulme Centre for Innovative Catalysis, Department of Chemistry, University of Liverpool, Liverpool, L69 7ZD, UK. E-mail: kozhev@liverpool.ac.uk

1 INTRODUCTION

The Friedel - Crafts aromatic acylation (Eq. 1) and related Fries rearrangement of aryl esters, *e.g.* phenyl acetate (Eq. 2; Ac = acetyl), catalysed by strong acids are the most important routes for the synthesis of aromatic ketones that are intermediates in manufacturing fine and speciality chemicals as well as pharmaceuticals.[1,2]

$$ \text{(1)} $$

$$ \text{(2)} $$

These reactions involve acylium ion intermediates that are generated from an acylating agent or aryl ester by interaction with an acid catalyst. For the Friedel – Crafts chemistry, present industrial practice uses acyl chlorides or acid anhydrides as acylating agents and requires a stoichiometric amount of soluble Lewis acids (*e.g.* $AlCl_3$) or strong mineral acids (*e.g.* HF or H_2SO_4) as catalysts, which results in a substantial amount of waste and corrosion problems.[2] The overuse of catalyst is caused by product inhibition – the formation of strong complexes between the aromatic ketone and the catalyst. In view of the increasingly strict environmental legislation, the application of heterogeneous catalysis has become attractive. In the last couple of decades, considerable effort has been put into developing heterogeneously catalysed Friedel – Crafts chemistry using solid acid catalysts such as zeolites, clays, Nafion-H, heteropoly acids, *etc.*,[2] zeolites being the most studied catalysts.[2-6] Likewise, the environmentally benign aromatic acylation with carboxylic acids (Eq. 3) instead of the anhydrides and acyl chlorides, resulting in the formation of water as the only by-product, has been attempted, mostly with zeolites as catalysts.[7-10] Despite the numerous studies, solid-acid catalysts have hitherto been proved efficient only for the acylation of activated arenes (*e.g.* anisole), whereas non-activated arenes remain beyond the reach of heterogeneous catalysis. The acylation of anisole with acetic anhydride using a zeolite catalyst has been commercialised by Rhodia.[2] Although relatively active catalysts, the zeolites (*e.g.* H-Beta) are deactivated during the acylation [6-8]. The main deactivation is deemed to be reversible; this is attributed to the strong adsorption of the acylation product on the catalyst, blocking access to the active sites. Another type of deactivation, which is irreversible, is caused by tar deposition on the catalyst surface (coking).

$$\text{(3)}$$

Heteropoly acids (HPAs) are promising solid acid catalysts for the Friedel – Crafts reactions.[11-15,17-20] They are stronger than many conventional solid acids such as mixed oxides, zeolites, *etc.* The Keggin-type HPAs typically represented by the formula $H_{8-x}[XM_{12}O_{40}]$, where X is the heteroatom (most frequently P^{5+} or Si^{4+}), x is its oxidation state, and M is the addenda atom (usually W^{6+} or Mo^{6+}), are the most important for catalysis.[11-16] They have been widely used as acid and oxidation catalysts for organic synthesis and found several industrial applications (for a recent comprehensive review, see the monograph[15]). The aim of this paper is to review recent studies on catalysis by HPA for the Friedel – Crafts acylation and related Fries rearrangement of aryl esters, mainly focusing on authors' results.

2 FRIEDEL-CRAFTS ACYLATION

2.1 Acylation by acyl chlorides and acid anhydrides

Izumi *et al.*[11] pioneered the use of heteropoly acids as catalysts for aromatic acylation. Silica-supported acids $H_4[SiW_{12}O_{40}]$ and $H_3[PW_{12}O_{40}]$ were found to effectively catalyse the acylation of p-xylene with benzoyl chloride. $Cs_{2.5}H_{0.5}[PW_{12}O_{40}]$ showed high efficiency in the acylation of activated arenes, such as p-xylene, anisole, mesitylene, *etc.*, by acetic and benzoic anhydrides and acyl chlorides. This catalyst provided higher yields of acylated arenes than the parent acid $H_3[PW_{12}O_{40}]$, the latter being partly soluble in the reaction mixture.[21]

Our recent study[20] showed that bulk and silica-supported $H_3[PW_{12}O_{40}]$ exhibit a very high activity in the acylation of anisole (Eq. 4) with acetic anhydride in liquid phase, yielding up to 98% *para* and 2 - 4% *ortho* isomer of methoxyacetophenone (MOAP) at 70 - 110°C and an anisole to acetic anhydride molar ratio AN/AA = 10-20 (Table 1).

$$\text{(4)}$$

The acylation of anisole appears to be heterogeneously catalysed; no contribution of homogeneous catalysis by HPA is observed. The $H_3[PW_{12}O_{40}]$ catalyst is reusable, although gradual decline of activity was observed due to the coking of the catalyst.

Importantly, $H_3[PW_{12}O_{40}]$ is almost a factor of 100 more active than the zeolite H-Beta, which is in agreement with the higher acid strength of HPA.[20]

Table 1 *Acylation of anisole with acetic anhydride (2 h)* [20]

Catalyst (amount, wt%)[a]	AN/AA (mol/mol)	T (°C)	Yield[b] (%) p-MOAP	o-MOAP
$H_3[PW_{12}O_{40}]$ (0.83)	10	70	67	c
$H_3[PW_{12}O_{40}]$ (0.83)	10	90	96	3.8
50% PW/SiO$_2$ (0.83)	10	90	88	4.0
40% PW/SiO$_2$ (0.83)	10	90	88	4.0
40% PW/SiO$_2$ (0.88)	20	110	98[d]	2.1
30% PW/SiO$_2$ (0.83)	10	90	82	3.7
20% PW/SiO$_2$ (0.83)	10	90	79	3.3
10% PW/SiO$_2$ (0.83)	10	90	78	3.5
$H_4[SiW_{12}O_{40}]$ (0.83)	10	70	70	c
40% SiW/SiO$_2$ (0.83)	10	70	61	c
$H_3[PMo_{12}O_{40}]$ (0.83)	10	70	0	0
40% PMo/SiO$_2$ (0.83)	10	70	2	0
$Cs_{2.5}H_{0.5}[PW_{12}O_{40}]$ (0.83)	10	90	44	1.5

a) The amount of catalysts per total reaction mixture. PW = $H_3[PW_{12}O_{40}]$, SiW = $H_4[SiW_{12}O_{40}]$ and PMo = $H_3[PMo_{12}O_{40}]$. b) Yield based on acetic anhydride. c) The yield of o-MOAP ca. 2-3%. d) Yield in 10 min.

Anisole acylation is first-order in acetic anhydride, the order in catalyst is 0.66, and the apparent activation energy is 41 kJ/mol in the temperature range of 70 – 110°C.[20] The reaction is inhibited by product because of adsorption of p-MOAP on the catalyst surface (Figure 1). Applying the Langmuir – Hinshelwood kinetic model, the ratio of adsorption coefficients of p-MOAP and anisole has been found to be 37 at 90°C. The activity of $H_3[PW_{12}O_{40}]/SiO_2$ increases with the HPA loading, passing a maximum at about 50% loading. Such behaviour may be explained as a result of increasing the HPA acid strength, on the one hand, and decreasing the HPA surface area, on the other, as the loading increases. It should be noted that the specific catalytic activity (per Keggin unit) of supported HPA is greater than that of bulk HPA. This demonstrates that the reaction occurs via the surface-type catalysis in terms of Misono's classification ("bulk vs. surface type").[12,13]

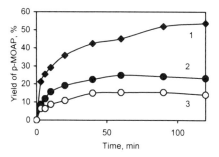

Figure 1 *Inhibition by product in the acylation of anisole (40% $H_3[PW_{12}O_{40}]/SiO_2$ (0.83 wt%), AN/AA = 10 mol/mol, 70°C, 2 h): (1) no p-MOAP added; (2) p-MOAP added initially, AA/p-MOAP = 2 mol/mol; (3) p-MOAP added initially, AA/p-MOAP = 1 mol/mol. Yields are based on acetic anhydride.*[20]

In contrast to anisole, the acylation of toluene with HPA is far less efficient than that with H-Beta. These results can be explained by the well-known strong affinity of bulk HPA towards polar oxygenates, which would lead to the preferential adsorption of acetic anhydride on HPA, blocking access for toluene to the catalyst surface. It appears that the hydrophobic zeolites with high Si/Al ratios less strongly differentiate the adsorption than the hydrophilic HPA and, therefore, are more suitable catalysts for the acylation of nonpolar aromatics like toluene.

2.2 Acylation by acids

The aromatic acylation with carboxylic acids (Eq. 3) instead of acid anhydrides and acyl chlorides has attracted interest, because it is an environmentally benign reaction, resulting in the formation of water as the only by-product. It has been attempted with zeolites and clays as catalysts.[7-10] Heteropoly acids have proved to be more active catalysts for this reaction.[17,18,22]

Castro *et al.*[17,18] reported that silica-supported 12-tungstophosphoric acid and its Cs^+ salts catalyse the acylation of toluene, p-xylene and m-xylene with crotonic acid. Some alkylation of aromatic compounds with crotonic acid also takes place. Heteropoly acid was found to be more active than zeolites HY and H-Beta in the acylation.

Recently, it was demonstrated that the acylation of toluene and anisole with C_2 – C_{12} aliphatic carboxylic acids in liquid phase is efficiently catalysed by $Cs_{2.5}H_{0.5}[PW_{12}O_{40}]$.[22] The acylation of toluene was carried out at a molar ratio PhMe/RCOOH = 50 and 110°C in the presence of *ca.* 10 wt% $Cs_{2.5}H_{0.5}[PW_{12}O_{40}]$ (Table 2). The reaction was clearly heterogeneous; it stopped when the catalyst was filtered off the reacting mixture. With acetic, propionic and butyric acids, the yield of acylated products was very low, though increasing in this series, similar to that observed for zeolites.[7,8] This may be due to the preferential adsorption of the lower acids on $Cs_{2.5}H_{0.5}[PW_{12}O_{40}]$, blocking access for toluene to the catalyst surface. The higher acids C_6 – C_{12} were more reactive in acylation, yielding 31 – 51% aromatic ketones. All three possible isomers, *ortho, meta* and *para,* were formed, the *para* isomers being the major products (55 – 73%), as expected. The reaction selectivity was virtually 100%. The yield increased in the series of acids from hexanoic to dodecanoic acid like in the reaction with CeY zeolite[7,8] and cation-exchanged montmorillonite.[9] $Cs_{2.5}H_{0.5}[PW_{12}O_{40}]$ (S_{BET}, 112 m^2g^{-1}) was a more efficient catalyst than the bulk $H_3[PW_{12}O_{40}]$ (S_{BET}, 7 m^2g^{-1}) (Table 2), which may be explained by a greater number of H^+ surface sites in $Cs_{2.5}H_{0.5}[PW_{12}O_{40}]$. The Cs^+ salt was also more active than the silica-supported HPA, 40% $H_3[PW_{12}O_{40}]/SiO_2$, which may be the result of the higher hydrophobicity of $Cs_{2.5}H_{0.5}[PW_{12}O_{40}]$,[12,13] favouring the adsorption of nonpolar reactants on the catalyst surface and making $Cs_{2.5}H_{0.5}[PW_{12}O_{40}]$ more resistant towards deactivation by co-product water compared to the more hydrophilic $H_3[PW_{12}O_{40}]$. After the reaction, $Cs_{2.5}H_{0.5}[PW_{12}O_{40}]$ could be easily separated by filtration and reused. Some catalyst deactivation was observed, though, which was probably caused by coking.

Table 2 *Acylation of toluene (100 mmol) with carboxylic acids (2.0 mmol) at 110°C, 48 h* [22]

Catalyst (g)	Acid	Yield (%)[a]	Product distribution (%)		
			para	*ortho*	*meta*
$Cs_{2.5}H_{0.5}[PW_{12}O_{40}]$ (1.0)	Hexanoic	31	55	37	8
$Cs_{2.5}H_{0.5}[PW_{12}O_{40}]$ (1.0)	Octanoic	47	72	22	6
$Cs_{2.5}H_{0.5}[PW_{12}O_{40}]$ (1.0)	Dodecanoic	51	71	22	7
$Cs_{2.5}H_{0.5}[PW_{12}O_{40}]$ (1.0)[b]	Dodecanoic	44	73	21	6
$H_3[PW_{12}O_{40}]$ (1.0)	Dodecanoic	14	60	23	17
40% $H_3[PW_{12}O_{40}]/SiO_2$ (2.5)	Dodecanoic	35	83	11	6

a) The yield of aromatic ketones based on carboxylic acid. b) A reuse of the above run. The catalyst was filtered off, washed with CH_2Cl_2, dried and rerun.

The most important advantage of $Cs_{2.5}H_{0.5}[PW_{12}O_{40}]$ catalyst is that it gives much higher productivity in aromatic ketones than the zeolite and clay catalysts reported so far, which may be attributed to the stronger acidity of $Cs_{2.5}H_{0.5}[PW_{12}O_{40}]$. Thus, for the acylation of toluene with dodecanoic acid, $Cs_{2.5}H_{0.5}[PW_{12}O_{40}]$ gives a 1.0% yield of ketone based on toluene (Table 2) which is three times that reported for CeY (0.31%)[7] and for Al^{3+}-montmorillonite (0.32%).[9] (For CeY, a 96% yield based on dodecanoic acid at PhMe/acid = 313 has been obtained (150°C, 48 h).[7] For Al^{3+}-montmorillonite, a 60% yield at PhMe/acid =187 (110°C, 24 h) has been reported.[9]) It should be noted, however, that CeY gives a higher selectivity to the *para*-acylation (94%)[7] than $Cs_{2.5}H_{0.5}[PW_{12}O_{40}]$, which may be the result of shape selective catalysis by the zeolite.

The acylation of anisole with C_2 – C_{12} acids was carried out under the same conditions as that of toluene, except a shorter reaction time (5 h).[22] The acylated anisole formed as the major product (*para/ortho* = 59:1 – 96:1 and no *meta* isomers) together with esterification products – methyl esters of carboxylic acids and phenol. No phenyl esters formed. The selectivity to esters increases from acetic to dodecanoic acid, reaching 40% for the latter. The acylation of anisole, in contrast to that of toluene, is most efficient with C_2 – C_6 acids, giving a 62 – 65% yield of acylated products and only 2 – 6% of methyl esters.

The acylation of anisole with HZSM-5 zeolite (Si/Al = 30) as a catalyst proceeds differently.[10] With C_2 – C_3 acids, at 120°C, PhOMe/acid = 4 and 20% HZSM-5, the phenyl esters are the main products; no methyl esters have been found. At 150°C and otherwise the same conditions, a 2:1 – 5:1 mixture of acylated anisole and phenyl ester forms at an 87 – 100% acid conversion. The conversion drops sharply for the acids higher than C_3, down to 0.6% for C_{12}, probably because of restricted access into zeolite pores. Thus $Cs_{2.5}H_{0.5}[PW_{12}O_{40}]$ is a more active as well as more selective catalyst than HZSM-5 for the anisole acylation.

3 FRIES REARRANGEMENT

The Fries rearrangement of aryl esters, *e.g.* phenyl acetate (Eq. 2) yields 2- and 4-hydroxyacetophenones (2HAP and 4HAP) and 4-acetoxyacetophenone (4AAP) together with phenol. Mechanisms for the formation of products have been discussed.[4] 2HAP, 4AAP and phenol are considered to be the primary products, 2HAP being formed by the intramolecular rearrangement of PhOAc, and 4AAP and PhOH by the self-acylation. In contrast, 4HAP appears to be the secondary product formed by the intermolecular acylation of phenol with PhOAc. Usually, the yield of phenol is greater than that of 4AAP, as part of PhOH results from the decomposition and/or hydrolysis of PhOAc that also produce ketene, acetic acid and acetic anhydride. Solvent plays a significant role in Fries reaction, polar solvents favouring the formation of the *para*-acylation products (4AAP and 4HAP).

Heteropoly acids, especially $H_3[PW_{12}O_{40}]$, have been demonstrated to be promising catalysts for Fries rearrangement of aryl esters.[23,24] The HPA-catalysed rearrangement of phenyl acetate (Eq. 2) occurs in liquid phase at $100 - 160°C$ (Table 3). One of important advantages of HPA, as compared to zeolites or mineral acids, is that the reaction can be carried out both homogeneously and heterogeneously. The homogeneous process occurs in polar media, for example, in neat PhOAc or polar organic solvents like nitrobenzene (PhNO₂) or o-dichlorobenzene that are commonly used for Fries reaction. All these media will dissolve $H_3[PW_{12}O_{40}]$ at elevated temperatures (*ca.* 100°C). On the other hand, when using nonpolar solvents, such as higher alkanes (*e.g.* dodecane) that will not dissolve HPA, the reaction proceeds heterogeneously over solid HPA catalysts. In the latter case, supported HPA, preferably on silica, is the catalyst of choice, as bulk HPA possesses a low surface area. The HPA catalysts are easily separated from the heterogeneous system by filtration and can be reused, though with reduced activity. The heterogeneous catalysis in the PhOAc – dodecane media was clearly proved by filtering off the catalyst from the reacting system, which completely terminated the reaction. In contrast, filtration did not affect the reaction course in homogeneous systems, *e.g.* PhOAc – PhNO₂.[23,24]

Table 3 *Fries rearrangement of phenyl acetate (2 h)*[a, 23,24]

Catalyst (wt%)	Solvent (PhOAc, wt%)	T (°C)	Conv. (%)	Selectivity (%)			
				PhOH	2HAP	4HAP	4AAP
$H_3[PW_{12}O_{40}]$ (0.60)	PhOAc (100)	150	5.5	49	5.2	5.6	40
$H_3[PW_{12}O_{40}]$ (3.0)	PhOAc (100)	150	19.2	52	5.7	15	28
$H_3[PW_{12}O_{40}]$ (3.0)	PhNO₂ (25)	150	45.8	52	12	24	12
$H_3[PW_{12}O_{40}]$ (0.60)	PhNO₂ (25)	130	21.0	46	7.8	18	27
$H_3[PW_{12}O_{40}]$ (0.60)	PhNO₂ (50)	100	10.5	55	5.1	10	29
H_2SO_4 (1.4)	PhNO₂ (25)	130	12.8	67	9.4	7.6	16
$H_3[PW_{12}O_{40}]$ (0.60)	dodecane (25)	130	3.1	69	8.0	0	23
40% PW/SiO₂ (1.5)	dodecane (25)	130	8.3	62	10	6.0	22
40% PW/SiO₂ (1.5)[b]	dodecane (25)	130	6.7	51	11	5.0	32
10% PW/SiO₂ (6.0)	dodecane (25)	130	11.8	66	8.0	9.6	16
40% PW/SiO₂ (3.3)[c]	dodecane (36)	160	18.0	66	11	8.2	14
H-Beta (1.3)[c,d]	dodecane (36)	160	9.3	38	32	6.4	24
$Cs_{2.5}H_{0.5}[PW_{12}O_{40}]$ (0.67)	PhNO₂ (25)	130	8.7	49	6.1	4.4	41

a) The reaction with $H_3[PW_{12}O_{40}]$ is homogeneous in PhOAc and PhNO₂ and heterogeneous in dodecane. b) Reuse of the above run. c) 5 h. d) Si/Al = 11.[4]

Strong inhibition of HPA-catalysed process with reaction products takes place both in homogeneous and heterogeneous systems like in anisole acylation. Addition of more HPA catalyst allows reaching a higher PhOAc conversion. Some irreversible catalyst deactivation is also observed.[23,24]

The total selectivity towards the sum of PhOH, 2HAP, 4HAP and 4AAP is over 98%. Some acetic acid and acetic anhydride are also formed. The homogeneous reaction is more efficient than the heterogeneous one because it makes less phenol and more acetophenones, the selectivity to the more valuable *para* acetophenones, 4AAP and 4HAP, being also higher. In terms of turnover frequencies, HPA is almost 200 times more active than H_2SO_4 in homogeneous reaction, as well as more selective to acetophenones. In heterogeneous systems, HPA is also two orders of magnitude more active than H-Beta zeolite, which is one of the best zeolite catalysts for this reaction.[6] However, H-Beta shows a higher total selectivity to acetophenones than HPA (Table 3). It should be pointed out that HPA in homogeneous systems gives a higher selectivity to *para* acetophenones 4AAP and 4HAP than H-Beta. The efficiency of solid HPA (at a constant loading) increases in the order $H_3[PW_{12}O_{40}]$ < 40% $H_3[PW_{12}O_{40}]/SiO_2$ < 10% $H_3[PW_{12}O_{40}]/SiO_2$ in which the number of accessible proton sites increases.

The insoluble salt $Cs_{2.5}H_{0.5}[PW_{12}O_{40}]$ is an efficient solid catalyst for the reaction in polar media such as $PhNO_2$ (Table 3). Although less active per unit weight than the homogeneous $H_3[PW_{12}O_{40}]$ or the solid catalyst $H_3[PW_{12}O_{40}]/SiO_2$, it is more selective to acetophenones than the parent HPA. The explanation of this may be that the less hydrophilic Cs^+ salt possesses stronger proton sites than the partially hydrated solid $H_3[PW_{12}O_{40}]$ or $H_3[PW_{12}O_{40}]/SiO_2$ that contained 4 – 6 H_2O molecules per Keggin unit.[24]

In contrast to silica-supported $H_3[PW_{12}O_{40}]$, the sol-gel $H_3[PW_{12}O_{40}]$ catalysts prepared by the hydrolysis of tetraethyl orthosilicate showed only a negligible activity in the Fries reaction of phenyl acetate, yielding mainly phenol with 92 – 100% selectivity. This may be explained by a weaker acid strength of the sol-gel catalysts due to strong interaction of the HPA protons with the silica matrix and the presence of relatively high amount of water in sol-gel catalysts.[23]

A continuous gas-phase catalytic acetylation of phenol with acetic anhydride to phenyl acetate followed by simultaneous Fries rearrangement to yield *ortho-* and *para-* hydroxyacetophenones over a silica-supported heteropoly acid has been described.[25] A complete conversion of phenol to phenyl acetate is achieved at 140°C. Upon increasing the temperature to 200°C, the phenyl acetate formed rearranges into hydroxyacetophenones with 10% yield and 90% selectivity to the *para* isomer.

The rearrangement of phenyl benzoate (PhOBz) occurs similarly to that of PhOAc (Eq. 2), yielding 2- and 4-hydroxybenzophenones, 4-benzoxybenzophenone (4BBP) and phenol together with benzoic acid.[23] The amount of phenol formed was found to be nearly equal to that of 4BBP, indicating that the hydrolysis of PhOBz is less significant in this case.

In p-tolyl acetate (p-TolOAc) rearrangement, the acylation in the *para* position is no longer possible. Hence the major products are 2-hydroxy-5-methylacetophenone (2H5MAP) and p-cresol (Eq. 5) together with acetic acid and acetic anhydride. A small amount of the *meta*-acylation product 3-hydroxy-6-methylacetophenone (3H6MAP) may also be formed. The homogeneous reaction with $H_3[PW_{12}O_{40}]$ in $PhNO_2$ or o-dichlorobenzene gives almost equal amounts of 2H5MAP and p-cresol, no 3H6MAP being formed.[23] The heterogeneous reaction with the Cs^+ salt gives 2H5MAP with a remarkably high selectivity of 82% and only 17% of p-cresol. A little of 3H6MAP (1.4%) is also formed. This indicates that the hydrolysis p-TolOAc is less significant with the Cs^+ salt than with $H_3[PW_{12}O_{40}]$, which is in agreement with the higher hydrophobicity of the Cs^+ salt.

(5)

4 CONCLUSION

The recent studies reviewed here demonstrate that HPA-based solid acids, including bulk and supported heteropoly acids (preferably $H_3[PW_{12}O_{40}]$) as well as acidic heteropoly salts (*e.g.* $Cs_{2.5}H_{0.5}[PW_{12}O_{40}]$), are active and environmentally friendly catalysts for the Friedel – Crafts acylation of aromatic compounds and related Fries rearrangement of aryl esters. These solid acids are superior in activity to the conventional acid catalysts such as H_2SO_4 or zeolites, which is in line with the stronger acidity of HPA. The HPA catalysts can be reused after a simple work-up, albeit with reduced activity. Similarly to zeolite catalysts, the HPA-catalysed acylations are inhibited by products because of strong adsorption of the products on the catalyst surface. Consequently, to achieve higher conversions a larger amount of the catalyst is needed or a flow technique should be applied. Adsorption of aromatic substrate and acylating agent on the catalyst, especially preferential adsorption of one of them (*e.g.* the acylating agent), can affect (inhibit) the activity of HPA catalyst in acylation. The irreversible deactivation (coking) of HPA catalysts in Friedel – Crafts reactions is an issue that needs to be addressed.

Acknowledgement. EPSRC and Johnson Matthey for support.

References

1 G.A. Olah, in *Friedel – Crafts and Related reactions,* Wiley – Interscience, Vol. I – IV, New York, 1963 – 1964. G.A. Olah, in *Friedel – Crafts and Related reactions,* Wiley – Interscience, New York, 1973.
2 P. Metivier, in *Fine Chemicals through Heterogeneous Catalysis*, (R.A.Sheldon and H. van Bekkum, Eds.), p. 161, Wiley – VCH, Weinheim, 2001.
3 E. G. Derouane, G. Crehan, C. J. Dillon, D. Bethell, H. He and S. B. Abd Hamid, *J. Catal.,* 2000, **194**, 410.
4 F. Jayat, M.J. Sabater Picot and M. Guisnet, *Catal. Lett.,* 1996, **41**, 181.
5 A. Vogt, H.W. Kouwenhoven and R. Prins, *Appl. Catal., A* 1995, **123**, 37.
6 M Guisnet and G. Perot, in *Fine Chemicals through Heterogeneous Catalysis*, (R.A. Sheldon and H. van Bekkum, Eds.), p. 211, Wiley – VCH, Weinheim, 2001.
7 B. Chiche, A. Finiels, C. Gauthier and P. Geneste, *J. Org. Chem.,* 1986, **51**, 2128.
8 C. Gauthier, B. Chiche, A. Finiels and P. Geneste, *J. Mol. Catal.,* 1989, **50**, 219.
9 B. Chiche, A. Finiels, C. Gauthier and P. Geneste, *J. Mol. Catal.,* 1987, **42**, 229.
10 Q. L. Wang, Y. Ma, X. Ji, H. Yan and Q. Qiu, *Chem. Commun.,* 1995, 2307.
11 Y. Izumi, K. Urabe and M. Onaka, *Zeolite, Clay and Heteropoly Acid in Organic Reactions,* Kodansha/VCH, Tokyo, 1992.
12 T. Okuhara, N. Mizuno and M. Misono, *Adv. Catal.,* 1996, **41**, 113.
13 M. Misono, *Chem. Commun.,* 2001, 1141.
14 I.V. Kozhevnikov, *Chem. Rev.,* 1998, **98**, 171.
15 I.V. Kozhevnikov, *Catalysts for Fine Chemicals. Vol. 2. Catalysis by Polyoxometalates*, Wiley, Chichester, England, 2002.
16 J.B. Moffat, *Metal-Oxygen Clusters. The Surface and Catalytic Properties of Heteropoly Oxometalates,* Kluwer, New York, 2001.
17 C. Castro, J. Primo and A. Corma, *J. Mol. Catal. A: Chem.,* 1998, **134**, 215.
18 C. Castro, A. Corma and J. Primo, *J. Mol. Catal. A: Chem.,* 2002, **177**, 273.
19 B.M. Devassy, S.B. Halligudi, C.G. Hedge, A.B. Halgeri and F. Lefebvre, *Chem. Commun.,* 2002, 1074.

20 J. Kaur, K. Griffin, B. Harrison and I.V. Kozhevnikov, *J. Catal.,* 2002, **208**, 448.

21 Y. Izumi, M. Ogawa, W. Nohara and K. Urabe, *Chem. Lett.,* 1992, 1987.

22 J. Kaur and I.V. Kozhevnikov, *Chem. Commun.,* 2002, 2508.

23 E.F. Kozhevnikova, J. Quartararo and I.V. Kozhevnikov, *Appl. Catal. A,* in press.

24 E.F. Kozhevnikova, E.G. Derouane and I.V. Kozhevnikov, *Chem. Commun.,* 2002, 1178.

25 R. Rajan, D.P. Sawant, N.K.K. Raj, I.R. Unny, S. Gopinathan and C. Gopinathan, *Indian J. Chem. Tech.,* 2000, **7**, 273.

SELECTIVE OXIDATION OF PROPANE ON $Cs_{2.5}H_{1.5}PV_1W_xMo_{11-x}O_{40}$ HETEROPOLYOXOMETALLATE COMPOUNDS

N. Dimitratos[1] and J. C. Védrine[1,2,*]

[1]Leverhulme Centre for Innovative Catalysis, Department of Chemistry, University of Liverpool, Oxford Street, L69 7ZD, Liverpool, United Kingdom
[2]Present address: Laboratoire de Physico-Chimie des Surfaces, ENSCP, 11 Rue P. & M. Curie, F-75005, Paris, France. E-mail: jacques-vedrine@enscp.jussieu.fr.
* Corresponding author

1. INTRODUCTION

The catalytic selective oxidation of light parafins to olefines or oxygenated compounds is an attractive way to utilise alkane feed. Propane oxidative dehydrogenation (ODH) to propene and propane selective oxidation to acrolein or acrylic acid have been extensively studied in the last twenty years both in academia and in industry. For ODH reaction basic catalyst and vanadium element appear to be two crucial parameters, as for VMgO.[1,2] For oxidation of propane to acrylic acid, MoNbTe(Sb)V mixed oxides appear to be the most promising catalysts.[3-5] Mo and V containing Keggin-type heteropolycompounds ($H_{3+x}PMo_{12-x}V_xO_{40}$) HPCs have also been studied[6-8] but for propane oxidation the selectivity to propene was observed to be low and carbon oxides were the main products.[9] In the latter case the stability of the compounds was the major problem and very often the compounds transformed to lacunary HPCs or even to a mixture of the constituting oxides. However it is known that alkaline salts of heteropolycompounds are thermally more stable.[10] Moreover, the acidity of such compounds is known to be particularly strong,[11] which could be detrimental to selective oxidation, although acidity is necessary to activate the poorly reactive alkane. As Mo-based heteropolyacids are better catalysts for oxidation reactions than their W counterparts[6], which are known to present a stronger acidity,[6,11] i.e. a higher ability for alkane activation, our work aimed at studying the effect of W substitution for Mo in the relatively stable $Cs_{2.5}$ salt of $H_4PV_1Mo_{11}O_{40}$ acid.

2. EXPERIMENTAL

2.1 Sample preparation

$H_4PV_1Mo_{11-x}W_xO_{40}$ and $Cs_{2.5}H_{1.5}PV_1Mo_{11-x}W_xO_{40}$ with x = 0, 1, 2, 4 and 6 heteropoly compounds have been prepared, according to the following procedures. For each acid form sample, calculated stoichiometric amounts of V_2O_5, MoO_3, H_2WO_4 and H_3PO_4 were dissolved in water upon stirring and heating at 80 °C under reflux for 24 h. The solid samples were obtained by evaporating water to dryness at 50 °C.

The $Cs_{2.5}H_{1.5}PV_1Mo_{11-x}W_xO_{40}$ salts were prepared by precipitation, upon adding desired amount of Cs_2CO_3 to the aqueous solution of the corresponding $H_4PV_1Mo_{11-x}W_xO_{40}$ samples. The precipitates were then dried by evaporating water solvent at 50 °C.

2.2   Thermogravimetric and surface area measurements

TGA and DTA measurements were performed using a Setaram TG-DSC 111 equipment. BET surface area values were determined at liquid nitrogen temperature using a Micromeretics ASAP 2000 equipment after outgassing the sample at 250 °C.

2.3   Infrared Spectrometry

The IR spectra were recorded with a Nicolet NEXUS FTIR spectrometer (spectral range down to 400 cm^{-1}, resolution 2 cm^{-1}) using diffuse reflectance mode. The samples were diluted (ca 2 wt%) and finely ground with dried KBr.

2.4   Catalytic testing

Catalytic study of propane oxidation was performed under atmospheric pressure in a stainless tube microreactor (2 cm o.d.) connected with a GC-MS apparatus, either at a temperature in the range 300-400 °C or at 340 °C by varying the contact time. To better control the reaction temperature, the thermocouple was installed within the reactor. The gas feed consisted of 40 vol% propane, 20% oxygen and He as balance. Total flow rates were in the range 7.5 to 30 cm^3 min^{-1} and the mass of the catalyst was 0.6g. Conversion and selectivities were measured after 2 h of reaction at each reaction temperature/contact time. No water was introduced in the feed.

3. EXPERIMENTAL RESULTS

The chemical composition of $Cs_{2.5}H_{1.5}PV_1Mo_{11-x}W_xO_{40}$ samples, prepared as described in §2.1 was determined using a Spectro CIROS SOP 120 ICP atomic emission spectrometer. They were found very close to those expected for all samples from stoichiometries used for the preparation.

3.1 Thermogravimetric and BET surface area data

Typical TGA and DTA curves are given in Figure 1 for $H_4PV_1Mo_9W_2O_{40}$ and $Cs_{2.5}H_{1.5}PV_1Mo_9W_2O_{40}$ samples taken as examples. From the shape of the curves, some information can be extracted about the loss of physisorbed and constitutional waters. Loss of physisorbed water happens in the 25-150 °C temperature range, while loss of constitutional water occurs in the 200-450 °C temperature range. Thus, from the amount of water released in the 200-450 °C temperature range, one can calculate the number of constitutional protons. The data given in Table 1 for the acid samples are very similar with the theoretical values of protons, supporting the assumption that W^{6+} was incorporated in the Keggin anion as substitute for Mo^{6+}.

Table 1. *TGA and BET data of $H_4PVMo_{11-x}W_xO_{40}$ samples (* measured as water weight loss between 200-450°C)*

Heteropoly compounds	endothermic peaks (°C)	exothermic peaks (°C)	H content per Keggin unit *	S.A (B.E.T) m^2/g
$H_4PVMo_{11}W_0O_{40}$	25-200, 200-440	440-470	3.9	4.3
$H_4PVMo_{10}W_1O_{40}$	25-200, 200-440	440-470	3.8	0.8
$H_4PVMo_9W_2O_{40}$	25-200, 200-445	445-470	3.9	4.3
$H_4PVMo_7W_4O_{40}$	25-200, 200-440	440-480	3.9	2.6
$H_4PVMo_5W_6O_{40}$	25-240, 220-430	430-510	4.0	2.2

Table 2. *TGA and BET data of $Cs_{2.5}H_{1.5}PVMo_{12-x}W_xO_{40}$ samples. (* measured as water weight loss between 200-450 °C. n.o means no exothermic peaks were observed)*

Heteropoly compounds	endothermic peaks (°C)	exothermic peaks (°C)	H content per Keggin unit *	S.A (B.E.T) m^2/g
$Cs_{2.5}H_{1.5}PVMo_{11}W_0O_{40}$	25-200, 200-450	n.o	1.6	63
$Cs_{2.5}H_{1.5}PVMo_{10}W_1O_{40}$	25-200, 200-450	n.o	2.5	78
$Cs_{2.5}H_{1.5}PVMo_9W_2O_{40}$	25-200, 200-460	n.o	2.6	81
$Cs_{2.5}H_{1.5}PVMo_7W_4O_{40}$	25-200, 200-455	n.o	2.8	65
$Cs_{2.5}H_{1.5}PVMo_5W_6O_{40}$	25-200, 200-450	450-480	1.8	89

It is worth noting that no exothermic peak was obtained on Cs salts compared to their acid analogues, showing their greater thermal stability. Note also that the content of protons, as calculated from the constitutional water evolved between 200 and 450°C, is reliable for the acid samples (close to 4) whereas it varies between 1.6 and 2.8 instead of 1.5 as expected for Cs salts. The TGA curves shown in Figure1 illustrate this phenomenon, in particular for Cs salts for which water loss occurred up to 600°C, with a small step between 450 and 500°C. This water loss between 200 and 600°C corresponds both to water adsorbed on Cs cations and to dehydroxylation of constitutional water, i.e. to the elution of protons. This is why the number of protons given in Table 2 is more than 1.5 as expected

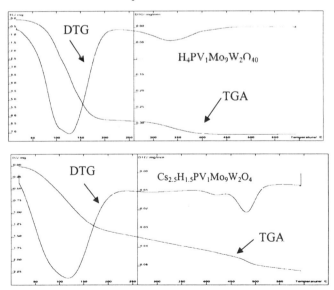

Figure 1   *TGA & DTG curves of $H_4PV_1Mo_9W_2O_{40}$ and $Cs_{2.5}H_{1.5}PV_1MO_9W_2O_{40}$ samples*

3.2   FT-IR study

The IR spectra of heteropolyoxometallates have been discussed and bands assigned previously.[12,13] The most relevant features are as follows: ν_{as}(P-O$_a$), (1080-1060 cm^{-1}), ν_{as}(Mo-O$_d$), (990-960 cm^{-1}), ν_{as}(Mo-O$_b$-Mo), (900-870 cm^{-1}) and ν_{as}(Mo-O$_c$-Mo), (810-760 cm^{-1}), where O$_a$ refers to the O atom common to PO$_4$ tetrahedron and a trimetallic group; O$_b$ to the O atom connecting two trimetallic groups; O$_c$ to the O atom connecting two MoO$_6$ octahedra inside a trimetallic group and O$_d$ to the terminal O atom. It appears that incorporation of W^{6+} in the acid form (0 to 6 per Keggin unit), ν_{as}(P-O$_a$) increases from 1063 to 1071 cm^{-1}, ν_{as}(Mo-O$_b$-Mo) from 867 to 877 cm^{-1}, ν_{as}(Mo-O$_c$-Mo) from 785 to 791 cm^{-1} and ν_{as}(Mo-O$_d$) from 962 to 973 cm^{-1}. When Cs$^+$ cations have replaced protons, these bands were little affected. In other words in both H$^+$ and Cs$^+$ forms, when W^{6+} was substituting Mo^{6+} in the Keggin structure, there was a general increase of the frequencies of the main bands, suggesting that incorporation of the W^{6+} was achieved and the strength of the M-O bonds was increased.

After catalytic testing in the 300-400 °C temperature range, the samples were again analysed by IR spectroscopy. It was observed that the Keggin structure was retained during catalytic reaction, with a small shift (1-3 cm^{-1}) to higher frequencies of the bands.

3.3   Catalytic results

3.3.1   Effect of reaction temperature

The effect of temperature was studied in the range of 300-400 °C under conditions described in the experimental section. The products detected were propene, acrolein (ACR), acetic (Acet.A) and acrylic (Acr.A) acids and CO$_x$, acrolein, acetic and acrylic acids being minor products.

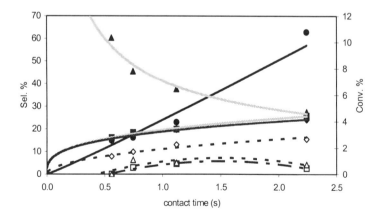

Figure 2   *Variations of propane conversion (●) and selectivities as a function of temperature for Cs$_{2.5}$H$_{1.5}$PV$_1$Mo$_9$W$_2$O$_{40}$ sample with C$_3$:O$_2$:He = 2:1:2, total flow rate 15 cm^3 min^{-1}. Selectivity to CO (◆), CO$_2$ (■), C$_3^=$ (▲), Acet.A (◇), Acr.A (□), ACR (△)*

Figure 2 shows the variations of selectivities and conversion versus reaction temperature for $Cs_{2.5}H_{1.5}PV_1Mo_9W_2O_{40}$ sample taken as an example. As expected, temperature increase was accompanied by an increase in propane conversion and a decrease in selectivities to propene, acetic and acrylic acids and an increase in CO_x. Only acrolein was favoured by the enhancement of temperature.

3.3.2   Effect of contact time

The effect of contact time was studied at 340 °C, by changing the contact time from 0.5 s to 2.5 s. Figure 3 shows the catalytic performance of $Cs_{2.5}H_{1.5}PV_1Mo_9W_2O_{40}$, taken as an example. Conversion increased linearly as expected, while selectivity to propene decreased. Selectivities to acetic acid, CO and CO_2 increased regularly while those to acrolein and acrylic acid reached a maximum, indicating that they were further oxidised to CO_x.

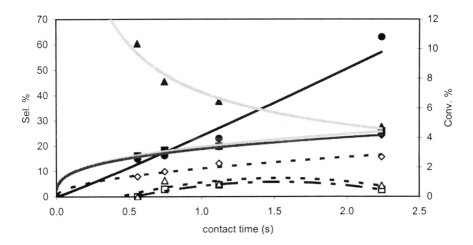

Figure 3   *Variations in propane conversion (●) and selectivities at 340°C as a function of contact time for* $Cs_{2.5}H_{1.5}PV_1Mo_9W_2O_{40}$ *sample with* $C_3:O_2:He = 2:1:2$. *Selectivity to CO (♦), CO_2 (■), $C_3^=$ (▲), Acet.A (◇), Acr.A (▯), ACR (△).*

Catalytic data were also compared at the same conversion, namely 5%, as shown in Figure 4. It can be seen that, with the introduction of W^{6+} to substitute Mo^{6+} in the Keggin anion, selectivity to propene dropped, while those in acetic acid and CO_x increased. Exception to this trend was the W_6 sample, which gave relatively high selectivities to propene and acetic acid, i.e relatively low selectivity in CO_x and other oxygenated compounds.

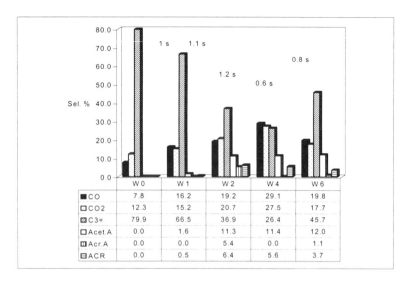

	W 0	W 1	W 2	W 4	W 6
■ CO	7.8	16.2	19.2	29.1	19.8
☐ CO2	12.3	15.2	20.7	27.5	17.7
▨ C3=	79.9	66.5	36.9	26.4	45.7
☐ Acet.A	0.0	1.6	11.3	11.4	12.0
▥ Acr.A	0.0	0.0	5.4	0.0	1.1
⊟ ACR	0.0	0.5	6.4	5.6	3.7

Figure 4.   *Variation in selectivities of $Cs_{2.5}H_{1.5}PV_1M_{11-x}W_xO_{40}$ samples at $340^{\circ}C$ and 5% propane conversion obtained at different contact times indicated on top of each bar diagram.*

3.3.3   Discussion

The catalytic properties of the $Cs_{2.5}H_{1.5}PV_1Mo_{11-x}W_xO_{40}$ heteropoly compounds followed the same general trend when W^{6+} content varied, namely a decrease in propene productivity, an increase in those in acetic acid and CO_x and a maximum in acrolein and acrylic acid for 2 W per Keggin anion. In order to explain these results we should keep in mind that molybdenum is able to insert oxygen into the activated hydrocarbon, while tungsten is only able to abstract protons from the molecule to yield the corresponding olefine.[14] Moreover, the metal-oxygen bond strength is progressively increased on increasing W^{6+} content as shown by the chemical shift increase observed in IR. This results in a lower tendency of the oxygen atom to participate in the oxygen insertion into propane and propene and therefore affects the catalytic properties in propane oxidation by giving more acetic acid. On the other hand, the lower negative charge on terminal oxygens lead the protons more free, i.e increases their acid strength.

3.3.4   Reaction pathways

The determination of the reaction network on $Cs_{2.5}H_{1.5}PV_1Mo_{11-x}W_xO_{40}$ heteropoly compounds can be deduced from the effect of contact time on propane oxidation shown in Figure 3. Extrapolation of the curves at zero contact time gives informations about the sequence of the formation of the products. Generally, primary products have nonzero intercept at zero contact time, whereas secondary or higher order products appear at a positive contact time.

The extrapolation of the curves at zero contact time shows that propene is the primary product. Acrolein and acrylic acid appear at non zero contact time which indicates that these compounds are secondary products. Their selectivities present a maximum with contact time, which shows that they are further transformed to CO_x. The variations of acetic acid formation show that it is a secondary product formed by another route than acrolein and acrylic acid. CO_x products show behaviour of typical end products, which are formed from the C-C cleavage of propane and the further total oxidation of acetic and acrylic acids. Thus, a reaction scheme can be proposed based on literature proposals[5,15] and our own previous work.[16,17] (Scheme 1).

Scheme 1   *Reaction pathways for propane oxidation on $Cs_{2.5}H_{1.5}PV_1Mo_{11-x}W_xO_{40}$ samples. The thick arrows indicate the major pathway.*

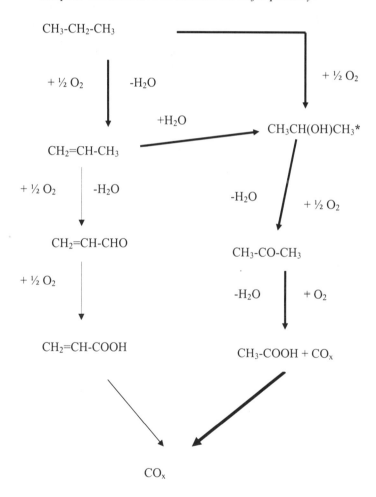

4 CONCLUSION

The present study has shown that W^{6+} element was incorporated in the Keggin anion substituting Mo^{6+} and has an appreciable effect on the catalytic performance of Keggin-type heteropolyacid Cs salts in propane oxidation. In general, increase of W^{6+} atoms in the Keggin structure leads to the enhancement of the formation of oxygenates, specifically of acrolein, acetic and acrylic acids. However, the above increase of formation of oxygenated products is also accompanied with an enhancement of CO_x products. Thus, the ratio of Mo/W atoms in the Keggin structure should be tuned in order to favour either the formation of propene pathway or the formation of oxygenated products.

The present study showed that the major reaction pathway on $Cs_{2.5}H_{1.5}PV_1Mo_{11-x}W_xO_{40}$ heteropoly compounds is the formation of acetic acid via acetone, while the formation of acrylic acid through acrolein is a minor reaction pathway. In both cases, the major primary product is propene. The formation of acetic acid could be assigned to the strong Brønsted acidity of the catalysts compared to that of MoNb(Sb)TeV mixed oxides[15,16] known to be the best catalysts for propane to acrylic acid in the presence of water vapour.

References

1. M.A. Chaar, D. Patel, H.H. Kung, *J. Catal.*, 1988, **109**, 463.
2. D. Siew Hew Sam, V. Soenen, J.C. Volta, *J. Catal.*, 1990, **123**, 417.
3. Mitshubishi Chem. Corp. *US Patent*, 5,472,925 (1995)
4. Toa Gosei Chem. Ind. Ltd., *JP Patent* 1013 7585 (1996)
5. M.M. Lin, *Appl. Catal. A: Gen.*, 2001, **207**, 1.
6. M.T. Pope, *Heteropoly and Isopoly Oxometalates*, Springler, Berlin !983
7. I.V. Kozhevnikov, Russ. Chem. Rev., 1987, **56**, 1417
8. M. Fournier, C. Feumi-Jantou, C. Rabia, G. Hervé, S. Launay, *J. Mater. Chem.*, 1992, **2**, 971.
9. R. Kringer ans L.S. Kirch, *European Patent* 0010902 (1979)
10. T. Okuhara, N. Mizuno, M. Misono, *Adv. Catal.*, 1996, **41**, 113.
11. N. Essayem, A. Holmqvit, P.Y. Gayraud, J.C. Védrine and Y. Ben Taarit, *J. Catal.*, 2001, **197**, 273.
12. C. Rocchiccioli-Deltcheff, M. Fournier, R. Franck and R. Thouvenot, *Inorg. Chem.*, 1983, **22**, 207
13. C. Rocchiccioli-Deltcheff and M. Fournier, *J. Chem. Soc., Faraday Trans.*, 1991, **87** 3913
14. W. Ueda and Y. Suzuki, *Chemistry Lett.*, 1995, 541.
15. M.M. Bettahar, G. Constentin, L. Savary, J.C. Lavalley, *Appl. Catal. A: Gen.*, 1996, **145**, 1.
16. E. K. Novakova, J. C. Védrine, E. G. Derouane, *J. Catal.* 2002, **211**, 226.
17. E. K. Novakova, J. C. Védrine, E. G. Derouane, *J. Catal.* 2002, **211**, 235

MULTIPHASE HYDROGENATION REACTORS - PAST, PRESENT AND FUTURE

E H Stitt[1], RP Fishwick[2], R Natividad[2] & JM Winterbottom[2],

1 Synetix, Johnson Matthey, PO Box 1, Billingham, Cleveland, TS23 1LB
2 School of Engineering, University of Birmingham, Edgbaston, Birmingham, B15 2TT

1 INTRODUCTION

Catalytic hydrogenation in multiphase systems has now been carried out commercially for over 100 years. The process is now used for a huge variety of reactions. Catalysts are predominantly heterogeneous, although the recent drive for enantiomeric selectivity in production routes for the pharmaceutical industry has prompted a significant growth in homogeneous catalysts, and going full circle immobilised homogeneous catalysts. Product values and volumes vary enormously; both over several orders of magnitude. Given this variety it is therefore maybe somewhat surprising that these reactions are carried out predominantly in one reactor type: the stirred tank reactor. And further that this type of reactor has been the staple of the industry for several hundreds of years.

2 MULTIPHASE CATALYTIC HYDROGENATION - AN OLD TECHNOLOGY

The hardening of oils and fats by hydrogenation is an important industrial process - with worldwide production of hardened oils in excess of four million tons. Product applications include edible oils, margarine, mayonnaise, frying fats, confectionary, cosmetics, tyres, plastics and many more. The application of this reaction is growing at a significant rate.

The invention of catalytic hydrogenation of tri-glycerides was made by Dr Wilhelm Normann in February 1901[1]. In his initial experiments he obtained the vapour phase hydrogenation of oleic acid over a calcined and reduced nickel catalyst. In subsequent work he used liquid phase processing and investigated supported catalysts.

The invention was patented in Germany a year later[2]. Normann the built a small catalyst preparation and hydrogenation unit in Warrington, England (on the site of the current Crosfield plant). In approximately 1912 his company, Ölwerke Germania, opened the first commercial catalyst manufacturing and fat hardening units at Emmerich in Germany; where Johnson Matthey to this day still manufactures fat hardening catalysts alongside the Uniqema fat hardening process plants. Photographs of reactors designed and built by Dr Normann, and of the actual plant at Emmerich, Fig.1, look disturbingly similar to modern

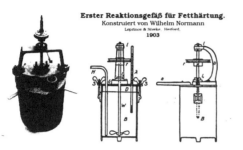

Figure 1: *Photograph and Drawing of Norman's Hydrogenation Reactor from 1903 and of the Ölwerke Germania Fat Hardening (Hydrogenation) Plant at Emmerich in 1912.*

reactors and chemical batch processing plants, with the notable exception that the belt drives have been replaced by overhead motors. Indeed later photographs and drawings from the Normann family archives include innovations such as different impellers and use of multiple impellers, dished end pressure vessels, jacket and internal cooling/ heating coils. This begs the question what improvements have been made since?

2 MULTIPHASE REACTOR DEVELOPMENT

The late Prof. Villermaux stated[3] that while the technology of Concorde has almost nothing to do with that of the Wright brothers or Bleriot, and that they would probably not be able to fly it, technical drawings of chemical processing taken from patents filed in the 1880s are remarkably similar to those still in use and being installed now. The observations on work of Normann above appear sympathetic this assertion.

Improvements have or course been made. Fixed internals such as baffles and draft tubes are commonplace. Many improvements and optimisations to impeller designs have been proposed, with the "optimum" design varying according to process conditions and phase ratios. Integration of stationary and rotating parts in impeller designs has also proven beneficial. The work on the optimisation of impeller designs and impeller configurations, as well as heat transfer provision in the context of Fig.1 is however essentially incremental in nature. The stirred tank is still "king".

Alternatives to the stirred tank have of course been proposed and used extensively for many years. Bubble column and trickle bed type reactors will be discussed below. But first however, let us consider the stirred tank in a little more detail.

3 MIXING AND MASS TRANSFER IN STIRRED TANKS

Mixing in stirred tanks is extremely complex; as complex as the overall design concept is simple. The situation gets more complicated when a second phase is introduced, never mind a third! There is a huge body of work and literature that has led to good understanding amongst specialists of the variable and input parameter effects. Design and operational optimisation however remains largely empirical.

Research and flow visualisation techniques allow detailed interrogation of mixing patterns and phase distributions in opaque multiphase systems. Recently developed

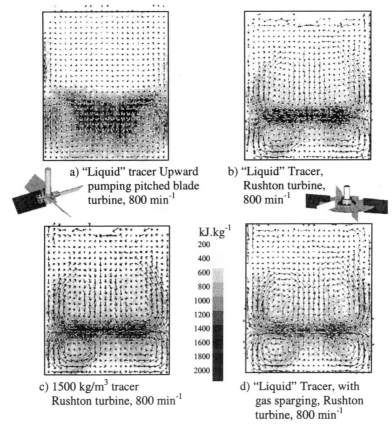

a) "Liquid" tracer Upward pumping pitched blade turbine, 800 min^{-1}

b) "Liquid" Tracer, Rushton turbine, 800 min^{-1}

c) 1500 kg/m^3 tracer Rushton turbine, 800 min^{-1}

d) "Liquid" Tracer, with gas sparging, Rushton turbine, 800 min^{-1}

Figure 2 Flow *patterns in a 100 mm stirred tank by PEPT (positron emission particle tracking) experiments*[5,6]. *Impeller speed 800 min^{-1} for all cases.*

radioactive methods such as tomography and particle tracking have rapidly increased the rate at which data can be acquired. The information that can be obtained from these methods is limited to Lagrangian traces of particle and fluid motion and time averaged global phase distributions at relatively coarse spatial resolutions (>500 µm). Well established photographic and velocimetry techniques are still required to probe bubble sizes, dispersed phase interactions and dynamics.

3.1 Liquid Flow Patterns

The overall mixing pattern in a stirred tank is a function of the impeller type, and the phase ratios. Figs 2 shows mixing patterns derived from particle tracking techniques[4] for (Fig. 2a) a pitched blade turbine[5] and (Fig 2b) a Rushton turbine[6] in the same vessel (100 mm diameter), with the same impeller diameter and rotation speed. It is evident that the overall patterns are very different. These patterns are in fact very time averaged and the actual trace of the "flow follower" particle indicates significant randomness and chaos in the temporally resolved flow. They also indicate that there are quiescent, low flow velocity regions, even in such a small and idealised vessel.

3.2   Particle Flow Patterns

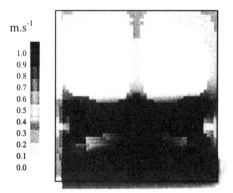

Figure 3   *Particle - liquid slip velocities for 100 mm diameter vessel, Rushton Turbine, 800 min^{-1}, particle diameter 250 μm, density ρ_p≡1,500 kg/m^3*

How then does this influence the flow of catalyst particles in the vessel and the liquid - solid mass transfer? Figure 2c shows the particle velocities, for a particle density of 1,500 kg/m^3 under identical conditions to Figure 2b. By subtraction of the temporally averaged liquid and particle velocities, the spatially resolved slip velocities can be obtained, Figure 3. This shows that the relative velocity of the liquid and particle (and hence the mass transfer) also varies considerably over the vessel, with the highest values confined to the impeller discharge region.

3.3   Effect of Gas

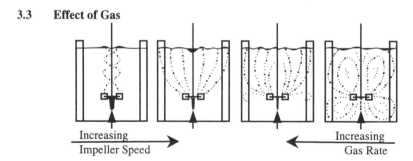

Figure 4   *Regimes of Bubble Flow in a Stirred Tank (after Nienow et al[7])*

The introduction of gas via a sparger leads to further complication and loss of momentum transfer from the impeller. The relative gas velocity and impeller speed essentially governs the nature and quality of gas dispersion[7] (Figure 4). The key learning here is that at low impeller speeds and/or high gas rates there is the possibility of the impeller "flooding", with consequent poor dispersion of the gas, and as it happens a significant loss of mixing energy imparted to the liquid phase. There is however more to it than this.

The velocity plot in Figure 2d is for the same vessel and impeller as Figure 2b, but now with gas sparging[6]. The velocities are severely diminished, and thus the overall mixing has deteriorated. In a multi-phase reaction where the need is to disperse the gas and suspend the solids catalyst particles, the gas will clearly impact on the solids dispersion. It is in fact relatively easy to allow a catalyst (and especially a dense catalyst such as a Raney nickel or high metal loading supported catalyst) to spend a significant proportion of the time on the bottom of the vessel in the quiescent zone underneath the impeller[5]. The situation is complicated further by the interaction of the dispersed gas and liquid phases. It is shown above that the gas influences the liquid and particle flows. Recent observations[8] have indicated that the presence of suspended particles influences the average bubble size and gas dispersion.

3.1   Scale Up

The problem in scaling up a stirred tank is that the key length-scales do not scale in similar fashion. The catalyst particle size, and thus the diffusional path lengths are essentially unchanged. The geometrical length of course may change through an order of magnitude or more. Stirred tanks of $30m^3$ and above are in fact used in the chemical industry for multiphase reactions. As was shown above, the momentum transfer away from the impeller results in dissipation. Essentially as the flow area increases with increasing diameter the fluid velocity decreases. While in scaling up a mixer "geometric similarity" is normally maintained, and impeller tip speed is commonly set to a similar value, the momentum and fluid velocities in the wall region are inevitably lower at the larger scale. Thus, the volume-averaged velocity is decreased and the average scale of turbulence will change, as a consequence. Micro-mixing will be less intense and thus transport processes within the liquid phase and inter-phase will be slower. The general non-uniformity of mixing will increase.

Despite all these complexities, the stirred tank is successfully used in many sectors of the chemical and process industries. It is flexible, of moderate cost and ultimately controllable. Many processes are batch, and poor scale up can always be countered by longer batch times or turning up the impeller speed. This can however never completely disguise losses in selectivity and reactivity arising from imperfections in the mixing and the resulting transport limitations.

4   ESTABLISHED ALTERNATIVES TO STIRRED TANKS

There are of course a number of well established alternatives to the stirred tank used in the large scale chemical industries[9], and strategies have been proposed for selection of the best reactor for a given multiphase reaction[10]. The nature of flow and contacting in these reactors differs significantly. Industrial applications are largely dominated however by a small number of these:

–   Trickle beds: almost universally used for hydroprocessing in refineries and widely used for hydrogenation in petrochemical plants. The liquid is allowed to trickle over a fixed bed (typically cylindrical or trilobe extrudate) catalyst in a gas phase continuum. The nature of contacting has been demonstrated clearly in recent flow visualisations using magnetic resonance imaging[11]. The scaling issues arise because of changes in superficial velocities to achieve the same residence time between a lab, pilot and production scale reactor. This leads to changes in liquid hold up, real liquid residence time, liquid film thickness and/or wetting efficiency of the catalyst.

– Slurry bubble columns: increasing use driven mainly by Fischer Tropsch synthesis, but also used for fat hardening and Air Product's slurry phase methanol process. The gas is sparged into the liquid continuum, or pseudo-homogeneous slurry phase. The gas tends to rise in the centre of the column, carrying liquid with it via an air lift type action. The liquid falls back to the bottom of the column in the annulus near the walls. Gas superficial velocities up to and above 0.5 m.s^{-1} are cited in the literature. The column normally includes a large number of vertical heat exchange tubes. These appear to have little impact on the overall flow pattern. The high velocities of the liquid leads to high heat transfer coefficients. The bubble column if thus favoured where heat removal and good mass transfer are at a premium. While the construction of the column is essentially simple, its hydrodynamics, design   and   scale   up   are   complex.

– Eductor reactors: of which the Buss loop is the commercially most successful. These have significant application in medium and fine scale reactions, especially with simple homogenous catalysis (eg. liquid acid or base catalysed). The gas liquid mixture is fed through a constriction that forces intense intermixing of the phases, and frequently initial bubble sizes of less than 1 mm. Catalyst attrition issues have thus far inhibited widespread use of these for slurry systems. The use of eductor systems with fixed beds has been demonstrated at the laboratory scale[12].

5 AN ALTERNATIVE APPROACH

The above reactors all rely on the use of a single vessel, where the overall dimensions of that vessel change in proportion as the scale of operation. This brings with it problems in scale up; different problems to the stirred tank but problems all the same. The common theme is that in scaling from the laboratory or pilot unit, the relevant length and time scales do not change in proportion.

The result is that, in moving a catalyst or a chemistry from the laboratory to the production scale, changes in apparent activity and selectivity of a catalyst are observed. This is the effect of changes in the mixing, mass and heat transfer as the reactor increases in size; and thus a change in the relative importance of the transport processes to the overall rate process. This results in changes in reactant concentrations on the catalyst surface.

An alternative approach, summarised effectively by Figure 5[13], has been proposed variously. The essential message from this representation is that although typical reactor residence times for multiphase hydrogenations are in the order of minutes to hours, the electronic and chemical interactions are very rapid (micro-seconds and less). This infers that the residence time in the reactor is taken up primarily by getting the reactive species around the vessel and to the active catalyst sites. If the mixing and diffusional distances were made smaller, then by inference the total time for reaction would be shorter.

There is a growing body of evidence in support of this hypothesis, especially for gas phase systems in channels in the order of 10-100 μm. For liquid phase processing it is probably not practical to go to these low dimensions. But can these gains be obtained at

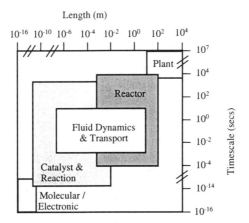

Figure 5 *Lengthscales and Timescales in Reaction Engineering (after Lerou[13])*

characteristic dimensions of a more pragmatic size: say in the order of a few mm? There is considerable work being carried out in academia and in industry in pursuit of this objective, largely using various embodiments of structured catalysts[14]. The best demonstration hitherto comes from the use of catalysed honeycomb.

6 MONOLITHS - AN EXAMPLE OF A STUCTURED CATALYST

The potential of monoliths to act as a catalytic support for multiphase reactions has been recognised for over 20 years[15]. Experimental data have demonstrated the rate benefits of this structured approach over, for instance, both the stirred tank[16] and the trickle bed[17]. These studies have used monolith channel dimensions in the order of 1-3 mm. The benefit is believed to arise from the slug (or Taylor) flow of the gas liquid dispersion through the monolith channels, wherein discrete gas bubbles in the liquid continuum essentially fill the channel cross section. This leads to a thin (20-100 μm) liquid film between the gas slug and the catalyst wall. Thus, by virtue of structuring the flow channels and by controlling the flow regime, the short diffusional path lengths have been achieved, resulting in the observed rate benefits.

Kinetic studies are indicating that there are not only rate advantages to be gained but also selectivity effects too. Winterbottom et al[18] compared kinetics for the hydrogenation of butynediol in slurry and monolith systems, for conditions that were ostensibly kinetically limited. They observed not only changes in reaction rate, but also changes in reaction order, specifically the reactant inhibition term. Based on kinetic modelling, they attribute this effect to changes in the surface coverage of the catalyst and adsorption mode.

7 CONCLUSIONS

Multiphase hydrogenation reactors have historically been dominated by the stirred tank, They are however hard to scale up, leading to loss of efficiency at larger scale. Many proprietary developments over the years have led to significant improvements. The trickle bed has been favoured in high volume refinery and petrochemical applications. For medium and fine scale operations, alternatives to the stirred tank are now being considered as a route to increased competitity. Reactors from larger scale and non-catalytic processing now implemented include trickle beds, bubble columns, and loop reactors. Structured reactors and catalysts are an exciting development that offers potential for a flexible, high efficiency approach for the future that may be close to commercialisation.

References

1   W. Normann, Tagebuch (*Laboratory Notebook*) 27 Feb 1901.
2   W. Normann, *German Pat.* 143,029, 14 Aug 1902.
3   J. Villermaux, "Future Challenges in Chemical Engineering Research", *Trans. I. Chem. E.*, 1995, **73**, 105-109.
4   Y.S. Fangary, M. Barigou, J.P.K. Seville, & D.J. Parker, "Fluid Trajectories in a Stirred Vessel of Non-newtonian Liquid Using Positron Emission Particle Tracking", *Chem. Eng. Sci.*, 2000, **55**, 5969-5979.
5   R. P. Fishwick, J.M. Winterbottom, E.H. Stitt, "Explaining Mass Transfer Observations in Multiphase Stirred Reactors: Particle-Liquid Slip Velocity Measurements using PEPT", *Catalysis Today*, 2003, *in press.*
6   R. P. Fishwick, J.M. Winterbottom, E.H. Stitt, "Effect of gassing rate on solid–liquid mass transfer coefficients and particle slip velocities in stirred tank reactors", Chem. Eng. Sci., 2003, **58**, 1087-1093.
7   A.W. Nienow, D.J. Wisdom & J.C. Middleton, *Proc 2nd Euro. Conf. Mixing*, April 1977, Pub. BHRA, Cranfield, pp F1-1 - F1-16.
    See also JC Middleton, pp 337 in N. Harnby, M. F. Edwards & A. Nienow, "Mixing in the Process Industries", 2nd Ed., Pub. Butterworth Heinemann 1992
8   A. Nienow, W. Bujalski & A .Etchells, "Some Effects of Particle Wettability in Solid-Gas-Liquid Systems", *AIChE Annual Meeting*, Indianapolis, 3-8 Nov. 2002.
9   P.L. Mills & R.V. Chaudhari, "Multiphase Catalytic Reactor Engineering and Design for Pharmaceuticals and Fine Chemicals", *Catalysis Today*, 1997, **37**, 367-404.
10 R. Krishna & S.T. Sie, "Strategies for Multiphase Reactor Selection", *Chem. Eng. Sci.*, 1994, **49**, 4029-4065.
11 L.F. Gladden, M.H.M. Lim, M.D. Mantle, A.J. Sederman and E.H. Stitt, "MRI Visualisation of Two-Phase Flow in Structured Supports and Trickle-Bed Reactors, *Catalysis Today*, 2003, *in press.*
12 J.M. Winterbottom, Z. Khan, A.P. Boyes and S. Raymahasay, "Catalytic Hydrogenation in a Packed Bed Bubble Column Reactor", *Catalysis Today*, 1999, **48**, 221-228.
13 J.J. Lerou & K.M. Ng, "Chemical Reaction Engineering: A Multiscale Approach to a Multiobjective Task", *Chem. Eng. Sci.*, 1996, **51**, 1595-1614.
14 E.H. Stitt, "Alternative Multiphase Reactors for Fine Chemicals: A World Beyond Stirred Tanks?", *Chem. Eng. J.*, 2002, **90**, 47–60.
15 W. Herrmann & C.T. Berglin, "Method in the production of hydrogen peroxide, *Eur. Pat.* EP 0102934 (to Eka AB, Sweden), March 1984 (Priority 8 Sept 1982).
16 S. Irandoust & B. Andersen, *Chem. Eng. Sci.*, 1988, **43**, 1983-1988.
17 R.K. Edvinsson & A. Cybulski, *Chem. Eng. Sci.*, 1994, **49**, 5653-5666.
18 J.M. Winterbottom, H. Marwan, E.H. Stitt & R. Natividad, "The Palladium catalysed hydrogenation of 2-butyne-1,4-diol in a Monolith Bubble Column Reactor", *Catalysis Today*, 2003, *in press.*

NOVEL SILICA ENCAPSULATED METALLIC NANOPARTICLES ON ALUMINA AS NEW CATALYSTS

Kai Man K. Yu and Shik Chi Tsang*

Surface and Catalysis Research Centre, Department of Chemistry, University of Reading, Whiteknights, Reading, RG6 6AD, UK, *E-mail: s.c.e.tsang@reading.ac.uk

1 INTRODUCTION

In applied heterogeneous catalysis the development of any novel catalysts or processes are not only dependent upon the performance (activity & selectivity) of catalytic system but also on the ease of catalyst separation from product. There have been many recent fundamental studies concerning the direct utilization of active but well-defined nanosized catalysts (homogenized heterogeneous catalysts) incorporating good control of catalyst size/ structure (site) /morphology. In particular, nano-sized metallic particles as catalysts are of a significant interest because the unusual high activity/selectivity for many reaction(s) associated with the small particles. The employments of micelles or reversed micelles stabilised by organic stabilizers in water/oil microemulsion for the template synthesis of the nanosized metals therein are well documented.[1,2] Studies of the catalytic activity of these organic appendage-stabilized nanoparticles in academic laboratories have also been widely investigated.[3,4] It is, however, technical issues such as separation and stabilization of these nano-catalysts are yet to be fully addressed. In addition, problems arisen from the use of organic stabilizers (interfere catalytic reactions, passivating the nanoparticle surface, isolate metal particle from desirable metal-support interaction or/and affect the crystallographic matching between metal and support, etc) limit them from real industrial practices. We have recently reported some new concepts towards developing these nano-catalysts with possible practical uses. That is the appliance of chemically-functionalized coatings (fullerene carbon[5], porous silica[6] & fluorinated compound coats[7]) onto nano-sized catalysts, rendering the nano-composite catalysts separable from product.

In this short paper, we report mainly on the synthesis and some characterisations of the nanosized metal cores (Pt and Pd) each stabilised with ultra-thin porous silica over-layers of controlled size and composition. These new but purely *inorganic* based nano-composite precursors, free from organic stabilizer, are shown to be colloidally stable against agglomerization in ethanolic solution. They can therefore be used as catalysts or catalyst precursors prior hosting them onto support materials. In this article, we also demonstrate that the new silica coated metallic nanoparticles anchored onto conventional γ-alumina display excellent catalysis for butane oxidation without encountering diffusion problems from the over-layers.

2 METHOD AND RESULTS

2.1   Unsupported Silogel@5% Pd and Pt nano-composites

Formation of microemulsion was carried out using a small quantity of ionic surfactant, CTAB (cetyltrimethylammonium bromide), and de-ionized water in excess dried toluene. Above critical micelle concentration (cmc) these three components form reversed micelles.[8,9] As the size of such micelle system is related to the ratio of water/surfactant (W) added, tailored amount of water to surfactant ratios in order to control the sizes of the nano-composites were investigated.

The preparation of materials was done at room temperature. Typically, Silogel@5%Pt nano-composite (Pt nanoparticle encapsulated in ultra-thin silica gel; 5wt% of Pt *w.r.t.* silica) was prepared as follows: (4.0442 g) CTAB was added to dry toluene (150 cm^3) under vigorous stirring where a suspension of CTAB in toluene was formed immediately. Then an aqueous solution of Pt precursor salt, $(NH_4)_2PtCl_6$, (0.0958 g in 4.347 cm^3 DI water) was added dropwisely to the suspension of CTAB in toluene and was stirred to create aqueous micelles in the bulk toluene for overnight. After that, 1.650 cm^3 of 1M NaOH solution was added and stirred for 2 hours. Reduction of platinum ion was accomplished, at ambient temperature by dropwisely adding excess hydrazine hydrate (0.5 cm^3). The solution turned to black in color but no precipitation was observed. This clearly implied that the reduction process does not significantly interfere the stability of the microemulsion. TEOS (tetraethyl orthosilica, 3.4637 g) was then added to the reaction mixture. Formation of the silica-gel coating onto the interface (the aqueous and toluene interface) of the micelle carrying the metal nanoparticle was expected to achieve since the NaOH previously added, is known to catalyze hydrolysis/condensation of the TEOS (nucleophilic substitution) to form a sol-gel coating. The composite colloids were allowed to age for 6 days at ambient conditions under a constant stirring. It was later found that adding an extra step involving the use of 5M HCl (5 ml in total mixed with ethanol) in the washing step produced less aggregation in the acidified & charged colloids (shown from the acoustic particle size analyzer). The prepared Silogel@Pt nanocomposites were then centrifuged, filtered using celite glass. It was noted that the filtrate became colorless indicating the composites were mostly retained in the residue after these treatments. The residue was then washed with excess toluene followed by hot ethanol. In order to extract all the trapped surfactant molecules from the nanocomposites the solid material was then exhaustively washed in refluxing ethanol overnight for two times. The final resulting nanocomposites were found dispersed in the ethanol as *meta-stable* colloids despite the absence of organic surfactant molecules, see Figure 3 (stable against precipitation for weeks). This indicates that the surface –OH on the silica sol-gel coating could stabilize the small particles against precipitation in the polar ethanol solution. The final product material was filtered and dried in 333 K vacuum oven and then calcined for 3 hours at 673K. The preparation of Silogel@5%Pd was carried out in the same manner but using corresponding amount of Pd precursor salt.

Supporting Silogel@Pt and Silogel@Pd nano-composites onto conventional γ-alumina was performed as follows. 2g of aluminium oxide (activated, acidic, std grade, ca. 150

mesh) was added to the same Silogel@Pt colloid solution described at above and stirred overnight. The colour of the resulting solution turned to light gray indicating that a substantial amount of Silogel@Pt nanoparticles was immobilized onto the support and produced Silogel@5%Pt-alumina. UV-vis spectroscopy was used to work out the nano-composite loading onto the support materials. The solid supported composites were then filtered, washed, surfactant extracted, dried as the same manner as the unsupported composites. Furthermore, FTIR was employed to characterize the Si-O-Al linkages in order to provide evidence for a successful impregnation of Siliogel@5%Pt onto alumina.[6]

Transmission electron microscopy (TEM) analysis was used to characterize particle size, structure, morphology and composition through direct imaging and elemental analysis of selected areas. A Philips CM20 microscope operating at 200 kV equipped with an Oxford Instrument EDS6767 energy dispersive X-ray analyzer (EDX) was used. High-resolution images were observed with a JEOL JEM-2010 transmission electron microscope operating at 200 kV. Sample was gently grinded, suspended in iso-propanol and placed on a carbon-coated copper grid after the evaporation of the solvent. Electron micrographs and EDX analyses of selected areas were taken.

X-ray powder diffraction (XRD) analysis of the samples was preformed with a Simen Instrument X-ray diffractometer using Ni-filtered copper $K\alpha 1$ of 1.54056Å radiation. Phase identifications of our samples were carried out by comparing the collected spectra with the published files from International Centre for Diffraction Data (JCPDS-1996). The average size of metal particle was calculated based on the corresponding metal (111) peak broadening using the Scherrer equation. Instrumental peak broadening has also been taken into account. Physical characterizations (porosity, surface area, pore distribution) of dried samples were carried using standard N_2 BET method at a range of partial pressure at 77K (Sorptometic 1990 from CE Instrument).

n-Butane catalytic oxidation (combustion) was conducted using our home-built microreactor whereby catalyst in powder form was placed. The microreactor was in form of a quartz tube with an internal diameter of 5 mm i.d. where 0.01g catalyst was packed and sandwiched between two silica wool plugs. All samples were used prior their pre-reduction at 300°C for 1 h under a 60 ml/min flowing stream of 5%H_2/Ar. The total gas flow rate used in typical reaction was 60 ml/min. The feed gas was composed of a 1:1 mixture from two BOC special gas cylinders (One is calibrated as 7,000 ppm n-butane in air and the other, the pure air cylinder), each delivered at 30 ml/min by two individual mass flow controllers. The catalytic test at a particular temperature was complete in an isothermal environment before bringing up to the next set of higher temperatures. After the catalyst was stabilized with the feed gas at the desired temperature for about 30 minutes, the product gases were analyzed by a Perkin Elmer AutoSystem XL gas chromatograph equipped with a methanator and a flame ionization detector (FID). A Carbosphere (Alltech) packed column (6 ft x 1/8 inch) was used to separate CO, CO_2 and n-butane with nitrogen as the carrier gas. Under typical conditions CO_2 was also exclusively produced from the total oxidation of n-butane with traces of CO and other hydrocarbons. The use of such a small amount of the composite catalysts in a fast reactant flow for butane combustion (the Gas hour space velocity, GHSV is estimated to be 73,000 h^{-1}) is to ensure a good heat management of the catalyst bed without engaging into any possible run-away reactions (highly exothermic).

Table 1   *Samples prepared using water/surfactant ratio, W = 30, unless stated*

Sample	N_2 BET /$(m^2 g^{-1})$	Average metal particle size (from XRD)
Silogel@5%Pt (W= 30)	193.85	4.8
Silogel@5%Pt (W= 30) with HCl treatment*	n.d.	3.5
Silogel@5%Pt on γ-alumina (W=30; Al/Si =1)	124.64	4.7
Silogel@5%Pd (W=30)	390.49	3.9
Silogel@16%Pd (W= 70)	n.d.	9.8

* an extra step, see experimental

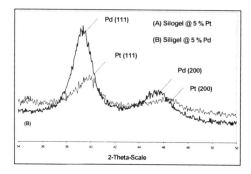

Figure 1   *XRD spectra (A) Silogel@5%Pt and (B) Silogel@5%Pd. The Pt {111} spacing and the Pd {111} spacing from XRD measurement were determined to be 2.26± 0.01Å (2θ of 39.85°) and 2.29 ± 0.01 Å (2θ of 39.30°), respectively*

Figure 1 shows typical XRD patterns of unsupported Silogel@Pd and Silogel@Pt samples prepared by our *method* using the same water/surfactant ratio, W=30. As reference to the published XRD data of Pd (JCPDS-05-0681) and Pt (JCPDS-04-0802) respectively, the two most intense peaks corresponding to (111) and (200) inter-planar separations of the noble metals were clearly visible from our samples though the peaks were very broad indicative of small particle size. There were no distinctive peaks detected due to the silica sol-gel coating. Using the water/surfactant ratio of W the average diameter of Pd in the Silogel@5%Pd sample was 3.9nm (Table 1). The Pt particle size recorded using the same W was 4.8 nm (Table 1). Our other results (unpublished) also showed average metal size of 4.5 ± 0.8 nm can be obtained within the experimental errors when W=30 ratio for 5% metal content was used. Adding an extra HCl pre-treatment during the preparation of Silogel@5%Pt (see experimental) could render the Pt nanoparticles a slightly smaller in size (3.5nm). It is noted that using non-ionic surfactant/water reversed micelle system (w= 6) Hanaoka *et al.*[10] were able to control the Rh particle to a similar size of 4-4.2 nm. In our case, by increasing our water/surfactant ratio to 70, an average size of 9.75 nm Pd particle with 16% metal loading was obtained (Table 1). Thus, our results agree with the observations that diameter of metal nanoparticles synthesized inside the micelle system depends on micelle diameter (the included material ≤ the aqueous core) which critically depends on water/surfactant ratio used.[11,12] The subsequent application of the silica-gel coating onto these micelle stabilized metal nanoparticle does not seem to seriously alter the metal particle size. It is however, noted that this technique could only indicate the average size of our metal cores (nor the size distribution) and gives no information on the silica sol-gel coating since no distinctive XRD peaks due to the silica was detected (indicative of its amorphous in nature).

Figure 2 (A) presents the typical TEM micrographs of the unsupported Silogel@5%Pt (acid pre-treated). It is noted that the average particle sizes as estimated by the TEM micrographs were in general consistent with that calculated from the XRD peak broadening. It is interesting to point out that most of the particle shape was nearly spherical. Figure 2(B) presents a particle size histogram for the typical Silogel@5%Pt. The result shows that the size distribution is unimodal giving an average particle size of 3.42 nm with a deviation of 0.43 nm which agrees reasonably closely with the metal nanoparticles of 3.5 ± 0.8 nm evaluated from the XRD data using the typical water/surfactant ratio. Closer examination of the micrographs reveals that the diameter distribution was extremely sharp suggesting our technique is able to tailor metal size approaching the mono-dispersity. The spherical particle shape and the sharp unimodal size distribution are consistent with that the previous observations that these nanoparticles are synthesized inside the micelles. From the contrast of the images in the micrographs the dense (dark) metallic clusters are clearly found all enclosed individually by a (light) sol-gel porous coating of ca. 2.50 ± 0.02 nm (Figure 2(A)) thickness depending on the W.

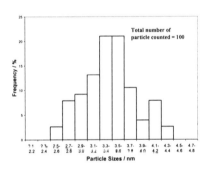

Figure 2   *Typical TEM images of (A) Silogel@5%Pt, scale bar= 5nm; (B) A size histogram of the Silogel@5%Pt (acid pre-treated) showing the Pt particle size distribution of 3.42 ± 0.43 nm*

As the key research was to investigate whether the mono-metal nanoparticles can be stabilized in solution by the silica sol-gel coating without any involvement of the organic surfactant molecules. Thus, extraction of surfactant molecules was achieved as described. Thermogravimetric analysis (TGA) was used to investigate the levels of organic moieties present in the samples by recording their weight changes (& heat changes in the DTA curves) upon their temperature ramping in air. Dried Silogel@5%Pt samples before and after the surfactant extraction (reflux in ethanol) were studied. It is noted from Figure 3A that the Silogel@5%Pt before the surfactant extraction shows a significant loss in weight (~22%) associated with an exothermic peak at around 220-280°C. (started at 200°C and finished at 300°C). There were further minor weight losses at 385°C(~1%, endothermic)

and 530°C (~0.5%). In order to identify these weight changes a sample based on a
physical mixture of surfactant (CTAB) and silogel was used for a comparative purpose
(Figure 3B). It was found that the physical mixture also produced a sharp loss in weight at
the same temperature regime (250-280°C) clearly indicative of this exotherm associated
with the combustion of surfactant molecules. The minor weight changes at the higher
temperatures were also visible (at 395°C and 520°C), which are attributed to the
dehydroxylation of silanol groups[13]. It is important to note that the Silogel@5%Pt, after
the ethanol reflux (Figure 3C), showed less than 1% drop in weight at 220°C hence a
majority of surfactant molecules can still be readily extracted despite the presence of the
porous sol-gel coatings (akin to extraction of surfactant from MCM-41 silicate samples).

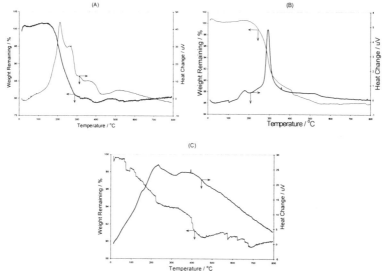

Figure 3   *TGA showing the organic contained: (A) Silogel@5%Pt prior to ethanol
treatment,   (B) a physical mixture of surfactant and silica gel, (C)
Silogel@5%Pt *extensively washed by ethanol*

2.2   Alumina supported Silogel@5% Pd and Pt nano-composites

Figure 4 (a-d) shows the XRD patterns of the Silogel@5%Pt nanocomposites before and
after their attachment to a high surface area γ-alumina support. Figure 4d shows clearly
the two broad peaks due to the (111) and (200) of Pt of the unsupported composites with
inner metal core of 4.8 nm. High surface area alumina does not produce any XRD feature
at around 40° (2θ) such that (111) Pt is chosen to study the loading effect. It was found
that despite a wide range of composite loading to the alumina (R = Al/Si ratios from 0.5 to
18) the inner Pt nanoparticle size retained more or less the same showing no extensive
metal aggregation due to a possible rupture of the coating upon attachment to the support.
This is in contrast to the organic surfactant stabilisers where metal aggregation is
commonly occurred on support[14].

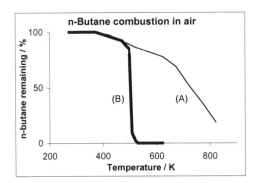

Figure 4   *XRD spectra showing different loading of the Silogel@5%Pt catalyst precursor onto γ-alumina (A) Al/Si=1, (B) Al/Si =2, (C) Al/Si=18, (D) Al/Si = 0*

2.3   Catalytic Reactions

Evaluation of the materials for catalytic n-butane combustion has been conducted. Typically, as shown in Figure 5, butane concentration (3,500 ppm) remained unchanged until the temperature reaching the light-off temperature where the butane was rapidly converted to CO_2. It was found that the samples with and without support were both active for catalytic combustion of n-butane in air. As shown in the Figure 5, the general light-off temperature range was at about 200-250 °C, which is amongst the most active metal catalysts tested under similar testing conditions.[15] Figure 5(A) shows the reaction curve displayed by the unsupported nano-composites, Silogel@5%Pt, as the catalyst. The apparent activation energy was calculated to be 25 kJ/mol. This low value strongly indicates that the reaction was in the diffusion-controlled regime. On the other hand, Figure 5(B) shows the reaction curve displayed by the same nano-composites when dispersed on alumina support (Silogel@5%Pt on γ-alumina). Despite of the dilution the supported catalyst apparently gave higher activity at elevated temperatures. The activation energy of this catalyst was calculated to be 380 kJ/mol, which is in the same order of magnitude of the apparent activation energy obtained over typical supported Pt catalysts. This clearly indicates that the Silogel@5%Pt on their own aggregates suffered a slow mass transfer because of the clustering of the nano-composites rather than any accessibility problems (porosity) of the individual composites. Hence, when these composites are dispersed the reaction could return to the chemical-controlled regime.

Although at this stage, the presented reactivity data over this type of metal composites are rather preliminary and far from complete, more detailed characterization (stability, structural & morphologic changes, if any, after the test) and comparison with conventional metal catalysts should be conducted. The oxidation reaction dependent on a chemical-controlled regime over these supported composites are clearly demonstrated. Detailed comparison of butane combustion activity of catalysts prepared by different methods, at different metal loadings with and without supports will be published elsewhere.

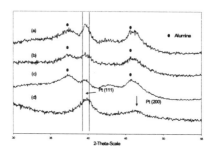

Figure 5   *A plot showing the decrease in butane concentration at increasing temperature over (A) Silogel@5%Pt material; (B) Silogel@5%Pt on alumina*

3   CONCLUSION

To summarise, we report a new technique for the preparation of defined metal particles using organic surfactant followed after chemical reduction. 3-5 nm mono-dispersed monmetallic nanoparticles of defined size and composition (Pt & Pd) using W=30 are achieved. We demonstrate that the organic surfactant molecules stabilizing these nanoparticles can be replaced with ultrathin sol-gel silica coating without leading to precipitation. Thus, extremely thin silica-gel coating is applied to these colloids based on hydrolysis/condensation of the sol-gel precursors at the micelle interface prior the removal of the surfactant. The purposes of the thin coating are to stabilize the metal clusters against agglomerization upon subsequent surfactant removal and to give surface functionality for subsequent anchoring to support materials. In this work, we also demonstrate that the extremely thin but porous layer of sol-gel coating on the metal nano-crystallites dispersed on support surface allows efficient mass transfers of molecules hence they are active catalysts for butane oxidation.

References

1   R.M. Crooks, M. Zhao, L. Sun, V. Chechik and L.K. Yeung, *Accounts Chem. Res.*, 2001, **34**, 181.
2   W.R. Moser, (Ed), *Advanced Catalysts and Nanostructured Materials – Modern Synthetic Methods*, 1st edn, Academic Press, London, 1996, Chapter 7.
3   M M. Maye, Y. Lou and C.J. Zhong, *Langmuir*, 2000, **16**, 7520.
4   M. Ikeda, T. Tago, M. Kishida and K. Wakabayashi, *Chem. Comm.*, 2001, **23**, 2512.
5   S.C. Tsang and V. Caps, a PCT patent application 0200259.0, 2003.
6   K.M.K. Yu, C.M.Y. Yeung, D. Thompsett and S.C. Tsang, *J. Phys. Chem.*, in press.
7   J. Zhu, A. Robertson, S.C. Tsang, Chem. Comm., 2002, 2044.
8   T. Li, J. Moon, A.A. Morrone, J.J. Mecholsky, D.R. Talham and J.H. Adair, *Langmuir*, 1999, **15**, 4328.
9   S. Papp and I. Dekány, *Colloid Polym. Sci*, 2001, **279**, 449.

10 T. Hanaoka, H. Hayashi, T. Tago, M. Kishida, K. Wakabayashi, *J. Colloid. Interf. Sci.* 2001, **235**, 235.

11 K. Torigoe, Y. Nakajima, K. Esumi, *J. Phys. Chem.* 1993, **97**, 8304.

12 P.A. Dresco, V.S. Zaitsev, R.J. Gambino, B. Chu, *Langmuir* 1999, **15**, 1945.

13 D. Hernandez, A.C. Pierre. *J. Sol-Gel Sci. Techn.* 2001, **20**, 227.

14 H. H. Ingelsten, J. C. Beziat, K. Bergkvist, A. Palmiqvist, M. Skoglundh, Q. H. Hu, L. K. L. Falk and K. Holmberg, *Langmuir*, 2002, **18**, 1811.

15 C. Bulpitt and S.C. Tsang, *Sensors and Actuators B*, 2000, **69**, 100.

STRUCTURE-TRANSPORT RELATIONSHIPS IN THE SURFACE DIFFUSION OF MOLECULES OVER HETEROGENEOUS SURFACES WITHIN POROUS CATALYSTS

Sean P. Rigby

Department of Chemical Engineering, University of Bath, Claverton Down, Bath, BA2 7AY, U.K.

1 INTRODUCTION

The surface diffusion of molecules can contribute up to ~90% of the transport flux within porous solids. This paper presents a model which relates the rate of surface diffusion to the properties of the surface. This model opens up the possibility of the control of the surface diffusion flux by the manipulation of the appropriate surface properties.

2 THEORY

The model discussed in this paper is a multi-fractal version of the so-called "homotattic patch" model frequently used for the description of adsorption of molecules on heterogeneous solid surfaces. The model was introduced in earlier work[1-3] by the author but will be developed further here. In the model the surface of a catalyst support is conceived of as being composed of a patchwork of many spatially extended zones within which the surface perceived by the molecules shows fractal geometry. A surface fractal dimension characterises the degree of heterogeneity within each patch and may differ between patches. An area of the surface is said to be homogeneous (or rather "homogeneously heterogeneous") if the heterogeneity in that area is characterised by a single fractal dimension. As will be shown below, the variations in the degree of heterogeneity between patches give rise to resultant variations across the surface in the heat of adsorption, and the Arrhenius parameters for the surface diffusivity.

Adsorption on an energetically heterogeneous surface can be described by the following equation:

$$n(T,P) = \int_{\phi_{min}}^{\phi_{max}} n(T,P,\phi) f(\phi) d\phi \qquad (1)$$

where $n(T,P,\phi)$ is the local adsorption isotherm on a particular homogeneous patch with adsorption energy ϕ, $f(\phi)$ is the energy distribution function, and $n(T,P)$ is the overall isotherm on the heterogeneous surface. According to Kapoor and Yang[4], a similar expression can be written for surface diffusivity on an energetically heterogeneous surface if the surface is assumed to consist of a series of parallel paths such that each path has uniform but different energy, and the surface flow is in the direction of these parallel paths:

$$D_s^{Het} = \int_{\phi_{min}}^{\phi_{max}} D_s^{Hom}(T,P,\phi)f(\phi)d\phi \tag{2}$$

where D_s^{Hom} is the surface diffusivity on a homogeneous patch with energy ϕ, and D_s^{Het} is the overall surface diffusivity on a heterogeneous surface. Conversely, if the surface consists of parallel, uniform patches of varying energies and the flow is perpendicular to the direction of the patches the surface diffusivity is given by[5]:

$$\frac{1}{D_s^{Het}} = \int_{\phi_{min}}^{\phi_{max}} \frac{f(\phi)d\phi}{D_s^{Hom}(\phi)}. \tag{3}$$

It is thus clear from eqs. (2) and (3) that, even when two surfaces possess the same probability density function for D_s^{Hom}, different spatial arrangements of patches will give rise to different values of D_s^{Het}. However, real surfaces are unlikely to be as regular as either of the two spatial arrangements discussed above. A real surface is likely to consist of a random or partially correlated network of patches. By analogy with the semiconductor system studied by Ambegaokar *et al.*[6] it is proposed that, where there is a relatively wide variation in diffusivity between patches, the overall diffusivity is likely to be determined by the diffusivity, D_c, possessed by a particular critical fraction of the adsorption sites. The reasoning behind this assertion is as follows. The adsorbed molecules can be considered as belonging to one of three groups:

(i)   a set of isolated "regions" of high diffusivity, with each region consisting of a group of adsorption sites with diffusivities rather greater than D_c;

(ii)   a relatively small number of sites with diffusivities of the order of D_c, which connect together the sites with high diffusivity to form a sample spanning pathway;

(iii)   the remaining sites possess diffusivities rather less than D_c.

It is thus clear that the sites with diffusivities $\sim D_c$ determine the overall surface diffusivity and, hence, surface diffusion is a percolation theory problem. The sites in category (i) could all have diffusivities set to infinity without greatly affecting the overall diffusivity. The diffusivity would be finite because the flux would have to pass through the regions with diffusivities $\sim D_c$. Whereas the sites with diffusivities rather less than D_c would make negligible contribution to the overall diffusive flux. Due to the commonly observed linear correlation between heat of adsorption and activation energy for surface diffusion, the adsorption sites of high diffusivity are likely to be those solid surface sites with relatively low heats of adsorption, and those within the multi-layers in regions where multi-layer adsorption had commenced. The sites of low diffusivity are likely to be predominantly those on the solid itself, which possess relatively high heats of adsorption, that would be first occupied in the adsorption isotherm. It is also conceivable, given particular spatial distributions of patches, that a critical surface diffusivity could exist on surfaces with relatively narrower spreads in diffusivity. The identities and populations of sites within categories (i)-(iii) could potentially change with surface coverage and thus the value of D_c could also change. Hence the overall measured surface diffusivity would change with surface coverage.

The influence of local surface structure on the values of the Arrhenius parameters for the surface diffusivity within a particular surface patch has been studied in previous work[1-3]. At the molecular scale, the surface diffusion mechanism can be modelled as a random walk, where the steps in the walk consist of hops between adsorption sites. The strong temperature dependence of the surface diffusivity that is generally observed in the literature, is consistent with the above mechanism and frequently exhibits an Arrhenius form. Thus the surface diffusivity, D_s, is given by:

$$D_s = D_0 \, exp \left(- E_D / RT \right) = \frac{\lambda^2}{4\tau} = \frac{\lambda_0^2 \, exp \left(- 2E_\lambda / RT \right)}{4\tau_0 \, exp \left(- E_\tau / RT \right)} \tag{4}$$

where λ is the jump length of the molecular hops, τ is the correlation time of the motion, D_0 is the pre-exponential factor and E_D is the activation energy. The correlation time, τ, for the molecular motion may be measured independently using deuteron NMR[1,2]. The correlation time has also been shown[1,2] to exhibit an Arrhenius-type temperature dependence. Therefore, as shown in eq. (4), both the jump length and correlation time exhibit Arrhenius behaviour, where λ_0 and τ_0 are the pre-exponential factors for the jump length and correlation time, respectively, and E_λ and E_τ are the respective activation energies.

In deuteron NMR studies[7] of the diffusion of benzene within zeolites it was found that the pre-exponential factor for the correlation time was inversely proportional to the number of available sites to which a molecule could jump. The pre-exponential factor is the limiting value of the correlation time as temperature tends to infinity. In previous work[1-3], the above findings for zeolites have been generalised to so-called "amorphous" structures which exhibit the property of self-similarity. It was assumed[1] that the pre-exponential factor for the correlation time was inversely proportional to the number of adsorption sites contained within a limiting length scale, R_∞. R_∞ is a parameter of the model determined by experiment[1]. If two different surfaces both exhibit fractal geometry, then it has been shown that the respective pre-exponential factors for the correlation times of the motion of the same chemical species on each surface are related by:

$$\frac{\tau_{01}}{\tau_{02}} = R_\infty^{\left(d_2 - d_1 \right)} \sigma^{\left(d_1 - d_2 \right)/2} = \frac{D_{02}}{D_{01}} \tag{5}$$

where the subscripts *1* and *2* refer to the two different surfaces, d is the fractal dimension of a surface within the current jump range of the molecule, and σ ($=r^2$) is the cross-sectional area of the diffusing molecule. If the Arrhenius expression for λ is independent of surface roughness and coverage[3], then the ratio of the pre-exponential factors for the correlation times is equal to the inverse of the ratio of pre-exponential factors for the surface diffusivities. It has been proposed that the heat of adsorption (or activation energy for the correlation time), E, at a particular site is made up of a constant component denoted E_S, and a component $n\varepsilon$ that is dependent on the number of nearest neighbour adsorption sites n. Therefore, for a fractal surface, E is given by[2]:

$$E = \left(E_S - \varepsilon\right) + \pi\varepsilon\left(\frac{R}{r}\right)^d \tag{6}$$

where R is a characteristic length scale, and the ratio R/r has a value of 1.5 if only nearest neighbour interactions are important. In a particular case, the specific values of E_S and ε will depend on whether eq. (6) refers to the heat of adsorption or the activation energy, and the particular surface-molecule pairing. Previously[2], it has been shown that if one of the surfaces in eq. (5) is a reference material (for which both d and τ_0 are known and constant), and the remaining value of d in eq. (5) is the same as that in eq. (6) then a compensation effect should be observed. This means that, for a set of data for diffusion of the same chemical species on different homogeneous surfaces with different fractal dimensions, the natural logarithm of either τ_0 or D_0 will be linearly related to the respective activation energy:

$$ln\, A = mE + c \tag{7}$$

where A represents either τ_0 or D_0, m and c are constants, and E is the relevant activation energy. However, as has been demonstrated previously[2], the compensation effect will only be observed if it is only nearest neighbour interactions that are important in determining the activation energy (and hence $(R/r) = 1.5$). It has also been shown[3] that the set of Arrhenius parameters for surface diffusivities obtained at different values of statistical, fractional surface coverage, θ, on the same heterogeneous surface will follow eq. (7) if the overall diffusivity at a particular value of θ is determined by a critical fraction of the adsorbed molecules. Where that critical fraction of molecules is the most recently adsorbed (as θ is increased) then a linear relationship between $ln\, D_0$ and the heat of adsorption will also be found. For heterogeneous surfaces, the parameters called d in both eqs (5) and (6) will be identical if the fractal dimension is constant over the length-scale range from the size of a single molecule up to the value of λ applying to the highest temperature at which the data was acquired.

It is proposed that the fractal dimensions appearing in eqs. (5) and (6) characterise the degree of heterogeneity of the potential energy surface perceived by a particular molecule. This heterogeneity may arise from both the underlying surface geometry and the chemical nature of the surface. It is conceivable that different chemical species may perceive differing degrees of heterogeneity depending on the extent and nature of their interactions with the surface. However, if two different chemical species perceive the same surface roughness and have similar strengths of interaction with the surface, then they are likely to occupy the same surface patches in the same order as θ is increased. Hence, at a particular value of θ, the values of the fractal dimension in both of the equations for the heat of adsorption (of the form of eq. (6)) for each molecule will be the same. In such circumstances, the above theory suggests that a linear relationship between the respective heats of adsorption for two different chemical species at the same value of θ will be observed such that:

$$\frac{E_1 - \left(E_{S1} - \varepsilon_1\right)}{\pi\varepsilon_1} = \frac{E_2 - \left(E_{S2} - \varepsilon_2\right)}{\pi\varepsilon_2} = \left(\frac{R}{r}\right)^d \tag{8}$$

where the subscripts *1* and *2* refer to the two different chemical species, and π, E_{Si} and ε_i are all constants. In addition, if, at identical values of θ for each type of molecule, the surface diffusion flux of both species is controlled by the same critical fraction of surface patches, each possessing the same fractal dimension, then the activation energies for the correlation times (and thus also for the surface diffusivities) of each species would also be expected to be linearly related. If the critical surface patches controlling the surface diffusion flux at a given value of θ are the same for two different chemical species, then the natural logarithm of the pre-exponential factors for the correlation times, and hence also for the surface diffusivities, will also be linearly related such that:

$$\frac{ln\left(\dfrac{D_{01}}{D_{0r1}}\right)}{ln\left(\dfrac{R_{\infty 1}}{r_1}\right)} = \frac{ln\left(\dfrac{D_{02}}{D_{0r2}}\right)}{ln\left(\dfrac{R_{\infty 2}}{r_2}\right)} = d \tag{9}$$

where the subscripts *1* and *2* refer to the two different chemical species, and the subscript *r* refers to the relevant parameters for the reference material. If the two species perceive the same surface on the test material then they will also perceive the same surface on the reference material. Hence, in deriving eq. (9) from eq. (5), it was assumed that $d_{r1} = d_{r2}$.

In previous work[3], it was suggested that, if the jump length of a particular molecule exceeded the characteristic size, ξ, of a surface patch, then a temperature-dependent deviation from linearity of the compensation effect plot (eq. (7)) would be expected. For the surface diffusion of benzene on a fumed silica it has been found[3] that there is a good linear relationship between *ln* τ_0 and the heat of adsorption, E_a, at the same fractional surface coverage. This was interpreted as evidence that the critical fraction of surface patches controlling the surface diffusion flux also belongs to the most recently adsorbed molecules. Hence, where a deviation from linearity of this plot is observed, then the fraction of molecules that have most recently been adsorbed (as surface coverage is increased), ψ_a, must be significantly more, or less, mobile than the fraction of molecules, ψ_c, adsorbed on the critical patches that actually control the surface diffusion flux. This reasoning is summarised in Table 1.

Table 1   *Influence of surface diffusion regime on the form of various data plots*

Characteristics of regime	*ln* D_0 vs. E_D plot	E_a vs. E_D plot	E_a vs. *ln* D_0 plot
$\psi_c = \psi_a; \lambda < \xi$	linear	linear	linear
$\psi_c = \psi_a; \lambda > \xi$	curve	linear	curve
$\psi_c \neq \psi_a; \lambda < \xi$	linear	curve	curve

3   RESULTS AND DISCUSSION

The experimental data discussed in this work were obtained from the results of single component surface diffusion and gas sorption studies for SO_2 and CF_2Cl_2 on Linde silica plugs reported by Carman and Raal[8]. This data has been re-analysed in the light of the new model. Figure 1 shows the plot of *ln* D_0 against E_D for both gases. It can be seen that both sets of data, for SO_2 and CF_2Cl_2, give good fits to eq. (7). Figure 2 shows a plot of values of E_D for CF_2Cl_2 against the corresponding values of the same parameter for SO_2, where each corresponding pair has been measured at the same value of θ. Figure 3 shows a plot of values of $ln\ D_0$ for CF_2Cl_2 against the corresponding values of the same parameter for SO_2, where each corresponding pair has been measured at the same value of θ. It can be seen that both these sets of data in Figures 2 and 3 give good fits to straight lines as expected, from eqs. (8) and (9), respectively, if both species perceive the same surface. As described previously[1], the fractal dimension for the surface patches occupied as θ approaches a statistical monolayer can be obtained using a fractal FHH analysis[9] of the relevant gas adsorption isotherm. The fractal dimensions obtained from a FHH analysis of the isotherms for SO_2 and CF_2Cl_2, are 2.38±0.02 and 2.37±0.02, respectively. The two values are thus identical within experimental error. Thus, when $\theta\sim1$, the values of d in the different versions of eqs. (5) and (6) for SO_2 and CF_2Cl_2 are the same. Hence, eqs. (8) and (9) will hold for $\theta\sim1$. The linear form of the data in Figures 2 and 3 shows that the relations in eqs. (8) and (9) hold at all values of θ. Hence SO_2 and CF_2Cl_2 perceive the same fractal dimension for patches occupied at values of θ other than ~1. Figure 4 shows a plot of $ln\ D_0$ against E_a for CF_2Cl_2. It can be seen that the plot is linear for the lowest values of θ but deviates from the initial straight line at higher values of θ. Figure 2 also shows a plot of the differential heat of adsorption for CF_2Cl_2 against the corresponding parameter for SO_2 at the same value of θ. It can be seen that there is a good fit of these data to a straight line.

Figure 1 shows that the compensation effect is occurring for the Arrhenius parameters of the surface diffusivities for both CF_2Cl_2 and SO_2, and hence, R/r is equal to 1.5 for both. Figures 1, 2 and 3 indicate that CF_2Cl_2 and SO_2 perceive identical degrees of surface heterogeneity and that the same surface patches are controlling the rate of surface diffusion of each species at a particular value of θ. However, Figure 4 suggests that for relatively high values of θ the most recently adsorbed CF_2Cl_2 molecules do not control surface diffusion. It is noted that a surface diffusivity for the surface patches containing the most recently adsorbed molecules for values of θ above where the deviation in linearity occurs in Figure 4 can be estimated. This is achieved by using the extrapolated line in Figure 4 to determine a corrected value of $ln\ D_0$ corresponding to a particular value of heat of adsorption. The equation fitted to the compensation effect plot can then be used to determine the corresponding activation energy, and hence the surface diffusivity at the experimental temperature can be calculated. It has been found that the surface diffusivities for the most recently adsorbed molecules calculated in this way are always higher than the measured values of diffusivity. This is what would be expected if these molecules do not limit the diffusive flux. The linearity of Figures 2 and 3 indicate that this would also be the case for SO_2 (even without the availability of sufficient data, equivalent to Figure 4, being present in the paper by Carman and Raal[8]). The data for E_a in Figure 2 suggest that, at a given value of θ, the most recently adsorbed molecules of CF_2Cl_2 and SO_2 adsorb in the same surface patches on the sample. In summary, the data discussed above suggests that the adsorption and surface diffusion properties of CF_2Cl_2 and SO_2 on Linde silica are similar to each other. However, the adsorption and surface diffusion properties of CF_2Cl_2 and SO_2 on Linde silica are different to those of benzene on aerosil reported previously[3]. For benzene on Aerosil, the most recently adsorbed molecules (as θ increases) are always the critical fraction determining the surface diffusion flux for values of θ below a statistical monolayer. The compensation effect plot for benzene on Aerosil is significantly more curved than that of CF_2Cl_2 and SO_2 on Linde silica. It was proposed that this was because, at the temperatures used in the benzene experiments, the value of λ exceeded the value of ξ of some of the surface patches.

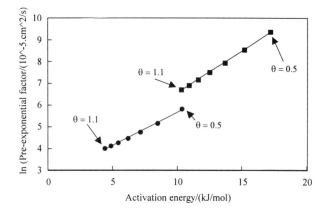

Figure 1 *Compensation effect plot for surface diffusivities of SO$_2$ (●) and CF$_2$Cl$_2$ (■).*

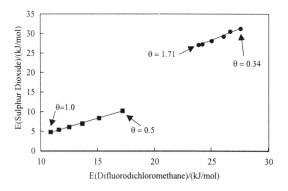

Figure 2 *A plot of the heat of adsorption (●) and surface diffusion activation energy (■) for SO$_2$ against the corresponding parameters for CF$_2$Cl$_2$ measured at the same value of θ.*

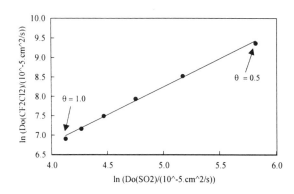

Figure 3 *A plot of ln D$_0$ for CF$_2$Cl$_2$ against ln D$_0$ for SO$_2$ at the same value of θ.*

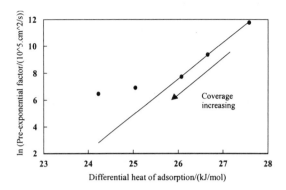

Figure 4 *Plot of ln D_0 against heat of adsorption for CF_2Cl_2 at various values of θ.*

3 CONCLUSIONS

A model has been presented that relates the degree of structural heterogeneity of a surface to the rate of surface diffusion upon that surface. The model has been shown to predict various relationships between the heat of adsorption, and the Arrhenius parameters for the surface diffusivity of molecules on a heterogeneous surface. The experimental data for the single component adsorption and surface diffusion of SO_2 and CF_2Cl_2 on a silica surface have been analysed in the light of the model, and good agreement has been found between the model predictions and experimental data.

References

1   S.P. Rigby, *Catal. Today*, 1999, **53**, 207.
2   S.P. Rigby, *Langmuir*, 2002, **18**, 1613.
3   S.P. Rigby, *Langmuir*, 2003, **19**, 364.
4   A. Kapoor and R.T. Yang, *A.I.Ch.E.J.*, 1989, **35**, 1735.
5   A. Kapoor and R.T. Yang, *Chem. Engng Sci.*, 1990, **45**, 3261.
6   V. Ambegaokar, B.I. Halperin and J.S. Langer, *Phys. Rev. B*, 1971, **4**, 2612.
7   P. Alexander and L.F. Gladden, *Zeolites*, 1997, **18**, 38.
8   P.C. Carman and F.A. Raal, *Proc. R. Soc. London, A* 1951, **209**, 38.
9   P. Pfeifer and K.-Y. Liu, *Stud. Surf. Sci. Catal.*, 1997, **104**, 625.

SUPPORTED SULFONIC ACID CATALYSTS IN AQUEOUS REACTIONS

S. Koujout and D.R. Brown

Centre for Applied Catalysis, Department of Chemical and Biological Sciences, University of Huddersfield, Huddersfield HD1 3DH, UK

1 INTRODUCTION

In the recent drive to prepare mesoporous solid acid catalysts for liquid phase processes, a popular approach has been to tether sulfonic acid groups to mesoporous silica supports.[1&2] The preferred method for functionalising has been to incorporate 3-mercaptopropylsilane in the synthesis gel for the mesoporous silica, and then, after precipitation and isolation, to oxidise the thiol group to sulfonic acid.[2] These materials have exhibited relatively high catalytic activity and the approach is emerging as one of the most successful for preparing acid forms of porous silica.

It is instructive to compare these new supported sulfonic acids with the well established polystyrene-supported sulfonic acids which have been used as acid catalysts in reactions such as MTBE synthesis for many years.[3-5] These sulfonated polymers are generally thought to be limited in application because of their relatively low acid strength and poor stability at reaction temperatures above 150 °C.[4&5] However, we and others have shown how sulfonated polystyrene resins with high levels of sulfonation (above the stoichiometric level of one sulfonic acid group per styrene unit) exhibit significantly higher acid strengths and catalytic activities than resins with the more usual, roughly stoichiometric, level of sulfonation.[5-11] An important aspect of this is that these higher acidities and activities are observed not only in dry resins, but also in the presence of water, where the resins operate in the fully hydrated, swollen state.[12&13]

The objective of the present work is to discover whether the same sort of acidity/activity enhancements seen for polymer-supported sulfonic acids can be obtained on more rigid inorganic supports. Acid strengths of the two types of catalyst have been measured in both the absence of solvent as molar enthalpies of ammonia adsorption, and in the presence of water through molar enthalpies of neutralisation with aqueous NaOH solution. Catalytic activities have been measured in water for the hydrolysis of ethylethanoate.

2 EXPERIMENTAL

2.1 Catalysts

2.1.1 Sulfonated Polystyrene Resins

The following macroporous resins were provided by Purolite International: (a) two specially prepared resins with less than stoichiometric levels of sulfonation (D4034 and SP21-51); (b) CT-175, a standard, normally sulfonated material and Amberlyst 15 (Rohm and Haas Co.) with which it is roughly analogous. Both are sulfonated at a level equivalent to approximately one sulfonic acid group per styrene or divinylbenzene monomer unit (stoichiometric sulfonation); (c) two "persulfonated" resins, based on CT175, but with higher levels of sulfonation (CT-275 and CT-375).

All resins were used in their H^+ forms. The sulfonic acid group concentrations were measured by Purolite, using a standard procedure of ion-exchange with Na^+ followed by aqueous titration with standard NaOH solution.[14] Water contents of the fully swollen resins were also determined by Purolite.

2.1.2 Sulfonated Mesoporous Silicas

Thiol-functionalised mesoporous silica molecular sieve HMS-SH was synthesized at room temperature from a gel containing 0.8 mol tetraethoxysilane (TEOS), 0.2 mol 3-mercaptopropyltrimethoxysilane (MPTS), 0.275 mol n-dodecylamine, 8.9 mol ethanol and 29.4 mol water. The amine was first dissolved in the alcohol-water mixture. Then the TEOS-MPTS mixture was added and stirred for 24 hours. The amine template was finally extracted from the as-synthesized HMS-SH with ethanol under reflux for 24 hours.[2]

The equivalent material based on the larger pore SBA-15 silica molecular sieve was prepared as follows. Pluronic 123 ($EO_{20}PO_{70}EO_{20}$, M_{av} 5800, Aldrich) (4 g) was dissolved with stirring in 125 g of 1.9 M HCl solution at room temperature. The solution was heated to 40 °C before adding 32.8 mmol TEOS. After 1 hour, 8.2 mmol thiol precursor MPTS was added to the mixture. The resultant solution was stirred for 20 hours at 40 °C, and then aged at 100 °C for 24 hours without stirring. The solid product was recovered by filtration and air-dried. The template was extracted with excess ethanol under reflux for 24 hours (1.5 g of as-synthesized material per 400 ml of ethanol).[15]

Materials with immobilised mercaptopropyl groups were oxidised with H_2O_2 in a methanol-water mixture. Typically 2.04 g of aqueous 35% H_2O_2 dissolved in three parts of methanol was used per gram of material. After 24 hours the mixture was filtered and washed with water and ethanol. The wet material was then re-suspended in 0.1 M H_2SO_4 solution for another 4 hours. Finally, the material was rinsed with water and dried at 60 °C under vacuum.[2&15]

The resultant materials were characterised by powder X-ray diffraction (XRD) and by nitrogen adsorption at 77K, using the adsorption isotherm to calculate BET surface area, and the desorption isotherm and BJH method to determine the pore size distribution. The concentrations of acid sites were measured by pH titration with standard NaOH solution, following exchange with excess NaCl solution. Water contents were measured by heating to constant weight at 150 °C.

2.2 Ammonia Adsorption Microcalorimetry

A Setaram C80 differential microcalorimeter coupled to an evacuable glass gas-handling system was used to monitor ammonia adsorption and associated enthalpies of adsorption. Catalyst samples (ca. 150 mg dry weight) were conditioned in the calorimeter at 100 °C under vacuum for two hours, with an empty reference cell. Successive pulses of ammonia (ca. 0.06 mmol) were introduced to the sample at 100 °C. Enthalpy changes associated with each dose were converted to molar enthalpies of adsorption and are expressed as functions of resin coverage. Further details of the technique have been reported previously.[5]

2.3 Aqueous Titration Microcalorimetry

A Setaram "Titrys" microcalorimeter was used for these experiments. This instrument is based on the C80 microcalorimeter described above, modified to allow continuous stirring of liquid samples. The pre-heated titrant is added to both sample and reference cells simultaneously using a programmable twin syringe pump.

In a typical experiment, 50 mg catalyst was suspended in 2 cm^3 water in the sample cell, with the same volume of water in the reference cell. Experiments were performed at 30 °C. The titrant, 0.100 mol dm^{-3} standard NaOH solution, was added to the sample cell in 0.20 cm^3 aliquots at one hour intervals, until neutralisation was complete. Water was added to the reference cell in the same way (adding NaOH solution introduces small errors due to dilution in the reference cell). The heat output was measured for each addition and the cumulative heat plotted against amount of added base. The gradient (which was essentially constant up to complete neutralisation in all cases) was calculated and is reported as the molar enthalpy of neutralisation in kJ mol^{-1}. Molar enthalpies of neutralisation were also measured for 0.10 mol dm^{-3} HCl solution and 0.50 mol dm^{-3} p-toluenesulfonic acid (p-TsOH) solutions.

2.4 Catalytic Activities

The silica-based catalysts were used in the powder form in which they were prepared. The resins were used as conventional beads and were also examined following grinding of the dried beads to fine powder. Hydrolysis of ethylethanoate was performed in a stirred 150 cm^3 batch reactor at 343 K with a water condenser. Ethylethanoate (16.9 mmoles) was dissolved in 50 cm^3 water. The temperature was raised to 343 K and the catalyst (0.2g) added. Samples (0.5μL) of solution were taken at regular intervals and analysed by GC. Reaction progress was followed through the peak area of ethanoic acid, measured against n-heptane internal standard. A sufficiently high stirrer speed was used to avoid diffusion (through the liquid) control of reaction rate.

3   RESULTS

The sulfonated silica catalysts were characterised by powder XRD. The HMS-SO₃H X-ray diffraction patterns were dominated by the 100 reflections corresponding to a d_{100} spacing of about 3.6 nm in each of the three samples. The SBA-15-SO₃H patterns also showed intense reflections from 100 planes with a spacing of 9.4 nm. In both cases the patterns were similar to those reported in the literature.[2&15]

From nitrogen adsorption data. the three HMS-SO₃H samples exhibited BET surface areas of 940 (HMS 1), 850 (HMS 2) and 750 m^2 g^{-1} (HMS 3), with pore volumes of 0.48, 0.34 and 0.31 cm^3 g^{-1} respectively. All three showed maxima in the pore size distributions at diameters of approximately 2.0 nm. The SBA-15-SO₃H exhibited a surface area of 680 m^2 g^{-1}, pore volume of 0.92 cm^3 g^{-1}, and showed a well-defined maximum in the pore size distribution around a diameter of 6.5 nm. These results are consistent with reported values.[11]

The concentrations of acid sites on all catalysts, measured by aqueous pH titration, appear in the Table, along with molar enthalpies of neutralisation in water, average molar enthalpies of ammonia adsorption in the absence of solvent, and catalytic data.

Table   *Acidity and catalytic data for sulfonated polystyrenes and sulfonated silicas*

Catalyst	Acid Site Concentration /mmol g^{-1}	$\Delta H^{\circ}_{ads.}$ [a] (NH$_3$) /kJ mol^{-1}	$\Delta H^{\circ}_{neut.}$ [a] (aq. NaOH) /kJ mol^{-1}	Initial Rate (TON)[c] /mol (acid-mol)$^{-1}$ min^{-1}
Polystyrene-SO₃H				
D4034	0.74	-105.4	-52.2	2.4
SP21/51	1.82	-110.3	-54.6	2.4
AMB-15	4.74	-112.7	-58.3	2.5
CT-175	4.9	-114.2	-58.6	3.9
CT-275	5.4	-119.2	-61.2	4.6
C1-375	5.56	-117.6	-60.9	4.5
Silica-SO₃H				
HMS 1	1.2	[b]	-54.6	2.7
HMS 2	1.31	[b]	-54.9	2.7
HMS 3	2.8	[b]	-54.5	2.7
SBA-15	1.15	[b]	-54.9	2.6
Aqueous Strong Acids				
0.1 mol dm^{-3} HCl	-	-	-52.8	
0.5 mol dm^{-3} *p*-TsOH	-	-	-52.4	

[a] from straight line fit in cumulative enthalpy vs coverage plot; 95% confidence limit = ± 1.0 kJ mol^{-1}
[b] straight line fit not possible, see Figure 1b
[c] 95% confidence limit = ± 0.02 mol (acid-mol)$^{-1}$ min^{-1}

3.1 Acidities

3.1.1   Acidities in the Absence of Solvent

Plots showing the molar enthalpies of ammonia adsorption at 100 °C as a function of acid site coverages for the dehydrated resins are shown in Figure 1a. Note that the profiles fall abruptly at coverages close to the sulfonic acid contents of the resin, showing that interaction with ammonia is essentially stoichiometric. An alternative way of showing this data is to plot the cumulative enthalpy of adsorption against coverage and measure the gradient of a best-fit line through points up to saturation. This gives an average value for the molar enthalpy of ammonia adsorption for each resin. These values appear in the Table. On the basis that molar enthalpies of ammonia adsorption can be used as indicators of relative acid strengths[16], both the tabulated data and the profiles in Figure 1a clearly show that the average strength of acid sites in dry resins increases as the concentration of acid sites on the resins is increased.

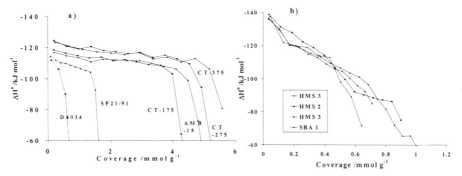

Figure 1   *Molar enthalpies of adsorption of ammonia on dry (a) sulfonated polystyrene and (b) sulfonated silica catalysts as a function of coverage.*

In Figure 1b similar plots are shown for the sulfonated silica catalysts. The profiles are very different to those of the resins. They do not show the abrupt decreases at coverages equivalent to the known concentrations of acid sites. In fact, all four catalysts exhibit a steadily falling differential enthalpy of adsorption as coverage increases. In each case, it drops below 80 kJ mol^{-1} (as an arbitrary indicator of the saturation of acid sites) at coverages well below the known concentrations of surface acid sites given in the Table, and there is only a weak correlation between the coverage at which it drops to this level and the acid site concentration.

The reason for this relatively inconclusive data on acidity of the dry sulfonated silicas is most likely that these materials, with high surface areas compared to the resins, interact with ammonia through physisorption at the same time as chemisorption on acid sites. Each point on the profile is a weighted average of the two types of interaction, with physisorption becoming increasingly dominant as coverage increases.

This makes direct comparison with data for the resins difficult. However, two observations can be made from these profiles. The first is that the acid strengths on all three HMS and on the SBA-15 supports appear to be similar, and these strengths seem to be independent of the concentration of acid groups, at least over the limited range of concentrations that is achievable with these functionalisation routes. The second is that, at very low coverages, the sulfonated silica catalysts appear to show higher acid strengths than the sulfonated resins, under these anhydrous conditions. Profiles for all four sulfonated silicas show molar enthalpies of ammonia adsorption of 135-140 kJ mol^{-1} for the first pulse of ammonia, compared to 125 kJ mol^{-1} for the most acidic sulfonated resin.

3.1.2 Acidities in the Presence of Water

A typical microcalorimeter output is shown in Figure 2 for an aqueous NaOH titration of a sulfonated silica. The molar enthalpy of neutralisation is essentially constant throughout the neutralisation and a precise value for the molar enthalpy of neutralisation can be obtained from the linear plot of cumulative enthalpy against added base.

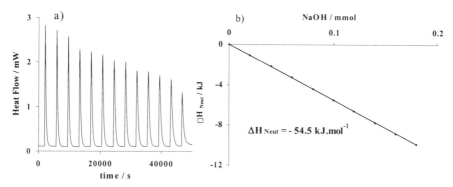

Figure 2 *(a) Output from Titrys microcalorimeter and (b) cumulative enthalpy vs. extent of neutralisation, for HMS 3-SO₃H on reaction with aqueous NaOH*

These average values for the molar enthalpy of neutralisation are shown in the Table, along with equivalent values for dilute HCl solution and dilute *p*-TsOH solution. Note that the molar enthalpies of neutralisation by strong aqueous base solution for the sulfonated silicas are very similar to those for the two strong acids in aqueous solution. The sulfonated resins, in comparison, show similar values for resins with low levels of sulfonation but dramatically higher (numerical) values for resins with high concentrations of acid groups.[7&13] The two most highly sulfonated resins show enthalpies of neutralisation six or more kJ mol^{-1} higher than those for the sulfonated silicas and the strong aqueous acids.

3.2 Catalytic Activities

Activities are reported in the Table as initial rates per mole of acid group, or initial turnover numbers. In the case of the resins, higher rates were recorded for the powdered than the bead resins, and the data shown is for the powdered materials. This suggests that, when in bead form, the reaction is largely under diffusion control (through the swollen gel). It is not possible to estimate the extent to which the reaction is still under diffusion control when the catalyst is powdered. However, the consistent pattern of rates for the sulfonated resins and the sulfonated silicas that appears in the Table suggests that kinetic control dominates in both catalysts under the conditions used.

The turnover numbers, or specific rates, of the sulfonated silicas are all similar and within experimental error of each other. This suggests that all acid sites are readily accessible, and diffusion rates to the sites are not rate limiting, in both HMS supported catalysts with relatively narrow pores, and SBA-15 where the pores are very much larger. On this basis the catalytic activities of the all acid groups in both sulfonated HMS and sulfonated SBA-15 are essentially constant.

The activities of the resins with low levels of sulfonation are similar to those of the sulfonated silicas, again suggesting that the acid groups are readily accessible and exhibit the same specific activity. A very significant difference is seen however, for resins with higher levels of sulfonation. Sulfonic acid groups on these catalysts are evidently very much more active than those on the silicas and those on the resins with acid concentrations.

4 DISCUSSION

The results show clearly that, in the presence of water, the acid catalytic activity of supported sulfonic acid groups is essentially the same on polymer and silica supports, except where the level of polymer sulfonation is high, when the sulfonic acid exhibits significantly enhanced activity. The trend in molar enthalpies of neutralisation with aqueous NaOH is similar. On silica supports, and polymer supports with low sulfonic acid concentrations, these enthalpies are very similar to those of strong mineral acid solutions. In contrast, resins with high levels of sulfonation show significantly higher molar enthalpies of neutralisation.

We have previously observed this elevated molar enthalpy of neutralisation with highly sulfonated resins, along with a similar enhancement in specific activity in another water-based reaction, the hydration of propene.[5&7] In this earlier work, we also studied the degree of dissociation of the sulfonic acid groups in hydrated resins, using FT-Raman spectroscopy.[13] We showed how highly sulfonated resins are largely undissociated even when fully hydrated, and that it is the undissocated acid groups that are responsible for the higher acid strength and higher catalytic activity of these catalysts. We extended an earlier model developed by Gates[12] to explain this enhanced acidity in terms of networks of interacting sulfonic acid groups, the formation of which depends on the flexibility of the polymer and the extent to which cross-linking brings neighbouring groups into proximity. The results here suggest that similar networks of sulfonic acid groups may not be able to form on sulfonated silicas, where equivalent enhancements in activity and acid strength are not seen. In fact, this is not altogether surprising in view of the rigidity of the silica supports, which might reasonably be expected to preclude the reconfiguring of adjacent acid groups that is possible with the flexible polymer chains in polystyrene resins.

Of course, only relatively low levels of sulfonic acid functionalisation have been studied on the silica supports in this work, and it could be argued that a fair comparison requires higher levels of sulfonation to be included. However, it is extremely difficult to functionalise silica at the sort of level achievable readily with polystyrene, so such a comparison is not possible.

A largely accepted view to date has been that the acid strength of sulfonated resins is low compared to many inorganic acid catalysts. Our data on dry catalysts, based on enthalpies of ammonia adsorption (Figure 1), still supports this view, showing that the strongest sites on sulfonated silicas exhibit significantly higher differential enthalpies of ammonia adsorption than sulfonated polystyrene. This is a reversal of the order of acid strengths seen in water, where the highly sulfonated resins show both higher acid strengths and higher specific activities than the sulfonated silicas. This reversal is not necessarily surprising however, as it is quite possible that without the leveling effect of excess water, the factors controlling acid strength on the two types of catalyst are different.

The overall conclusion of this work concerns acidity and activity in the solvent water. It seems that, in aqueous reactions, sulfonated polystyrene catalysts offer two important advantages over sulfonated silicas, firstly that higher surface acid concentrations can be made available, and secondly that the acid strength and catalytic turnover number of the acid groups can be very much higher. Indeed, the specific mechanism which gives rise to the enhanced acid strengths of highly sulfonated resins in water may give these materials unique catalytic properties in water. This may have important implications for the development of more benign solvent systems for fine chemicals synthesis.

References

1   W.M. Van Rhijn, D.E. DeVos, B.F. Sels, W.D. Bossaert, and P.A. Jacobs, *J. Chem. Soc., Chem. Commun.* 1998, 317.
2   W.D. Bossaert, D.E. DeVos, W.M. Van Rhijn, J. Bullen, P.J. Grobet and P.A. Jacobs, *J. Catal.*, 1999, **182**, 156.
3   M.A. Harmer and Q. Sun, *Appl. Catal. A: Gen.*, 2001, **221**, 45.
4   A. Chakrabarti and M.M. Sharma, *React. Polym.*, 1993, **20**,1.
5   M. Hart, G. Fuller, D.R. Brown, C. Park, M.A. Keane, J.A. Dale, C.M. Fougret and R.W. Cockman, *Catal. Lett.*, 2001, **72**, 135.
6   C. Buttersack, H. Widdecke and J. Klein, *React. Polym.*, 1987, **5**, 181.
7   M. Hart, G. Fuller, D.R. Brown, J.A. Dale and S. Plant, *J. Molec. Catal. A: Chemical*, 2002, **182-183**, 439.
8   F. Ancillotti, M.M. Mauri and E. Pescaraollo, *J. Catal.*, 1977, **46**, 49.
9   K. Jerabek, J. Odnoha and K. Setinek, *Appl. Catal.*, 1987, **37**, 129.
10   J.H. Ahn, S.K. Ihn and K.S. Park, *J. Catal.*, 1988, **113**, 434.
11   K. Jerabek and K. Setinek, *J. Mol Catal.*, 1987, **39**, 161.
12   R. Thornton and B.C. Gates, *J. Catal.*, 1974, **34**, 275.
13   S. Koujout, B.M. Kiernan, D.R. Brown, H.G.M. Edwards, J.A. Dale and S. Plant, *Catal. Lett.*, 2003, **85**, 33.
14   C E Harland, *Ion Exchange: Theory and Practice*, 2nd edn., Royal Society of Chemistry Monograph, London, 1994, Chapter 4, p. 73.
15   D. Margolese, J. A. Melero, S.C. Christiansen, B. F. Chmelka, G. D. Stucky, *Chem Mater.*, 2000, **12**, 2448.
16   P.J. Parrillo, R.J. Gorte and W.E. Farneth, *J. Am. Chem. Soc.*, 1993, **115**, 12441.

Pt/H-MOR AND Pt/H-BEA CATALYSTS WITH VARIOUS Pt CONTENTS
AND BIMETALLIC PtPd/H-MOR, PtPd/H-BEA, PtIr/H-MOR AND PtIr/H-BEA
CATALYSTS WITH VARIOUS SECONDARY METAL CONTENTS
FOR THE HYDROCONVERSION OF n-HEXANE.

A.K. Aboul-Gheit, S.M. Abdel-Hamid and A.E. Awadallah

Process Development Division, Egyptian Petroleum Research Institute, Nasr City,
P.O. Box 9540, Cairo 11787, Egypt ; E-mail:<aboulgheit2000@hotmail.com>

1   INTRODUCTION

The hydroisomerisation of n-paraffins in the low-boiling naphtha fraction of crude oils
is the reaction of interest for increasing octane number. There is a great demand for
isoparaffins in the reformulated gasoline pool, due to the impact of environmental
regulations which exclude completely alkyl-lead additives and limit olefins, benzene
and total aromatics. Chlorine promoted Pt/Al_2O_3 catalysts were used commercially for
n-paraffin hydroisomerization [1] until the recent use of Pt/H-MOR catalysts [2] which are
thermally stable, resist heteroatom poisoning, and are strongly acidic and hence do not
require Cl- injection (corrosive). However, a more recent development of isomerization
catalysis introduces Pt/BEA and mazzite catalysts [3,4].

Monometallic Pt/H-BEA and Pd/H-BEA and various bimetallics including Pt plus
Pd, Ni, Cu or Ga have been examined for n-hexane hydroisomerization [4]. The
monometallic catalysts show similar activities and isomer selectivities as the bimetallic
PtPd catalyst, but in presence of sulphur, the bimetallic catalyst is more resistant to
poisoning [5]. Blomsma et al.[6] have investigated the hydroconversion of n-heptane on
catalysts containing Pt and Pd on H-BEA and USY zeolites and found that bimolecular
reactions are suppressed and the hydrogenolysis activity eliminated; hence the
performance of hydroisomerization is greatly improved. These authors attribute these
beneficial effects to better intimacy and balance of the metal/acid functions.

Combination of Pt and Ir, though not widely exploited commercially, has attracted
attention. It has many similarities to Pt. Pure Ir metal is inactive for isomerization, but
more active than Pt for hydrocracking [7]. PtIr catalysts have gained importance in
naphtha reforming [8]. Dees and Ponec [9] and Rice and Lu [10] report that PtIr catalysts have
higher activities than Pt catalysts. Rasser et al.[11] claim that the main function of Ir is the
suppression of surface carbiding.

In the present work, monometallic catalysts containing 0.35 up to 0.50wt%Pt on H-
MOR or H-BEA zeolites, as well as bimetallic catalysts containing Pd or Ir of 0.05 up to
0.20-0.25wt% combined with a fixed content of 0.35wt%Pt on H-MOR or H-BEA,
have been tested for n-hexane hydroconversion.

2   EXPERIMENTAL

2.1. Catalysts

Two series of monometallic catalysts (Pt/H-MOR and Pt/H-BEA catalysts), each containing 0.35, 0.40, 0.45, 0.50 and 0.55 wt%Pt on either support as well two series of bimetallic catalysts containing 0.35wt%Pt plus 0.05, 0.10, 0.15, 0.20 (or 0.25) wt% Pd or Ir, on H-MOR and H-BEA zeolites were prepared. The catalysts containing various Pt contents on either support were prepared via impregnating solutions containing the requisite quantities of chloroplatinic acid together with citric acid for obtaining optimum dispersion of the metal [12]. The bimetallic combinations (Pt+Pd and Pt+Ir) supported on either zeolite were prepared via successive impregnations; the first using H_2PtCl_6 followed by drying overnight at 110°C then calcination for 4 h at 530°C. The second impregnation used solutions of $Pd(NO_3)_2$ or H_2IrCl_6, respectively. These catalysts were again dried and calcined as above. Before testing the catalysts in the hydroconversion runs, they were reduced *in situ* at 500°C for 8 h.

2.2. Metals Dispersion (Fraction of Metal Exposed) in the Catalysts

Dispersion of Pt, as well as its combinations with Pd or Ir, in H-MOR and H-BEA zeolites were determined by H_2 chemisorption using a pulse technique similar to that of Freel [13].The calcined catalysts were heated in the chemisorption furnace at 500°C for 3h in a H_2 flow of 50 cc/min then in a N_2 flow of 30 cc/min for 2 h   (degassing). The furnace was shut off and the catalyst was cooled to room temperature. H_2 was then pulsed into the N_2 carrier until saturation (appearance of H_2 peaks equivalent to non-chemisorbed pulses). The H_2 uptake was calculated as hydrogen atoms adsorbed per total metal atoms on the basis of 1:1 stoichiometry [14].

2.3 Temperature Programmed Desorption (TPD) of Ammonia for Acid Site Strength Distribution using Differential Scanning Calorimetry (DSC)

The procedure adopted by the authors [15,16] to follow the desorption of presorbed NH_3 on the catalysts was used. NH_3 adsorption was first carried out in a silica tube furnace after evacuation at 1.33×10^{-3} Pa whilst heating at 500°C and subsequent cooling to 50°C (under vacuum). NH_3 gas was passed on the catalyst at 50cc/min flow rate. The catalyst was measured in a DSC-30 unit (Mettler TA-3000) with a gold sensor using Al crucibles. NH_3 desorption was then carried out in N_2 purge gas at a flow rate of 50 cc/min. The heating rate was 10 K/min. and full-scale range was 25 mW.

   Two endothermic peaks appeared in each desorption thermogram; a low temperature peak representing weak acid sites and a high temperature one representing strong acid sites.   The enthalpy value corresponded to the number of sites (peak magnitude), whereas the peak temperature compares site strength. Only the strong site peak is of interest in the present work. Figure 1 gives the DSC thermograms obtained for NH_3 desorption from H-MOR and H-BEA zeolites. Evidently, H-MOR possesses higher sites number and strength. Similar TPD thermograms are obtained for the metal loaded zeolite catalysts. Both acid site number and strength decrease with increase of metal content. The TPD data obtained are given in Tables 1 and 2.

Table 1 *Ammonia desorption enthalpy (ΔH_d) and peak temperature for catalysts containing 0.0-0.55wt%Pt supported on H-MOR and H-BEA zeolites*

Wt % °C	H-MOR Zeolite ΔH_d, Jg^{-1}	Peak temperature, °C	H-BEA Zeolite ΔH_d, Jg^{-1}	Peak temperature,
Unloaded	125	550	105	525
0.35 Pt	105	522	85	500
0.40 Pt	96	516	77	490
0.45 Pt	90	510	73	485
0.50 Pt	84	504	69	481
0.55 Pt	77	498	65	478

2.4 Hydroisomerization Technique and Analysis

A micro-reactor tube jacketed with an electrically thermostated copper block heater was installed at the injection port of a gas-chromatograph (Sigma-3 Perkin Elmer). The reactor always contained 0.20 g of a catalyst whereupon H$_2$ carrier gas of 20cc/min passed.

Table 2 *Ammonia desorption enthalpy (ΔH_d) and peak temperature for catalysts containing 0.35wt%Pt*
Plus Pd or Ir supported on H-MOR and H-BEA zeolites

Wt % Pd	H-MOR Zeolite ΔH_d, Jg^{-1}	Peak temperature, °C	H-BEA Zeolite ΔH_d, Jg^{-1}	Peak temperature, °C
0.00 Pd	105	522	85	500
0.05 Pd	98	519	80	492
0.10 Pd	92	513	76	489
0.15 Pd	85	507	72	485
0.20 Pd	79	500	68	481
0.25 Pd	74	495	--	---
0.05 Ir	95	515	76	488
0.10 Ir	89	509	73	485
0.15 Ir	82	502	70	482
0.20 Ir	75	496	66	479

The feed injected was always 1.0 μl of AR n-hexane. Reaction temperatures from 500°C downward (25°C intervals) were used. Eluted products passed directly to a GC column of length 3.0 m and internal diameter 5 mm packed with 10wt% OV-101 on Chromosorb W of 60-80 mesh. The column temperature was maintained isothermally at 60°C.

3 RESULTS AND DISCUSSION

n-Hexane isomerises to four branched isomers possessing higher octane numbers. These isomers are formed at relatively lower temperatures (250-350°C) using the catalysts under study mostly at very high selectivities. Typically, isomers increase with temperature to reach a maximum, beyond which they decline via a further increase of temperature by accelerated hydrocracking. Since n-hexane rearranges and cracks via a carbenium ion mechanism on the same catalytic acid sites, their relative occurrence is controlled by catalyst bifunctionality. Dehydrocyclization of n-hexane to benzene also occurs at higher temperatures.

In the past decade, Pt/H-MOR was considered as the beginning of a new family of catalysts for the hydroisomerization of C_5/C_6 n-paraffins.[2] More recently, Pt/H-BEA has been used. [3,4]. In this study, catalysts containing H-MOR and H-BEA zeolites loaded with 0.35-0.55 wt% Pt as well as 0.05-0.25 wt% Pd or Ir combined with 0.35 wt% Pt are tested for n-hexane hydroconversion. It is known that commercial hydroisomerisation and catalytic reforming catalysts normally contain 0.35–0.40 wt%Pt.

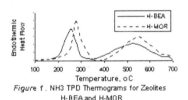

Figure 1. NH3 TPD Thermograms for Zeolites
H-BEA and H-MOR

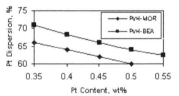

Figure 2. Pt Dispersion in H-MOR and H-BEA

3.1 Effect of Pt and PtPd or PtIr on their Dispersion in H-MOR and H-BEA Zeolites

A successive increase of Pt content in either Pt/H-MOR or Pt/H-BEA produces a continued decrease of Pt dispersion in an almost parallel fashion (Figure 2). Pt dispersion in H-BEA is always higher than in H-MOR at corresponding Pt contents, which may be attributed to the larger surface area (700 vs. 500 m^2g^{-1}) and larger pore volume (0.234 vs. 0.209 cc g^{-1}) of H-BEA than H-MOR. Very close values of surface area and pore volume of these zeolites to ours are given by other authors.[4] The 0.35-0.50wt %Pt/H-MOR catalysts acquire metal dispersions of 66-60%, respectively, whereas the 0.35-0.55wt%Pt/H-BEA catalysts acquire dispersions of 71- 62%, respectively.

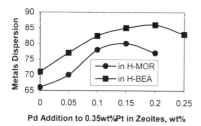

Figure 3 Effect of Pd Addition on Metals
Dispersion

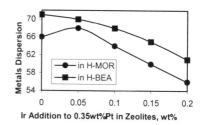

Figure 4 Effect of Ir Addition on Metal
Dispersion

On the other hand, dispersion of the combined metals (Pt+Pd) or (Pt+Ir) in either H-MOR or H-BEA zeolites, primarily containing 0.35wt%Pt, is found to depend largely on the second metal added and the zeolite type. Successive additions of Pd to Pt in H-BEA results in gradual improvement of dispersion of PtPd up to 0.20wt%Pd, where a maximum dispersion of 86% is attained. Further addition to 0.25wt%Pd reduces dispersion (Figure 3). Also, successive addition of Pd to Pt in H-MOR, increases dispersion up to a maximum of 80% at 0.15wt%Pd, then declines on adding 0.20wt%Pd. Such a decline beyond the maximum may be attributed to blockage of the zeolitic channels. The achievement of maximum dispersion at a higher Pd content in H-BEA than in H-MOR is evidently attributed to the larger surface area and pore volume of H-BEA. Jacobs and coworkers [6] observe dispersion improvement via combining Pd *with Pt*. Their TPR data indicate that this improvement is due to a catalytic effect of Pd on Pt reduction where rapid "fixation" of PtPd in highly dispersed form occurs. We may assume that their improved dispersion in the zeolites is most probably attributed to their similar valence states, i.e.; 2 and 4, which affect similar oxidation states of both metals during the calcination of the metals precursors followed by the formation of more dispersed metal crystallites during the reduction step.

On the other hand, successive addition of Ir to Pt/H-BEA continually decreases dispersion of the bimetal (Figure 4). However, the first addition of Ir (0.05wt%) to Pt in H-MOR gives little increase of dispersion, beyond which dispersion decreases again linearly with increasing Ir. The PtIr dispersion and acidity of the PtIr/H-BEA catalysts (Figure 4 and Table 2) don't differ significantly from those of the Pt/H-BEA catalysts (Figure 2 and Table 1). Dispersion of Ir crystallites in Ir/H-BEA may not vary significantly from those of Pt in Pt/H-BEA. Previously, similar findings by the authors [17] show that dispersion of the metal in a 0.35wt%Pt/Al$_2$O$_3$ catalyst equals that in a 0.35wt%Ir /Al$_2$O$_3$, and that bimetal dispersion in a catalyst containing 0.35wt%Pt + 0.35wt%Ir/Al$_2$O$_3$ equals that obtained for Pt dispersion in a 0.70wt%Pt/Al$_2$O$_3$ catalyst. The close periodic position of Pt and Ir in the 3rd period of group VIII, and hence the similarity of their electronic structure, could play a role in acquiring similar dispersions of the two metals. Ir (at. wt. = 192.2) directly precedes Pt (at. wt. = 195.1) in the periodic system of elements. Nevertheless, on H-MOR, very low Ir (0.05wt%) may catalyse dispersion of Pt agglomerates at hidden situations in channels.

3.2 Maximum Iso-Hexanes Production

3.2.1 Using Monometallic Pt/H-BEA and Pt/H-MOR Catalysts

The maximum production of isohexanes is considered the practical parameter of choice upon which refiners base their selection of operating conditions in light naphtha isomerization.

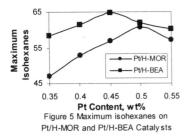

Figure 5 Maximum isohexanes on Pt/H-MOR and Pt/H-BEA Catalysts

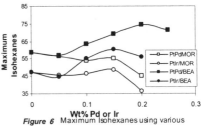

Figure 6 Maximum isohexanes using various Pd or Ir additions to Pt on Zeolites

Using *monometallic catalysts* (Pt/H-BEA and Pt/H-MOR), maximum isohexanes production (Figure 5), evidently differs with Pt content variation and the type of zeolite support. This maximum is always larger on the H-BEA supported catalysts than using the H-MOR supported ones. The increase of Pt in the H-BEA zeolite from 0.35wt% up to 0.45wt% has linearly increased the maximum iso-hexanes from 58.5% up to 64.8%, beyond which they decline in product by a further increase of Pt. However, the increase of Pt in H-MOR from 0.35wt% up to 0.50wt%Pt also increases the maximum production of iso-hexanes, almost linearly, from 47.1% up to 60.6%, beyond which maximum iso-C6s decline. So, a*bove 0.50wt%Pt on H-MOR, a blockage of the channels could take place* whereby the maximum isohexanes value decreases. However, *above 0.45wt%Pt on H-BEA the decline of iso-C6s can not be assumed to be due to blockage of channels*, because the channels of H-BEA are wider than in H-MOR beside its tri-directional arrangement which is easier for diffusion processes than in the uni-directional channels of H-MOR. Hence, it can be assumed that the attainment of the highest iso-hexanes maximum using *the 0.45wt%Pt/H-BEA catalyst is attributed to the attainment of the proper balance between the metal and the acid sites.* The somewhat lower Pt content, giving highest maximum isomers using H-BEA seems adequate for balancing the relatively lower acid site number and strength in H-BEA compared to H-MOR. Guisnet et al.[18] show that higher loadings with Pt >0.30wt%, decrease maximum isomerisation by partial blockage of the MOR channels with Pt crystallites.

Evidently, no correlation can be withdrawn between the trend of Pt dispersion (Figure 2), and that of maximum isohexanes production (Figure 5), indicating that this reaction may be *"structure insensitive"* or *"facile"*. Moreover, the acid sites density and strength which decrease with increasing Pt content (Table 1) also do not correlate with maximum isohexanes production.

3.2.2 Using Bimetallic PtPd and PtIr/H-BEA and PtPd and PtIr/H-MOR Catalysts

Combinations of PtPd and PtIr supported on H-BEA zeolite give a significant increase of maximum isohexanes production with increasing the second metal (Figure 6) showing structure sensitivity in case of the PtPd/H-BEA catalysts as correlated with the metal dispersion data in Figure 3. Moreover, on H-MOR, the increase of Pd or Ir seems to exhibit some gradual decrease of the isohexanes maximum production (Figure 6), which may also show some compatibility with the behaviour towards metal dispersion (Figure 4) as well as with acid sites strength distribution (Table 1). Hence, using the PtPd and PtIr/H-MOR catalysts, the isomerization maxima can be considered structure sensitive or "demanding".

Structure sensitivity may be controlled by the catalytic support rather than by the metals. This may be substantiated by the authors[17] previous findings concerning Pt, Ir and PtIr/alumina catalysts, discussed above. Practically, it is found that successive addition of Pd to Pt causes isohexanes maximum to occur at a lower temperature of 275°C, whereas Ir addition causes this maximum to occur at 300°C. Although Pd incorporation is more economic from the view of reaction engineering, Ir is quite resistant to sintering on a number of metal oxides and zeolite supports in presence of H_2[19].

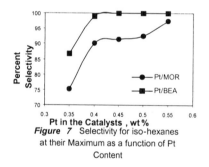

Figure 7 Selectivity for iso-hexanes at their Maximum as a function of Pt Content

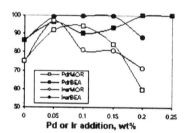

Figure 8 Selectivity at Maximum isohexanes via Pd or Ir addition

3.2.3 Hydroisomerisation Selectivity at Maximum Isohexanes Production

Using the monometallic Pt/H-BEA and Pt/H-MOR catalysts (Figure 7), isohexanes selectivity at their maximum production generally increases with Pt content. It is lowest using the 0.35wt%Pt, (87.0% and 75.0%, respectively). At 0.40wt%Pt, it increases to 98.5 and 90.0%, respectively, and then remains almost unchanged in a plateau-like fashion as a function of Pt content. However, using the 0.55wt%Pt/H-MOR catalyst, again isohexanes selectivity increases to 97.5%. This behaviour is incompatible with Pt dispersion and acidity; hence, isohexanes selectivity at maximum isomers production on these catalysts is structure insensitive (facile).

Using the bimetallic catalysts (Figure 8), the selectivity is generally higher on the H-BEA supported catalysts than on the H-MOR supported ones. Furthermore, correlation of Figures 6 and 8 shows that higher isohexanes maximum yields associated with higher selectivities are   attained using the H-BEA supported catalysts. Evidently, the first addition (0.05wt%) of the secondary metal gives a sharp increase of selectivity, beyond which they behave differently. Again, beyond 0.15wt% addition, a sharp drop of selectivity is attained, except for the PtPd/H-BEA catalyst, which continues 100% selectivity up to 0.25wt%Pd. The superiority of these catalysts (Figure 8) may be primarily attributed to high PtPd dispersion (Figure 3).

3.3. Hydrocracking of n-Hexane

3.3.1 Using Monometallic Pt/H-MOR and Pt/H-BEA

The catalysts containing successively increased Pt contents starting with 0.35wt% up to 0.55wt% (increments = 0.05wt%) are compared for their hydrocracking activities at a reaction temperature of 350°C (Figure 9). Using the Pt/H-MOR catalysts, hydrocracking significantly exceeds that attained using the corresponding Pt/H-BEA catalysts, most probably due to their higher acid site density and strength (Table 1). Hence, using the catalysts containing 0.35-0.50wt%Pt on H-MOR, at least 96.5% of hydrocracked products are obtained, but this decreases to 88.7% at 0.55wt%Pt. Correspondingly, hydrocracked product range between 60.7 and 69.4% using the Pt/H-BEA catalysts. Apparently, hydrocracking of n-hexane on the latter catalysts is not controlled by their Pt content, Pt dispersion or even the catalyst acidity. Their hydrocracking activities may be controlled by the balance of their acid/metal bifunctionality which governs the competing reactions occurring together with hydrocracking.

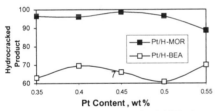

Figure 9 Effect of Pt content in Pt /H-MOR and Pt/BEA Catalysts on n -Hexane Hydrocracking

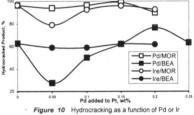

Figure 10 Hydrocracking as a function of Pd or Ir addition to 0.35wt %Pt onZeolites

3.3.2 Hydrocracking using Bimetallic Catalysts

On these catalysts, hydrocracking is compared, also at a temperature of 350°C, using the bimetallic PtPd/H-MOR, PtPd/H-BEA, PtIr/H-MOR and PtIr/H-BEA catalysts, where Pt content is always 0.35wt% (Figure 10). The mordenite supported catalysts are more active in hydrocracking reactions than the H-BEA supported ones. However, Pd imparts higher activity than Ir on H-MOR. At all Pd additions (0.05-0.20wt%), hydrocracking is >95 to 100% using the PtPd/H-MOR catalysts. There may be a synergistic effect (promotion) of Pd on Pt on the H-MOR zeolite (compare 0.55wt%Pt/H-MOR, Figure 9, and the 0.35wt%Pt+0.20wt%Pd/H-MOR, Figure 10). The synergistic effect between Pt and Pd is manifested in improving the metal dispersion (Figure 3). However, using the PtIr/H-MOR catalysts, the hydrocracked product is not significantly different from those produced on the PtPd/H-MOR catalysts, except at the first addition of 0.05wt%Ir, where hydrocracked product amounts to 78.8%.

On the other hand, addition of 0.05wt%Pd to Pt in H-BEA has sharply diminished hydrocracked product from 62.8 down to 27.5%. Such inhibition of hydrocracking is an economical accomplishment for light-naphtha hydroisomerisation catalysts. However, the hydrocracking activity is found to increase again via successive increase of Pd up to 0.20wt% where a maximum of 76.8% is attained. The decline at 0.25wt%Pd, may be attributed to blockage of the zeolitic channels by the metals. However, on addition of Ir to Pt/H-BEA from 0.05-0.20wt%, hydrocracking does not change significantly (62.8 to 59.1%).

3.4. Benzene Formation via Dehydrocyclisation

Using all the catalysts under study (mono- or bimetallic), benzene is the only aromatic compound detected suggesting that no other mechanism except dehydrocyclisation can occur during benzene formation in this study. This reaction is endothermic. On monometallic and bimetallic catalysts it is not very significant since maximum production using the most active catalyst (PtPd/H-BEA, Fig. 12) does not exceed 19.0%.

3.4.1 Using the monometallic Pt/MOR and Pt/H-BEA catalysts

Using the monometallic catalysts (Figure 11) benzene production is studied at two reaction temperatures (425 and 500°C) because there is an overlap of its formation activity over the Pt/H-BEA and Pt/H-MOR catalysts at 500°C. At 425°C, this reaction is normally more enhanced on the Pt/BEA than on Pt/H-MOR catalysts at all Pt contents due to the higher hydrophilicity of the former (Figure 11). However, at 500°C the higher loading of Pt (0.45-0.55wt%) seems to increase the relative hydrophilicity of Pt/H-MOR catalysts to be more active than the corresponding Pt/H-BEA ones for benzene production.

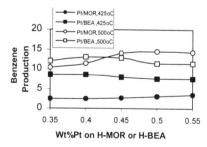

Figure 11 Benzene Production, %

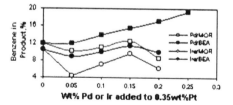

Figure 12 Benzene Production at 500o C as a function of Pd or Ir addition to 0.35wt Pt

3.4.2 Using Bimetallic PtPd/H-MOR, PtPd/H-BEA, PtIr/H-MOR and PtIr/H-BEA Catalysts

The PtPd combinations on either zeolite produce more benzene than the corresponding PtIr combinations at all temperatures (Figure 12). As Pd addition increases in Pt/H-BEA, benzene increases almost linearly up to 0.25%Pd, whereas on the three other catalysts, benzene increases with increasing the secondary metal from 0.05 up to 0.15wt%, then declines at 0.20wt% addition. This is attributed to partial blockage of the channels of the zeolite support. Blockage in H-MOR channels is a common phenomenon, since these channels are of unidirectional arrangement as indicated above. However, in the H-BEA channels, the same occurs with the addition of Ir in presence of Pt. The benzene molecule is a large one (around 6.3A) and may encounter significant blockage in the PtIr loaded channels of H-BEA. The significantly larger atom of Ir (at. Wt= 192.2) restricts benzene production on the 0.20wt%Ir+0.35wt%Pt/H-BEA catalyst, whereas the smaller atom of Pd (at. Wt = 106.4) does not cause such restriction.

4 CONCLUSION

Increased Pt in either Pt/H-MOR or Pt/H-BEA continues decreasing Pt dispersion. This dispersion in H-BEA is always higher than in H-MOR due to the larger surface area and pore volume of H-BEA. Adding Pd to Pt in *H-BEA* and *H-MOR* improves PtPd dispersion up to 0.20wt% Pd in the former zeolite and up to 0.0.15wt% in the latter. Reduction beyond a maximum dispersion can be attributed to blockage of the zeolite channels. Addition of Ir to Pt in both zeolites decreases dispersion.

Maximum isohexanes production is always larger on the H-BEA supported catalysts than on H-MOR supported ones. Above 0.50wt%Pt on H-MOR, a blockage of the channels could take place whereby the maximum isohexanes value decreases. However, above 0.45wt%Pt on H-BEA the decline of iso-C6s cannot be due to blockage of channels, because the channels of H-BEA are wider than in H-MOR beside its tridirectional vs. unidirectional channels arrangement (easier for diffusion). Hence, highest iso-hexanes maximum on 0.45wt%Pt/H-BEA is attributed to proper balance between metal and acid sites. Maximum isomer production seems structure insensitive on all catalysts. Generally, at maximum isomer production, isomerisation selectivity is best at 0.40-0.55 wt%Pt in monometallics and 0.05-0.15 wt% Pd or Ir in bimetallics. Hydrocracking is higher on H-MOR supported catalysts than on H-BEA (contrary to isomerisation). However, Pd introduction to Pt in H-BEA shows a varying trend. Benzene production is higher on the more hydrophilic H-BEA containing catalysts. The most active are the PtPd/H-BEA catalysts, whereas the least active are the PtIr/H-MOR catalysts.

References

1. I.E. Maxwell and W.H.J. Stork, *Studies in Surface Science and Catalysis*, 1990, **58**, 571.
2. P.J. Kuchar, J.C. Bricker, M.E. Reno and S.R. Haizmann, *Fuel Process Technology*, 1993, **35**, 183.

3. L.-J. Leu , L.-Y. Hou, B.-C. Kang, C. Li, S.-T. Wu and J.-C.Wu, *Appl. Catal.*, 1991, **69**, 49; F. Fajula, Boulet, B. Cop, V. Rajacfanova, F. Figueras, T. Des Courieres; in L. Guczi, F. Solymosi, P. Telenyi, eds., *10th International Congress on Catalysis*, Budapest, Elsevier, 1993, 1007.

4. J.-K. Lee and H.-K. Rhee, *Catal. Today*, 1997, **38**, 235.

5. J.-K. Lee and H.-K. Rhee, *J. Catal.*, 1998, **177**, 208.

6. E. Blomsma, J.A. Martens and P.A. Jacobs, *J. Catal.*, 1997, **165**, 241.

7. J.P. Brunelle, R.E. Montarnal and A.A. Sugier, *6th ICC, London, Proceed.*,1976 (G.C.Bond et al., eds.), Chem. Soc., London, Vol. 2, (1977), p. 844.

8. V. Ponec and G.C. Bond, Catalysis by Metals and Alloys, *Studies in Surface Science and Catalysis*, Vol. 95, Chap. 6, Elsevier, Amsterdam, 1995.

9. J.M. Dees and V. Ponec, *J. Catal.*, 1998, **115**, 347.

10. R.W. Rice and K. Lu, *J. Catal.*, 1982, **77**, 104.

11. J.C. Rasser, W.H. Bindroff and J.J.F. Scholten, *J. Catal.*, 1979, **59**, 211.

12. A.K. Aboul-Gheit, *J. Chem. Technol. Biotechnol.*, 1979, **29**, 480.

13. J. Freel, *J. Catal.*, 25 (1972) 193

14. R.M. Fiedorow, B.S. Chahari and S.E. Wanake *J. Catal.*, 1978, **51**, 193

15. A.K. Aboul-Gheit, *Thermochim. Acta.*, 1991, **191**, 233.

16. A.K. Aboul-Gheit, *Solid State Ionics*, 1997, **101-103**, 893.

17. A.K. Aboul-Gheit and S.M. Abdel-Hamid,; *6th International Symposium on the Scientific Bases for the Preparation of Heterogeneous Catalysts*; Louvain La-Neuve, 1994; *Studies in Surface Science and Catalysis*, (G. Poncelet, J. Martens, B. Delmon, P.A. Jacobs and P. Grange, eds.), Vol. 91, Elsevier, Amsterdam, 1995, p. 1131.

18. M. Guisnet, V. Fouche, P.A. Belloum, J.P. Bournomille, C. Traverse, *Appl.Catal.*, 1991, **71**, 283.

19. N.D. Triantafillou, S.E. Deucsch, O. Alexeev, J.T. Miller and B.C. Gates, *J. Catal.*, 1996, **159**, 14.

COMPARISON OF THE ACID PROPERTIES ON SULPHATED AND PHOSPHATED SILICA-ZIRCONIA MIXED OXIDE CATALYSTS

J.A. Anderson[*], B. Bachiller-Baeza and D.J. Rosenberg

Surface Chemistry and Catalysis Group, Division of Physical and Inorganic Chemistry, University of Dundee, DD1 4HN, Scotland, UK

1 INTRODUCTION

Although zirconia is widely used both as a catalyst and as a catalyst support, two component oxides such as silica-zirconia display certain advantages in terms of both the thermal stability and the acidic properties. One method of obtaining homogeneously mixed oxides containing two or more components involves sol-gel methods. The use of sol-gel chemistry to prepare mixed oxides introduces a high degree of flexibility as their final properties can be deliberately influenced by manipulation of their preparation parameters. Millar *et al* and Anderson *et al,*[1-5] have employed such methods to prepare silica-zirconia aerogels and the influence of the degree of mixing and the Si/Zr ratio on the density and strength of generated acid groups have been determined.

An alternative method of modifying the acid properties of a solid oxide is by use of dopant ions such as sulphate. The enhanced surface acidity and possible superacid catalytic properties resulting from sulphate addition is most widely recognized in the case of neat zirconia[6-8]. However, the surface acid properties of transition metal promoted and alumina containing sulphated zirconias have also been reported[9,10] as has the influence of sulphate on the acidic properties of silica-titania[11] and silica-zirconia[12] mixed oxides. Incorporation of sulphate often involves wet impregnation whereby the mixed oxide is contacted with an aqueous solution of sulphuric acid. However, this method has been reported to induce segregation at the surface of both silica-zirconia[13] and silica-titania[11] mixed oxides as a consequence of hydrolysis of the Si – O – M bonds leading to extraction of the M cation producing an amorphous sulphated zirconia or titania being deposited on silica. Sulphation may also have an effect on the crystallization of zirconia[14], delaying this process to higher temperatures. When a mixed oxide is sulphated in such a manner, the sulphur containing species are thought to be exclusively associated with the non – silica component of the mixed oxide[15] and thus a sulphated silica – zirconia mixed oxide may be envisaged as sulphated zirconia dispersed within a silica lattice.

This work involves the modification of silica–zirconia (33 mol% zirconia) *via* sulphation and phosphation processes and a comparison of the resultant acidic properties. Mixed oxides were prepared by the sol-gel method which allowed homogeneous mixing of the two component oxides at room temperature. Procedure involved pre-hydrolysis of TEOS with nitric acid and water followed by the addition of zirconium isopropoxide. S : Zr and P : Zr ratios of 0.20 : 1, 0.25 : 1, and 0.30 : 1 were employed whereby the addition of modifiers involved impregnation of a previously calcined silica–zirconia mixed oxide with sulphuric/phosphoric acid. This method should ensure that all of the sulphate/phosphate species which are retained after calcination were present at the surface

2 EXPERIMENTAL

2.1 Sample preparation

An unmodified mixed-oxide and a number of phosphated and sulphated 33 mol% zirconia-silica mixed oxides were prepared by modifying the method described by Yoldas[16]. Tetraethyl orthosilicate, TEOS (Silibond 90 wt.%) was combined with water, propanol (Aldrich), as a solvent and nitric acid (Aldrich) used as a hydrolysis catalyst. The above were c ombined i n o verall r atios o f, 1 : 1 .2 : 1 .5 : 0 .2. T he r eagents w ere s tirred u nder nitrogen for 2 h pre-hydrolysis time, after which, zirconium isopropoxide (Aldrich 70 wt.%) diluted 10:1 in propanol was added such that the ratio of Si^{4+}: Zr^{4+} was 2 : 1. After a further hour the final amount of hydrolysis water was added drop wise, the final water : metal cation ratio was 2.6 : 1. All samples gelled within approximately 3 days. Propanol was then exchanged for ethyl acetate *via* Soxhlet extraction for 5 h, the ethyl acetate then subsequently removed using super critical drying. Initially the sample was left for 12 h in super critical CO_2 followed by a 30 min period of flushing every 2 h until no further ethyl acetate was detected in the effluent (typically after 5 flushes). Samples were then transferred to a tube furnace and calcined in flowing air at 873 K for 6 h.

Phosphate and sulphate modifiers were incorporated by the addition of appropriate amounts of 0.01 M sulphuric or phosphoric acid to a pre-calcined aerogel followed by further calcination at 873 K. Samples are labeled as X-SiZr (y) where X refers to either sulphated (S) or phosphated (P) samples, and y refers to the mole ratio of sulphate/phosphate relative to zirconium in the preparation method. For comparison purposes, samples of zirconia and sulphated zirconia were also prepared. This was achieved *via* precipitation from zirconium isopropoxide (Aldrich 70 wt.%). The same H^+: H_2O : Zr^{4+} : propanol ratios were employed as used during the preparation of the mixed oxides. A sulphated zirconia, prepared by the use of sulphuric acid as hydrolysis catalyst was prepared for comparative purposes and had a nominal S:Zr ratio of 0.30 : 1. A further sample was prepared where segregation of components was induced by thermal treatment by calcination at 1373 K for 6 h.of the non-treated SiZr (0)

2.2 Characterization

BET surface areas were measured using a multi-point Coulter SA 3100 instrument with data collected over the P/P$_0$ range of 0.02-0.2. Adsorption of N_2 at 77 K was carried out after outgassing the samples at 573 K. BJH pore distributions were determined using 45 data points over a full adsorption desorption isotherm.

Surface acid densities were estimated using pyridine adsorption monitored by combined thermogravimetric and infra-red spectroscopic techniques[5]. Thermogravimetric analyses were carried out using a PC controlled CI microbalance attached to a conventional vacuum line fitted with rotary and diffusion pumps. Approximately 100 mg of sample as a fragmented disc (prepared as per FTIR experiments, see below) was outgassed for 2 h at 573 K, then exposed to 1 Torr pyridine and cooled to 373 K. A further 0.5 Torr of pyridine was introduced and the system allowed to reach equilibrium over 30 min. After this period, the sample was heated under vacuum to 423 K for 2 h, then at 473 K for 2 h while the mass was monitored continually at 3 s intervals throughout the experiment. The infra-red experiments were carried out in identical fashion using *ca* 80 mg of sample pressed into a 2.5 cm diameter discs at 0.10 tons cm^{-2}. Spectra were recorded after the initial evacuation of the sample at 573 K and then after following exposure to pyridine and outgassing at 423 and 473 K. Due to the weaker adsorption on the phosphated samples, data was also collected after 373 K outgassing.

The mass of pyridine remaining adsorbed on the sample at the two (or three) adsorption temperatures, in combination with the integrated areas underneath the bands due to the 19b ring vibrations of pyridine adsorbed at Lewis and Brønsted sites (*ca*. 1450 and 1540 cm^{-1}, respectively) at the corresponding temperatures allowed calculation of the Brønsted and Lewis absorption coefficients. The number of Lewis and Brønsted sites could be calculated by fitting this data to the equation (1):

$$n_T = \frac{A_L C_d}{\varepsilon_L m} + \frac{A_B C_d}{\varepsilon_B m} \tag{1}$$

where n_T is the total number of micromoles of pyridine per gram sample adsorbed at each temperature, A is the integrated absorbance (cm^{-1}) of FTIR bands due to pyridine on each site, C_d is the cross sectional area (cm^2) of the pressed disc, ε is the absorption coefficient (cm $\mu mole^{-1}$) for pyridine at each site and m is the mass (g) of the pressed disc. Combined FTIR-gravimetric experiments were conducted between 3 and 5 times for each sample to ensure that the values obtained were reliable.

Sulphur contents were determined using a Leco – CHNS – 932 – determinator. Samples were weighed within a silver capsule and then the encapsulated sample dropped into a furnace when the components were combusted in an oxygen excess environment. The sulphur content, measured as SO_2 was then determined by IR absorption.

X-ray photoelectron spectra were obtained using a VG Microtech Multilab electron spectrometer using the Mg Kα radiation (1253.6 eV) from a twin-anode in the constant energy analyzer mode with a pass energy of 50 eV. The pressure in the analysis chamber was maintained at 5 x 10^{-10} m Bar. The binding energy and Auger kinetic energy scale were set by assigning a value of 284.6 eV to the C 1s transition. The accuracy of the binding energy and Auger kinetic energy values were 0.2 and 0.3 eV, respectively.

3 RESULTS

All samples prepared by the current methods gave gel times of around 3 days, somewhat longer than those achieved using previous preparation methods[5] but one which gave a higher degree of reproducibility amongst different batches. All mixed-oxide samples calcined at 873 K were X-ray amorphous. The 1373 K calcined sample gave peaks at 30, 34.5, and 50°, 2θ (Cu Kα radiation) indicating the presence of tetragonal zirconia. These peaks were also apparent in the sample of sulphated zirconia. Table 1 includes some of physical characteristics of the sulphated samples. The mixed-oxide aerogel gave a value of 317 m^2g^{-1}, somewhat greater than the value reported (283 m^2 g^{-1}) for the 33 mol% Zr sample prepared by our previous method[5]. Sulphation of the aerogel followed by further calcination led to surface area loss of around 30% (Table 1). The measured sulphur contents indicate t hat n ot a ll s ulphate w as r etained b y t he s amples a nd w as p robably l ost a s S O_2 during the calcination treatment.

Table 1   *Physical characteristics of sulphated samples*

Sample	BET (m^2g^{-1})	Pore volume $(cm^3\,g^{-1})$	Nominal sulphur content (wt%)	Measured sulphur content (wt%)
SiZr (0)	317	0.99	0	0
Segregated	16.7	0.09	-	-
SiZr (0.2)	230	0.87	2.44	1.07
SiZr (0.25)	212	0.82	3.00	1.84
SiZr (0.30)	208	0.87	3.56	1.84
ZrO$_2$(sulphated)	186		-	-

Table 2   *Physical characteristic of phosphated samples*

Sample	BET (m^2g^{-1})	Pore volume (cm^3g^{-1})	Nominal phosphate content (wt%)
SiZr (0)	317	0.99	0
Segregated	16.7	0.09	0
SiZr (0.20)	312	1.02	7.2
SiZr 0.25)	340	1.00	8.9
SiZr (0.30)	296	0.93	10.5

The physical characteristics of the phosphated samples are shown in Table 2. Samples all gave BET areas in the region *ca.* 296–340 m^2 g^{-1} with pore volumes between 0.9 and 1.1 cm^3 g^{-1}. Unlike the sulphated samples where loss of sulphur *via* SO$_2$ is possibly, the absence of v olatile p hosphorus o xides w ould s uggest t hat a g reater p roportion o f p hosphate t han sulphate was retained following calcination. Samantaray and Parida[24] found no weight loss due to phosphate removal by heating phosphated TiO$_2$-SiO$_2$ samples at temperatures up to 1423 K. The greater retention of phosphate than sulphate was confirmed by the measurement o f s urface a tomic r atios b y t he u se o f X PS. T hese r esults a re p resented i n Table 3 and bring together surface atomic ratios for Si/Zr and S/Zr for the sulphated samples (2nd/3rd columns) and Si/Zr and P/Zr ratios for the phosphated samples (4th/5th columns). The most significant difference was the variation in the Si/Zr ratio from 2.6 for SiZr (0) to 3.81 for the segregated sample. There were also significant differences observed when the modification was carried out by either phosphation or sulphation. For example the P-SiZr (0.3) gave a Si/Zr ratio of 2.0 compared to 3.48 for the S-SiZr (0.3) sample indicating major changes in surface composition follow sulphation. Different levels of phosphation had a far less significant effect on the Si/Zr ratios amongst samples. The ratio of sulphur to zirconia also displayed significant changes between various samples, for example t he S -SiZr (0.3) s ample h ad a S /Zr r atio o f 0 .2 c ompared w ith 0 .15 for S -SiZr (0.25). T his d ifference was s ignificant g iven t hat t he e lemental a nalysis s howed t he t wo samples to contain identical amounts of sulphur. The phosphated samples had a P/Zr ratio which decreased as the nominal amount of phosphate was increased suggesting the possibility of poorer dispersed species at the higher loadings.

Table 3   *XPS surface atomic ratios*

Sample	Si/Zr	S/Zr	Si/Zr	P/Zr
SiZr (0.30)	3.48	0.20	2.00	0.660
SiZr (0.25)	3.28	0.15	2.03	0.600
SiZr (0.20)	-	-	2.11	0.520
SiZr (0)	2.63	-		
Segregated	3.81	-		
ZrO$_2$ (sulphated)	-	0.27		

Combining FTIR and gravimetric measurements, allowed absorption coefficients to be calculated for both forms of adsorbed species (Tables 4 and 5). These were of a similar magnitude and range for both the pyridinium ion (0.7 to 1.66 cm µmol^{-1} at 1540 cm^{-1}) and Lewis bound pyridine (1.00 to 1.57 cm µmol^{-1} at 1450 cm^{-1}) on the sulphated samples. The absorption coefficients allowed calculation of the surface densities of both types of acid site as detected by pyridine (Tables 4 and 5). Surface densities as determined by amount of adsorbate retained at 423 K were lower for the sulphate-free sample employed here than for an equivalent 33 mol% sample reported previously[5], again emphasizing the sensitivity of acid site formation to the exact nature of the prepared mixed-oxide. Unlike results for the sulphated silica-zirconia samples, absorption coefficients for the pyridinium ion did not vary significantly across the series for the phosphated mixed-oxides (0.50 to 0.62 cm µmol^{-1} at 1540 cm^{-1}). Likewise, values for the Lewis bound pyridine showed a narrow range of coefficients (1.02 –1.05 cm µmol^{-1} at 1450 cm^{-1}) for the phosphated samples, these being significantly lower than the value obtained for the untreated parent mixed-oxide. All absorption coefficients determined were highly consistent irrespective of the out-gassing temperature range which was used to calculate the values. The extent of Lewis acid formation was much less sensitive to the amount of phosphate added than the amount of sulphate added. Increased Brønsted acidity was achieved by increasing the sulphate levels whereas no clear trend was observed on phosphation.

Table 4   *Infrared and Gravimetric data for pyridine adsorption on sulphated samples*

Sample	Conc[a]	Conc[b]	ε_{1540}[c]	ε_{1450}[c]	$n_{Br\o nsted}$[d]	n_{Lewis}[d]
SiZr (0)	198	0.625	0.53	1.78	0.217	0.161
	(111)	(0.352)			(0.096)	(0.116)
SiZr (0.20)	185	0.804	1.66	1.00	0.088	0.395
	(120)	(0.522)			(0.049)	(0.266)
SiZr (0.25)	174	0.820	1.18	1.11	0.121	0.374
	(115)	(0.544)			(0.069)	(0.258)
SiZr (0.30)	148	0.711	0.7	1.57	0.239	0.191
	(91)	(0.437)			(0.133)	(0.134)

[a,b] Based on mass retained after evacuation of pyridine at 423 K (and 473 K). [a]µmoles of pyridine per gram sample. [b]µmoles of pyridine per m^2 sample. [c]IR absorption coefficient (cm µmol^{-1}). [d]N^0 of acid sites per nm^2 (i.e. N^0 of pyridine molecules retained per nm^2 after evacuation at 423 K (473 K)).

Table 5　　*Infrared and Gravimetric data for pyridine adsorption on phosphated samples*

Sample	$Conc^a$	$Conc^b$	ε_{1540}^c	ε_{1450}^c	$n_{Br\o nsted}^d$	n_{Lewis}^d
	373/ 423/ 473	*373/ 423/ 473*			*373/ 423/ 473*	*373/ 423/ 473*
SiZr (0)	- / 198 / 111	- / 0.625 / 0.352	0.53	1.78	- / 0.217 / 0.096	- / 0.161 / 0.116
SiZr (0.20)	276 / 157 / -	0.89 / 0.50 / -	0.50	1.04	0.157 / 0.072 / -	0.375 / 0.232 / -
SiZr (0.25)	270 / 156 / -	0.79 / 0.46 / -	0.62	1.02	0.110 / 0.046 / -	0.37 / 0.230 / -
SiZr (0.30)	287 / 166 / -	0.97 / 0.56 / -	0.60	1.05	0.175 / 0.064 / -	0.41 / 0.27 / -

[a,b]Based on mass retained after evacuation of pyridine at 373, 423 and 473 K. [a]µmoles of pyridine per gram sample. [b]µmoles of pyridine per m^2 sample. [c]IR absorption coefficient (cm µmol⁻¹). [d]N⁰ of acid sites per nm^2 (i.e. N⁰ of pyridine molecules retained per nm^2 after evacuation at 373, 423 and 473 K)

All phosphated samples, irrespective of the preparation procedure employed, showed reduced Brønsted site densities compared with the parent, non-phosphated mixed-oxide (Table 5, column 6). In sharp contrast, Lewis site densities were always higher for the phosphate containing samples. Phosphated samples displayed a trend not too dissimilar from that obtained for sulfated samples where the gain with respect to the parent mixed oxide in Lewis sites as a function of increased phosphate loading is mirrored by a corresponding loss in Brønsted sites. This mirror effect was also apparent at the highest phosphate loading where a partial recovery of Brønsted sites was matched by a loss in Lewis sites. The consequence was that the total number of acid sites remained fairly constant across the series.

4 DISCUSSION

A non-sulphated sample which was subsequently treated with sulphuric acid solution clearly lead to textural changes in the mixed oxide, resulting in loss of pore volume and consequently, BET surface area (Table 1) for the sulphated samples. However, surface Si/Zr ratios of the sulphate treated mixed oxides were higher (3.28 and 3.48) than the sample prior to sulphation (2.63). The two apparently conflicting facts can only be reconciled if extraction of zirconia to the surface was accompanied by agglomeration, of zirconia to form small X-ray amorphous particles, i.e. acid treatment has a similar but less dramatic effect as the thermal treatment in that both induce surface segregation of zirconia to form a particulate zirconia phase. It was previously argued[5], that the generation of surface acid groups depends on the surface arrangement of the two component oxides rather than being related[17] to the degree of hetero-linkage formation throughout the structure, so ignoring for the moment the influence of retained sulphate groups on acidity,[18-23] extraction of zirconia to the surface and the formation of 3D zirconia phases from the 2D mixed surface layer of the mixed oxide should lead to significant modification to the acid sites present. Lewis acid density in non-sulphated silica-zirconia aerogels shows a monotonic increase with mol% zirconia[5] so extraction of zirconia to the surface by sulphation would be expected to enhance the number of these centers. This is confirmed in Table 4 where S-SiZr (0.2) and S-SiZr (0.25) show Lewis acid densities that are *ca.* 2.5 times greater than the parent sulphate-free mixed oxide. This increased Lewis acid density as a consequence of zirconia extraction from the bulk was obtained despite a decrease in the surface Zr/Si ratio (Table 3) of 0.38:1 to 0.3:1 resulting from sulphation. This can only be rationalized if Lewis sites are formed only in silica-free zirconia regions rather than at exposed zirconia sites in well-mixed regions of the surface. The latter are more likely to be the source of Brønsted acid sites leading to predict that loss of Brønsted acidity by phase separation in the surface layers would lead to enhanced Lewis acidity and *vice versa.*

Unlike t he c hanges i n Brønsted a nd Lewis a cid site d ensities i nduced b y changing surface composition of silica-zirconia by a change in molar ratios[5], changes in acid site densities induced here by sulphuric acid treatment may additionally result from acidity induced by retention of sulphate type species. Samples S-SiZr(0.25) a nd S-SiZr(0.3-ex) retain (Table 4) identical amounts of sulphur yet display widely different acid type and densities which might suggest that the amount of retained sulphur (sulphate) plays a minor role in acid site generation and that the nature of the treatment and the final surface composition of the component oxides was more significant. Although the measured sulfur concentration was identical for S-SiZr(0.25) and S-SiZr(0.3), the surface S/Zr ratio was lower for the former (Table 3). This results from the higher surface Zr concentration for S-SiZr(0.25) than S-SiZr(0.3). (Note direct comparison of surface ratios is possible here due to the almost identical (212 and 208 m^2g^{-1}) surface areas exhibited by these 2 samples). The greater Zr:Si ratio for S-SiZr(0.25) than S-SiZr(0.3) would be expected[5] to generate a surface with a greater Lewis acid density for the former, which is consistent with the findings (Table 4). Maintaining the density of surface sulfate, but increasing the S : Zr ratio as a result of reduced surface Zr for S-SiZr(0.3) lead to a loss in Lewis acid sites as these become covered by sulphate species, which may themselves be the source of proton acidity. The lower surface Zr levels for S-SiZr(0.3) detected by XPS (Table 3) are most likely the consequence of larger 3D amorphous clusters rather than reduced levels of extracted Zr from the bulk mixed oxide.

The phosphated samples on the other hand displayed little textural changes (Table 2) with the pore volumes and surface areas similar to those of the non-treated sample. XPS ratios (Table 3) showed that there was a decrease in the surface Si/Zr ratios compared to the non-treated mixed oxide. Within the series there was no significant trend with the degree of phosphation with all Si/Zr ratios of around 2 compared to 2.6 for the non-treated sample. This observation indicates more exposed zirconia at the surface of the phosphated samples and thus a reversal of the trend observed by sulphation. This may suggest that a more highly dispersed form of zirconia existed on the surface of the mixed oxide following p hosphation. However, i t i s a lso p ossible t hat w hereas s ulphate s pecies w ere predominantly associated with the zirconia species for sulphated samples, phosphate species m ay h ave p artially covered t he s ilica s urface. It i s c lear t hat s ignificantly m ore phosphate that sulphate was retained by the materials (Table 3 and the reduced P/Zr ratios with increasing nominal loading would suggest that 3-D phosphate species were present. Polyphosphate species were created[24] for loadings above 7.5wt% phosphate on TiO_2-SiO_2.

The lack of significant variation of Brønsted or Lewis acid sites as a function of phosphate loading would be consistent with the generation of samples where aggregation of phosphate units occurred. If, as in the case of sulphated samples, Lewis acidic sites were only generated at exposed zirconia sites, then an absence of any influence of phosphate loading would be consistent with the invariance in the Zr/Si ratios (Table 3). If the values at the maximum coverage on Lewis sites are compared (373 K outgassing for P-SiZr and 423 K for S-SZr then there is a reasonable degree of agreement for the 0.20 and 0.25 samples in both series indicating that the density of sites was similar. However, given that the outgassing temperatures required to retain these quantities of base were different, this would suggest that inductive effects by the sulphate/phosphate groups on the exposed Zr cation sites may have existed but that the former had the greater influence. The strength of the Brønsted acid sites generated by sulphation was also greater than that obtained by phosphation as shown by the outgassing temperatures chosen to obtain a specific density of retained base. The XPS ratios showed that there was much more phosphate retained at the surface compared to an analogous sulphated sample (S-SiZr (0.25) S/Zr = 0.15 *cf.* P-SiZr (0.25) P/Zr = 0.60). The large surface concentration of phosphate combined with the increase in Lewis acidity with a corresponding decrease in Brønsted acidity compared to the non-treated sample for all phosphated samples suggests that Brønsted acid sites existing on the mixed oxide may have been destroyed by the adsorption of phosphate species. As a consequence of the high amount of retained phosphate, these were poorly dispersed and may have existed on the surface as pyrophosphate type units where only the terminal units expose P-OH groups. While these terminal P-OH groups, may generate new Brønsted acid sites, such groups appear to act as much weaker acids than those obtained by sulphation.

References

1 Miller, J.B., and Ko, E.I., *Catal. Today* 1997,**35**, 269.
2 Miller, J.B., Rankin, S.E., and. Ko, E.I., *J. Catal.,*1994, **148**, 673.
3 Anderson, J.A., and Fergusson, C.A., *J. Non-Cryst. Solids,* 1999, **246**, 177.
4 Anderson, J.A., and Fergusson, C.A., *J. Mat. Sci. Letts.,* 1999, **18**, 1075.
5 Anderson, J.A., Fergusson, C.A., Rodríguez-Ramos, I. and Guerrero-Ruiz, A., *J. Catal.,*2000, **192**, 344.
6 Morterra, C., Cerrato, G., Emanuel, C., and Bolis, V., *J. Catal.,* 1993, 142, **349**.
7 Morterra, C., Cerrato, G., Pinna, F. and Signoretto, M., *J. Phys. Chem.,*1994, **98**, 12373.
8 Clearfield, A., Serrette, G.P.D., and Khazi-Syed, A.H., *Catal. Today,* 1994, **20**, 295.
9 Moreno, J.A., and Poncelet, G., *Appl. Catal.A,,* 2001, **210**, 151.
10 Hua, W., Goeppert, A., and Sommer, J., *J. Catal.,* 2001, **197**, 406.
11 Jung, S.M., Dupont, O., and Grange, P., *Appl. Catal. A.,* 2001, **208**, 393.
12 Barthos, R., Lónyi, F., Engelhardt, J., and Valyon, J., *Topics in Catal.,* 2000, **10**, 79.
13 Lopez, T., Tzompantzi, F., Navarrete, J., Gomez, R., Boldú, J.L., Muñoz, E., and Novaro, O., *J. Catal.,* 1999, **181**, 285.
14 Miller, J.B., and Ko, E.I., *The Chem. Eng. Journal,* 1996, **64** , 273.
15 Morrow, B. A., McFarlane, R.A., Lion, M., and Lavalley, J.C., *J.Catal.,* 1996, **158**, 116.
16 Yoldas, B.E., *J. Non-Cryst. Solids,* 1980, **38**, 81.
17 Miller, J.B., and Ko, E.I., *Catal. Today* 1997, **35**, 269.
18 Lónyi, F., Valyon, J., Engelhardt, J., Mizukami, F., *J. Catal,* 1996, **160**, 279.
19 Aguilar, D.H., Torres-Gonzalez, L.C., Torres-Martinez, L.M., Lopez, T. and Quintana, P., *J. Solid State Chem.,* 2000, **158**, 349.
20 Navío, J.A., Colón, G., Macías, M., Camplelo, J.M., Romero, A.A., and Marinas, J.M., *J. Molec. Catal.,* 1998, **135**, 155.
21 Morterra, C., Cerrato, G., Pinna, F., and Meligrana, G., *Topics in Catal.,*2001, **15**, 53.
22 Signoretto, M., Pinna, F., Strukul, G., Cerrato, G., Morterra, C., *Catal. Letts.,* 1996, **36**, 129.
23 Morterra, C. and Cerrato, G., *Phys. Chem., Chem., Phys.,*1999, **1**, 2825.
24 Samaantaray S.M. and Parida, K., *Appl. Catal.,* 2001, **220**, 9.

DEACTIVATION OF THE Pd-La/SPINEL CATALYST FOR THE PREPARATION OF 2,6-DIISOPROPYL ANILINE

Jiang Ruixia [1], Xie Zaiku [2], Zhang Chengfang [3] and Chen Qingling [4]

[1]Institute of Chemical Technology, East China University of Science and Technology, Shanghai 200237, P. R. China
[2]Shanghai Research Institute of Petrochemical Technology, Shanghai 201208, P. R. China.

1 INTRODUCTION

2,6-diisopropylaniline (2,6-DIPA) is an important intermediate in fine chemical industry. It is applied widely in the preparation of pesticide, medicine, plastics, dye and food additive. The method to prepare 2,6-DIPA has been developed in two different ways: alkylation of aniline and amination of 2,6-diisopropylphenol (2,6-DIPP). The selectivity of 2,6-DIPA on aniline in the alkylation was very low. The gas-phase amination method to prepare 2,6-DIPA was reported by BASF Corp, in which the catalyst Pd/spinel was used. Reaction temperature was 200°C, and the overall yield of 2,6-DIPA was more than 90% at 200°C[1,2]. Liu et al.[3] prepared Pd catalyst supported on magnesia-alumina spinel. On this catalyst, under the conditions of 200°C and LHSV=0.1h^{-1}, the conversion of 2,6-DIPP was more than 90%, and single pass selectivity of 2,6-DIPA on 2,6-DIPP was more than 70%. The overall run time of this catalyst was 300h. The gas amination is an ideal method to prepare 2,6-DIPA for its high conversion and high selectivity, but the reactive space velocity and the catalyst stability still need to be improved. Most of the published reports emphasized the preparation and evaluation of catalyst. The studies on the deactivation of catalyst are still very few.

In the present paper, the Pd-La/spinel catalyst for this reaction was developed. On this catalyst, the activity and selectivity were high, especially the stability was increased greatly, and it was up to 480 hours under the conditions of 220°C and LHSV=0.3h^{-1}. But the high initial activity decreases along with time, which impairs the catalyst practice in industrial applications. For this reason, it is important to understand the deactivation mechanism. This work concerns an investigation into the coke formation on the Pd-La/spinel catalyst in the gas-phase amination of 2,6-DIPP. Furthermore, the correlation between coke deposition and combustion was analyzed. The results provide fundamentals not only to develop new catalysts but also to optimize the reactor operation parameters.

2 RESULTS AND DISCUSSION

2.1 The catalytic performance of the Pd-La/spinel catalyst for amination

The catalytic performance of the Pd-La/spinel catalyst for gas-phase amination of 2,6-DIPP is provided in Figure 1. In the run time of 480h, the conversion of 2,6-DIPP and the selectivity of 2,6-DIPA dropped from 98.5% and 88.9% to 61.8% and 69.3% respectively. The yield of 2,6-DIPA decreased from 87.5% to 42.8%.

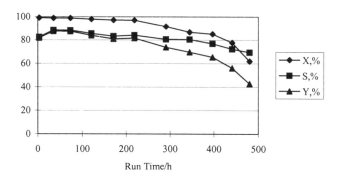

Figure 1 *The catalytic performance of the Pd-La/spinel catalyst for amination* *Reaction conditions: T=220 □, LHSV=0.3 h⁻¹, H₂=30 ml/min.*

2.2 BET studies

After reaction, the catalyst was cooled down in a hydrogen stream. The catalyst was taken out and characterized. The coke content on the deactivated Pd-La/Al$_2$O$_3$ catalyst after reaction was 28%. The color of catalyst turns from light gray to black during the reaction process. Characterization of fresh and coked catalysts shows a significant decrease in the catalyst BET specific area and pore volume values (Table 1). The decrease of BET area was up to 41.4%, and the decrease of pore volume was up to 7.1%. The decrease of BET area values measured for the coke catalyst is strictly related to the catalyst deactivation, since the deposition of organic materials prevents any further access of molecules to the active sites. According to the most probable pore diameter in Table 1, one can also conclude that coke deposits on the catalyst surface, the active sites are covered by coke, and then the catalyst deactivates.

Table 1 *Physical properties of different catalysts*

Catalyst	BET area m^2/g	Pore volume cm^3/g	Average pore diameter nm
Fresh *Pd-La/spinel* catalyst	172.4	0.3983	9.23
Coked *Pd-La/spinel* catalyst	101.1	0.3702	12.65

2.3 XRD studies

The nature of the coke was studied by X-ray diffraction. Fig. 2 shows the spectra of the fresh catalyst (a) and the coked catalyst (b). Compared to that of the fresh catalyst, there is a coke signal in the XRD pattern of the coked catalyst. Fig.3 shows the diffraction patterns of carbon extracted from the coked catalyst. The difference between the XRD patterns of the fresh and coked catalysts and the XRD patterns of the separated coke indicate that the coke probably consists of unorganized aromatic systems and of alkyl chains and alicyclic moeties joined to a polynuclear aromatic system. In conclusion, the carbonaceous matter possibly exists in a pseudo-graphitic structure.[4]

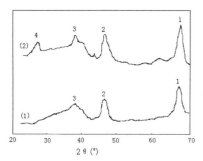 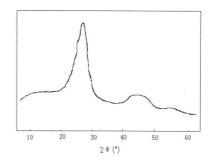

Figure 2 *X-ray diffraction patterns of different Pd-La/spinel catalysts*
(1) Fresh catalyst (2) Coked catalyst
1,2 and 3, γ-Al$_2$O$_3$ signals; 4, coke signal

Figure 3 *X-ray diffraction pattern of coke extracted from the coked catalyst*

2.4 DTA and TG studies

The DTA and TG-DTG profiles of the fresh and coked catalyst are given in Fig. 4 and 5. The DTA curve of the fresh catalyst did not have any peaks, which is in accordance with their thermogravimetric analysis. The weight loss was mainly observed below 200°C, produced by water desorption. The DTA curve of the coked catalyst showed two exothermic peaks. The former sharp peak at 242°C was due to the coke on the metal of the catalyst. Another peak appeared at 324°C was attributed to the combustion of the coke deposited on the acidic sites of the support.[5] The TG curve shows that the release materials took place at three different temperature ranges as listed in Fig. 5. The three distinct weight loss steps suggested that these deposits generated during the amination of 2,6-DIPA consisted of different types of compounds, which are dissimilarly bonded to the catalyst. The first weight loss step at 82°C was characteristic for desorption of physisorptive water; The second weight loss, 0.02745mg, at 203°C was due to the combustion of the coke on the metal; the third one, the largest weight loss, 0.08685mg, at 335°C was attributed to the burning of the coke deposited on the acidic sites of support. The above results indicate that the coke was predominantly deposited on the support, and it can be gasified below 450°C.

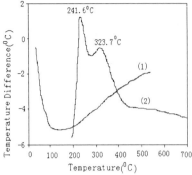

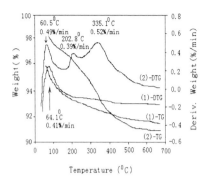

Figure 4 *DTA curves of different Pd-La*
/spinel catalysts
(1)Fresh catalyst (2) Coked catalyst

Figure 5 *DTG and TG curves of different*
Pd-La/spinel catalysts
(1) Fresh catalyst (2) Coked catalyst

2.5 FT-IR studies

In this present work, the coke was isolated by dissolving the catalyst in hydrofluoric acid so it does not modify the coke. The coke liberated from the coked catalyst shows some absorption bands in the range between 750 and 4000 cm^{-1} (Fig.6). The bands at 3415cm^{-1} and 3235cm^{-1} were assigned to the amine group. The band at 1326cm^{-1} was denoted to flexion vibration of C-N bonds in Ar-NH$_2$. These two absorption bands at 2878cm^{1} and 2965cm^{-1} were caused by the symmetric and asymmetric flexion vibration of the C-H bonds associated to the CH$_3$. The band at 1469cm^{-1} was assigned to C-H wagging vibration in CH$_3$ groups. The bands at 1636cm^{-1} and 865cm^{-1} were produced by polycyclic aromatics. The bands at 1400cm^{-1}, 1084cm^{-1}, 1124cm^{-1} and 965cm^{-1} were characteristic of the vibration absorption of C=C bonds in 1,2,3-sustituted phenyl groups. The bands at 1365 cm^{-1} was attributed to isopropyl groups. The band at 1183 cm^{-1} was produced by the vibration of C-O(H) bonds in Ar-OH. Those results confirmed the presence of aromatic and aliphatic rings, alkyl groups, hydroxyl groups and amine groups in the coke sample.

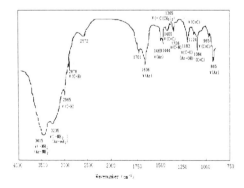

Figure 6 *FTIR spectrum of the coke liberated from the coked Pd-La/spinel catalyst*

2.6 Discussion

The catalyst used for the gas-amination has been studied on its preparation only. Most in-depth researches on the catalyst for amination focus on the synthesis of aniline from phenol and ammonia. Watanabe et al[6] study on the gas-amination of phenol on TiO_2-SiO_2 etc oxide catalyst. The reaction proceeds through the phenol adsorption on the acid sites of catalyst. Sites of moderate acid strength were most effective in the reaction. Catalysts having highly acidic sites were poisoned by phenol or aniline. Katada et al[7] proposed a reaction mechanism of the amination of phenol on the zeolite catalysts such as FAU, BEA, MFI and MOR. He noted that the rate-determining step is desorption of aniline. The stronger acid site seems to show the lower activity, because of strong adsorption of aniline. In case of MOR, it is moreover suggested that the strongly adsorbed aniline is converted into heavy molecules.

The product 2,6-DIPA is a large basic molecular material, so it adsorbs on the acid catalyst support easily but desorbs with difficulty. Furthermore, it can generate larger molecules that form carbon deposits on the catalyst through polymerization, which result in catalyst deactivation.

3   CONCLUSION

It can be concluded from the catalytic performance of Pd-La/spinel catalyst and the characterization results of fresh and coked catalysts that the catalytic performance decreased along with the amination. The coke formation is the main reason for catalyst deactivation. Both XRD and IR spectra indicate that shows that the coke contains aromatic and aliphatic rings, alkyl groups, condensed ring aromatic hydrocarbon, and amine groups.

References

1   G. Nobert, J. Peter, H. Leopold, EP 53819, 1982
2   G. Nobert, H. Leopold, F. Wolfgang, et al., DE 3425629, 1986
3   Z. L. Liu, W. W. Liao, W. N. Tan, et al., *Shiyou Huagong, Chin Petrochem Tech*, 1999, **28**, 585
4   N. S. Figoli, J. N. Belttramini, C. A. Querini, et al., *Appl. Catal.* 1986, **26**, 39
5   C. L. Li, O. Novaro, E. Munoz, et al., *Appl. Catal.* 2000, **199**, 211
6   Y. Watanabe, N. Nojiri, *Nippon Kagaku Kaishi* 1974, 540
7   N. Katada, S. Iijima, H. Igi, et al., Progress in Zeolite and Microporous Materials 1997,**105**, 1227

Mn-CONTAINING THERMOSTABLE MULTICOMPONENT OXIDE CATALYSTS OF LOW-CONCENTRATION METHANE MIXTURE OXIDATION IN AIR

N.M. Popova, K.D. Dosumov, Z.T. Zheksenbayeva, L.V. Komashko, V.P. Grigoriyeva, A.S. Sass, and R.K. Salakhova

D.V. Sokolsky Institute of Organic Catalysis and Electrochemistry, MES RK
142 Kunayev St., 480100, Almaty,
Tel.: 91 58 08, 91 64 73. Fax: 7(3272) 91 57 22

1 INTRODUCTION

The paper presents data on development of Mn oxide catalysts for selective oxidation of lean methane mixtures with air to produce CO_2 and generate heat. To obtain catalysts, new approaches to the synthesis of polyoxide materials based on Mn were adopted. Catalysts were modified by doping with La, Ce, Ba and Sr nitrates which were deposited from solutions onto the stabilized $2\%Ce/\theta\text{-}Al_2O_3$ support (of surface area 100 m^2/g; and pellet diameter 4-5 mm). By varying the components of the impregnation mixture, it was possible to optimize the chemical composition and ratio of elements in the multi-component catalysts (at Ba:Sr:La:Ce:Mn = 1:1:1:7:10 ratio). The catalyst composition conformed to the oxide stoichiometry in the perovskite structure.

2 METHODS AND RESULTS

2.1 Methods of Investigation

The activity of catalysts was determined for the oxidation of 0.5% CH_4 in air using a flow-type apparatus operating at a volumetric flow rate of $10 \cdot 10^3 h^{-1}$ in the 673-973K temperature range. The analysis of reactants and products was performed using a JIXM-72 gas chromatograph equipped with a katharometer detector. During investigation of the effect of process variables, the volumetric flow rate was varied from 10 to $20 \cdot 10^3 h^{-1}$, O_2 concentration from 2 to 20%, and CH_4 concentration from 0.2 to 4%. The catalysts were tested for thermal stability by heating them in air at 873K for 1 h (standard treatment), and then for 5 h successively at 1073, 1273, 1373, and 1473 K.

Powder x-ray diffraction (XRD), emission spectrum analysis, electron microscopy (EM) with micro-diffraction, BET, UV-visible diffuse reflectance spectroscopy, temperature-programmed desorption of O_2 (TPD), temperature-programmed reduction with

H_2 (TPR) and temperature programmed oxidation with O_2 (TPO); were used to characterise the chemical and phase composition of catalysts, their morphology, particle size, surface area, as well as the oxygen adsorption activity and reactivity.
In the following, the abbreviation REE stands for rare earth element and AEE for alkaline earth element.

2.2. Methane Oxidation over 7% MnREEAEE/Ce/θ-Al$_2$O$_3$.

The oxidation of 0.5% CH_4 in air at a volumetric flow rate of $10 \cdot 10^3$ h^{-1} over Mn catalysts begins at 773 K, reaching 80-98% conversion to CO_2 at 973 K [1]. Investigation of the effect of contact time, concentration of reagents - CH_4 (0.2-4.0 vol.%) and oxygen (2-20 vol. %.) - shows that the maximum conversion of CH_4 (90-98%) occurs over the range of 10-20% O_2, 0.5-4.0% CH_4, at 973K and a volumetric flow rate of $10 \cdot 10^3$h^{-1} [2]. Heating Mn-containing catalysts to 1373K does not negatively affect the CH_4 oxidation rate. It is only after heating at 1473K that there is a slight decrease (by 10%) in the degree of conversion, and specific rates of conversion, of CH_4 at 973 and 773K (Fig.1) [3].

These observations suggest that Mn-containing catalysts with additions of Ce, La, Ba, and Sr possess high thermal stability during methane combustion. To elucidate the reasons for this stability and to interrelate the state of elements and oxygen in the catalysts and their activity, catalysts were characterised by a variety of physical-chemical methods.

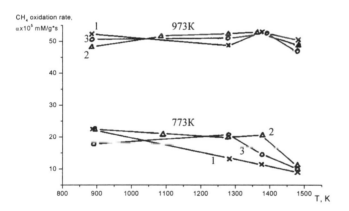

Figure 1. Influence of the temperature of pre-treatment in air on the specific oxidation rates of 0.5% CH_4 over MnREEAEE/2%Ce/θ-Al$_2$O$_3$ (1) with Pt (2) and Pd (3) addition.

2.3 Investigation of Mn Catalysts by Physical-Chemical Methods.

The application of XRD, EM, and UV- visible diffuse reflectance [4-6] methods to characterise the initial MnREEAEE/2%Ceθ-Al$_2$O$_3$ catalyst (pre-treated at 873K) demonstrates the presence of CeO_2 crystallites, X-ray amorphous Mn_2O_3 clusters and other elements (d=30-40 Å). According to XPS data, cerium is present in the sample both in the 4+ oxidation state (E_{bind}=891.1 eV) and, after heating at 1170K (20h), in the Ce^{3+} state (with a binding energy of Ce$_{3d5/2}$=887.0eV).

UV - visible spectra of the MnREEAEE/Al$_2$O$_3$ catalysts prior to reaction show a strong absorption above 400 nm, and a maximum at 340-350 nm [6]. The latter absorption band is typical for the $Mn^{3+}+O^{2-} \rightarrow Mn^{2+} +O^-$ charge transfer band in oxide clusters and indicates the formation of a dispersed Mn_2O_3 phase in the catalysts. Increasing in the catalyst pre-treatment temperature to 1170 K, decreases the absorption over 240-320 range (maximum at 320-340 nm) and reduces the absorption intensity in the range above 500 nm. This indicates that there is a decrease in dispersion and concentration of the Mn_2O_3 active phase on the surface. These results show the difficulty of manganese incorporation into Al_2O_3 in the presence of REE and AEE, unlike Mn/Al_2O_3 for which Mn^{4+} formation occurs (as evidenced by absorption band over 400-530 nm range).

From the thermal desorption of oxygen from the Mn $REEAEE/\gamma+\theta-Al_2O_3$ [7] material, it can be inferred that $\beta-Mn_2O_3$ clusters form. These clusters decompose at 973-1173K generating Mn_3O_4 and then MnO. Addition of La, Ce, Ba, and Sr elements enhance the release of O_2 from Mn_2O_3, lowering its temperature. The emergence of weakly bound oxygen (T_M=473-673 K) and excess of stoichiometry of the amount of oxygen released during Mn_2O_3 decomposition is characteristic of Mn compound catalysts, this being due to formation of new polyoxide compounds of ABO_3 type with stoichiometric oxygen.

The results obtained demonstrate that when the mixture of Mn oxides, REE, and AEE compound oxides is heated to 1170 K, ABO_3 perovskite type oxides addition to Mn_2O_3 form on the catalyst surface, due to an increase in dispersion and close interaction. The possibility of the formation of perovskites and kurnakovites on the surface at 1130 K in the system under study, was pointed out by Rode [8] and others [9]. During heating the concentration of CeO_2 crystals in Mn catalysts increases, and above 1273K the $\theta-Al_2O_3 \rightarrow \alpha-Al_2O_3$ phase transformation becomes rapidly intensified. This results in a slight decrease (from 50-60 to 40-50 m^2/g) of the overall surface area of the materials (Fig. 2).

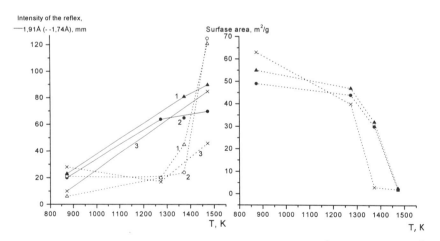

Figure 2. Relation between intensities of reflexes of CeO_2 (1.91 Å), $\alpha-Al_2O_3$ (1.74 Å) and overall surface of MnREEAEE/2%Ce/$\theta-Al_2O_3$ (1), with addition of Pt (2) and Pd (3) after heating in air successively for 5 h - CeO_2, ---αAl_2O_3, ... surface.

On pro-longed heating of Mn REEAEE/2%Ce/$\theta-Al_2O_3$ catalysts in air, starting from 1273K, in addition to the formation of fine oxide particles (30-40 Å), interaction occurs between Mn and other elements with $\theta-Al_2O_3$ (despite its protection by cerium). This gives rise to coarser dense particles with signs of faceting (70-100 Å) relating, according to their diffraction patterns, to aluminates of different composition, $LaAlO_3$, $SrLaAlO_4$, $MnAl_2O_4$, $BaSrAl_2O_7$, and hexaluminates $LaMnAl_{11}O_{18}$, $LaMnAl_{11}O_{19}$ (2,80; 2,64; 2,52 Å reflections) [4]. The amount of Mn hexaluminates formed (judging by the intensity of the 2.80 Å reflexion) as aggregates of semitransparent plate-like crystals of prismatic shape, increases on increasing the heating temperature to 1473 K and incorporation of promoters.

To confirm the formation of Mn hexaluminates in the MnREEAEE catalysts, counter synthesis of Mn hexaluminates was conducted by the method of deposition of hydroxide mixtures by ammonia both in the presence and absence of supports, with subsequent drying at 393 K for 15 h and heating at 1373 K for 2 h. The x-ray diffraction patterns of the Mn compound catalysts are shown in Fig. 3. For Ba, Sr, Mn, La, and Ce mixtures over 2%Ce/θ-Al_2O_3 the diffraction pattern of heated samples shows reflections from CeO_2, θ- and α-Al_2O_3 and at 2.80; 2.9; 1.99; and 1.49 Å corresponding, according to [10-12], to hexaluminate in addition to reflections from β-MnO_2 and α-Mn_2O_3. In the process of deposition of Ba, Mn, and La hydroxides by ammonia without a support, Mn hexaluminate (2.80; 2.52; 2.64; 3.0; 1.98; 2.45; and 2.00 Å), an amorphous phase and low intensity reflections corresponding to β-MnO_2 are evident following heating at 1373K.

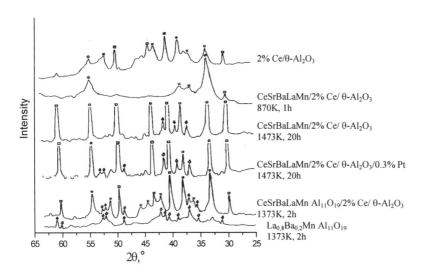

Figure 3. X-ray pictures of 7-9.5% Mn catalysts after drying, heating in air, o-CeO_2, □ – αAl_2O_3, x –θ-Al_2O_3, ◊ - hexaluminate.

The involvement of part of the surface Mn_2O_3 and other elements in the interaction with support during heating is indicated by a considerable decrease in the peak of oxygen release peak due to decomposition of Mn_2O_3 on the thermal desorption curve of samples heated at 1473 K.

The data obtained are not related to volatilization of manganese and other elements from the catalyst at high temperatures in oxygen-containing media. This is evidenced by the data on their content in initial and calcined catalysts obtained by method of emission-spectral analysis.

Figure 4a shows TPR spectra of unsupported $LaBaMnAl_{11}O_{19}$ hexaluminate, which was synthesised by the interation of metal nitrates with ammonia (T_{heat}=1373 K), and 7% MnREEAEE over 2% Ce/θ-Al_2O_3. Comparison of the spectra shows them to be identical: H_2 absorption range is 473-773 K, 773-1173 K, T^1M=673-773 K, T^2M=873 K, and T^3M=1173 K. Similar TPR spectra were obtained in work [13] for $BaMn_xAl_{12-x}O_{19}$ hexaluminate. The authors explain the obtained data by H_2 interaction with oxygen released from hexaluminate structure due to $Mn^{3+} \rightarrow Mn^{2+}$ reduction with simultaneous formation of oxygen vacancies in the catalyst.

The adsorption of oxygen on reduced hexaluminate, $LaBaMnAl_{11}O_{19}$, and 7% MnREEAEE/Ce/θ-Al_2O_3 (Fig. 4b) starts from 373 K.

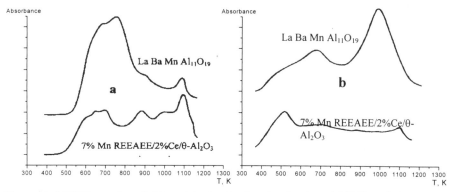

Figure 4. (a)TPR spectra of Mn catalysts after heating in air at 1373 K and (b) TPO after H_2 reduction to 1173 K.

3   CONCLUSION

Thus the processes occurring on heating MnREEAEE catalysts are beneficial : up to 1273 K, REE interact with θ-Al_2O_3 to form Ce(La)Al_2O_3 preventing the formation of $MnAlO_4$ aluminates; at 873-1173 K there occurs additional dispersion of Mn_2O_3 clusters and formation of Mn perovskites with REE and AEE. At 1373-1473 K, in addition to Mn_2O_3 clusters and perovskites, $MnLaAl_{11}O_{19}$ hexaluminates form, the percentage of which increases on heating and Pt and Pd promotion [4,5]. All three types of manganese compounds formed are active in the oxidation of CH_4 to CO_2, and this is an obvious reason of stable activity of Mn catalysts. The high activity of hexaluminates of Mn especially when they contain La, Ba, and Sr, has been pointed out in a number of papers.

According to TPD data, oxygen in Mn_2O_3, solid solutions, MnREEAEE, shows low desorption activation energy (80.2-98.2 kJ/mole) and readily interacts with reducing agents. Heating of Mn compound catalysts at 1173 K decreases the activation energy of M-O bond break processes by 10-20 kJ/mole. Under these conditions, as known for Mn perovskites, $LaMnO_3$, the concentration of oxygen vacancies rises increasing the activity in CH_4 oxidation reaction.

The results obtained point to the real possibility of practical application of Mn REEAEE/Ce/θ-Al_2O_3 catalysts for utilization of CH_4 lean mixtures in catalytic heat generators. As to its activity and thermal stability, the catalyst developed is not inferior to well-known analogs using perovskites, aluminates, and manganese hexaluminates.

The investigation was supported by the International Scientific and Technical Center (Grant K-270).

References

1 N.M.Popova, Y.S. Marchenko, Z.G. Zheksenbayeva, and K. Dosumov, Theses of International Conference "21 Century Problems of Catalysis", June 2000, Almaty, 94.
2 N.M. Popova, L.A. Sokolova, G.R. Kosmamabetova, Z.T. Zheksenbayeva, Report of International Seminar "Oxidation of Substances on Solid Catalysts", 22 May 2000, Novosibirsk, 248-253.
3 G.R. Kosmamabetova, N.M. Popova, Z.A. Marchenko, K. Dosumov, Z.T. Zheksenbayeva, Proceedings of International Symposium "Combustion and Plasmochemistry", Almaty, 2001, 217.
4 L.V. Komashko, Z.T. Zheksenbayeva, N.M. Popova, K. Dosumov, Izv. MES RK, NAS RK, Ser. Khim., 2002, **6**, 68-72.
5 V.P. Grigoriyeva, N.M. Popova, Z.T. Zheksenbayeva, A.S. Sass, R.K. Salakhova, K. Dosumov, Izv. MES RK, NAS RK, Ser. Khim., 2002, **5**, 63-69.
6 N.M. Popova, G.R. Kosmamabetova, L.A Sokolova, Izv AN RK, Ser. Khim., 2000, **6**, 23.
7 N.M. Popova, G.R. Kosmamabetova, L.A. Sokolova, Z.T. Zheksenbayeva, K.D. Dosumov, Zh.Phiz. Khim. (Rus), 2001, 75, **1**, 39-44.
8 E.Y. Rode, Manganese Oxides, Published by AS USSR, M., 1952, 397.
9 V.D. Parkhomenko, A.S. Katashinsky, G.N. Serdyuk, Ecotechnology and Resource Saving, 2000, **6**, 9-14.
10 Y. Groppi, A. Belloni, E. Fronconi, Catalysis Today, 1996, 403-407.
11 M. Machida, A. Sato, et al., Catalysis Today, 1995, **26**, 239-245.
12 M. Machida, K. Eguchi,H.J. Arai, J. Catalysis, 1990, **123**, 477-485.
13 P. Artizzu-Duart, J.M. Millet N. Guilhaume E. Garbowski and M. Primet, Catalysis Today, 2000, **59**, 163-177.

CATALYSTS BASED ON FOAM MATERIALS FOR NEUTRALIZATION OF GAS EMISSIONS

A.N. Pestryakov[1,*], V.V. Lunin[2], N.E. Bogdanchikova[3]

[1]Fritz-Haber-Institut der MPG, D-14195 Berlin, Germany
[2]Moscow State University, Chemistry Department, Moscow 119899, Russia
[3]Centro de Ciencias de la Materia Condensada, UNAM, Ensenada 22800, Mexico

1 INTRODUCTION

Foam materials or high-porosity cellular materials represent a very promising type of structural material.[1-5] An elementary cell of the foam metal has the form close to a pentadodecahedron that forms a three-dimensional network-cellular open-porosity frame with the cell packing resembling the densest packing of spheres. This structure ensures a higher degree of bonding and rigidity of the entire structure of the foam metal and anisotropy of the mechanical properties. The geometrical parameters (average diameter) of the elementary cells of the frame may be set in the range 0.5-5.0 mm. Completed foam materials have a higher fraction of the free volume, 80-98 %. It is thus possible to produce foam metals with a very low volume density, 0.1-0.3 g/cm^3. Foam metals are easily formed into the blocks of any configuration that is rather convenient for preparation of catalytic neutralizers.

The foam materials have a number of interesting, even unique properties, which may be utilized in various areas, especially in heterogeneous catalysis. However, until recently, insufficient attention has been paid to catalysts based on foam materials in comparison with conventional catalytic systems - granulated and block-honeycomb. Some previous studies showed high efficiency of the foam catalysts in a number of processes including deep oxidation of organic compounds.[6-15] Due to their particular structural properties the foam catalysts exhibited high gas permeability, mechanical strength, and good contact of the reaction flow with the catalyst surface. Any corrosion-resisting metal or alloy - Cu, Ni, Fe, Cu-Ni, Ni-Cr, steel - can be a foam support stock. Foam ceramics can also be used as support of the catalytically active phase.

The aim of the present study is the investigation of activity of the catalysts based on foam materials (foam metals and foam ceramics) in the processes of deep oxidation of volatile aromatic compounds and neutralization of gas emissions of enameled wire production.

---

*Permanent address: Tomsk Polytechnic University, Dept. of Organic Chemistry, Tomsk 634034, Russia

2 EXPERIMENTAL

The catalysts were prepared by thermal or chemical deposition of active cover based on Pt and Pd (using chloride precursors) or complex oxides of transition metals (using nitrate precursors) on the foam-metal or foam-ceramics support from aqueous solutions of the metal salts. The foam metal (Nichrome or steel) was coated with the layer of γ-Al$_2$O$_3$ intermediate support in order to increase the catalyst surface area. Content of Pt or Pd in the catalysts accounted for 1 wt.%, concentration of the transition metal oxides amounted to 10-12 wt.%.

The laboratory experiments were carried out in a flow catalytic system under the following operating conditions: temperature - 150-450°C; concentration of the oxidized substance in gas-air mixture – 0.5 vol.%; flow rate - 0.5 l/min; volume of the catalyst 10 cm^3.

The most active foam catalysts were tested in the process of neutralization of gas emissions of furnaces for wire enameling. Waste gases contained mixture of vapors of different (mainly aromatic) organic compounds in amount of 10-50 g/m^3. Temperature of the waste gases was 380-450°C. The existing catalytic neutralization system (2 % Pd on Nichrome wire) did not give the necessary degree of purification (72-80 % conversion only).

3 RESULTS AND DISCUSSION

The initial material for the foam catalyst may be any metal or alloy capable of existing, under normal conditions, in the metallic state (Cu, Ag, Ni, W, steel, Nichrome, etc.), or ceramics of various composition. Corrosion-resisting Nichrome and stainless steel are preferable materials of the foam support for deep oxidation processes. Foam ceramics is the most stable to corrosion; however, it is brittle and can be broken under hard operating conditions.

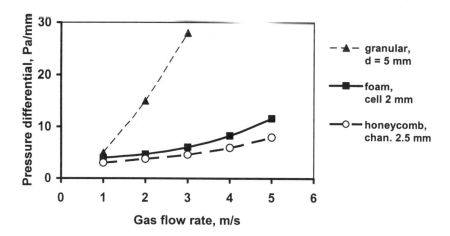

Figure 1 *Gas permeability of the catalysts*

An important parameter of the catalysts is the value of the specific surface which equals, for pure foam metals, approximately 0.01-0.1 m/g. For the majority of catalytic processes, taking place in the kinetic region, this is insufficient to produce active catalysts. Therefore, a number of methods have been developed for increasing the specific surface by direct deposition on the foam support of a catalytically active phase (up to 3-8 m^2/g),[7] or using a layer of the intermediate support γ-Al$_2$O$_3$ (up to 20-50 m^2/g).[8] Low-surface foam metals may be used in high-temperature external diffusion processes.

The high porosity of foam materials results in their satisfactory gas permeability (Fig.1). Gas-dynamic measurements show that the specific pressure losses in filtration of gases through foam catalysts are comparable with the characteristics of block-honeycomb specimens and considerably lower than the parameters of the granulated catalysts. At the same time, at almost all flow rates of the gas flow, typical of catalytic processes, movement of the gas in the foam catalysts takes place under the conditions of separation vortex bypass flow, characterized by high flow turbulence. This greatly improves the contact of the reaction gas mixture with the surface of the foam catalysts and increases its operating efficiency, whereas in block-honeycomb catalysts part of the molecules slips in the laminar mode without contact with the active surface of the catalyst. High gas permeability of the catalyst is very important for the studied process of neutralization of gas emission of the wire enameling as the furnace require minimal resistance to the gas flow.

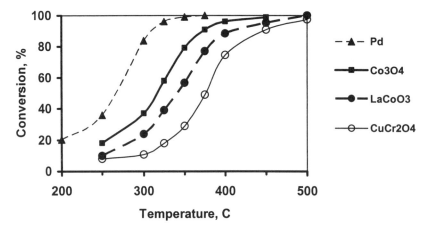

Figure 2 *Activity of the catalysts supported on foam ceramics in toluene oxidation*

Catalytic tests showed the highest activity of Co-containing samples among the oxide catalysts (Fig. 2) that is in good accordance with the literary data. Cobalt oxide, however, has low thermostability. Much higher stability, close to conventional CuCr$_2$O$_4$, was observed for the LaCoO$_3$ sample.

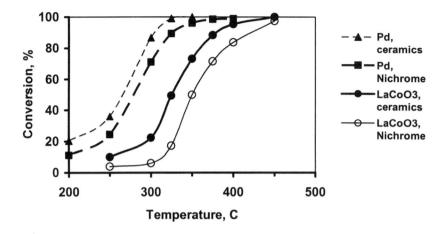

Figure 3 *Catalytic activity of the foam samples in xylene oxidation (foam ceramics contains γ-A₂O₃ layer).*

Testing the catalysts with different supports and methods of the active phase deposition revealed the better properties of the samples based on foam ceramics (Fig. 3). This is caused by higher surface area of this support. Foam ceramics has maximal amount of the intermediate layer $\gamma\text{-Al}_2O_3$ (up to 35 wt.%), whereas Nichrome contains only 15-18 wt.%. The samples without the intermediate support have much less surface area (5-8 m^2/g).

Differences in catalytic activity of the samples have no major importance in our case as at the temperatures typical for the studied real process (400-450°C) all the catalysts exhibit m aximal c onversion. T he m ain p arameter i n t his c ase i s s tability o f t he c atalyst work during the neutralization process. Aromatic compounds contained in the waste gases caused significant coke deposition on the catalyst surface. Tests of the catalysts in enameling furnaces showed that the highest coke deposition was observed for ceramic support, the least one – for Nichrome without intermediate layer (Table 1). As expected, Pt and Pd samples were more resistant to the coking as compared with oxide ones. However, all the catalysts can be easily regenerated by calcination at 500°C.

Table 1 *Period of the catalyst active life (more than 80 % conversion) in the neutralization system of enameling furnaces*

Active phase	Foam support	Period, days
Pd	Nichrome, without layer	94
Pd	Ceramics, γ-Al₂O₃ layer	86
LaCoO₃	Ceramics, γ-Al₂O₃ layer	75
CuCr₂O₄	Ceramics, γ-Al₂O₃ layer	62

Foam Nichrome seems to be a better support for Pt and Pd catalysts, whereas foam ceramics may be recommended for oxide catalysts. Foam ceramics are much cheaper and easier to prepare as compared with foam metals; however, it is brittle and can not operate under hard mechanical loads. For the studied process ceramics may be used as the operating conditions are not too hard.

4 CONCLUSIONS

1. The experiments showed high efficiency of the catalysts based on foam materials for the processes of deep oxidation of volatile aromatic compounds.
2. Foam catalysts with intermediate layer γ-Al_2O_3 exhibit the best catalytic activity in deep oxidation reactions.
3. Foam Pd/Nichrome and $LaCoO_3$/ceramics catalysts can be recommended for neutralization of gas emissions of enameled wire production.

5 REFERENCES

1. J. Banhart, *Progress in Materials Science,* 2001, **46**, 559.
2. P.S. Liu and K.M. Liang, *J. Mater. Sci.*, 2001, **36**, 5059.
3. C.S.Y. Jee, N. Ozguven, Z.X. Guo and J.R.G. Evans, *Metal. Mater. Trans. B,* 2000, **31**, 1345.
4. M. F. Ashby, A. Evans, N.A. Fleck, L.J. Gibson, J.W. Hutchinson, H.N.G. Wadley, *Metal foams: a design guide,* Butterworth-Heinemann, Oxford, UK, 2000, 251 pp.
5. A. Bhattacharya, V.V. Calmidi and R.L. Mahajan, *Int. J. Heat and Mass Transfer,* 2002, **45**, 1017
6. A.N. Pestryakov, A.A. Fyodorov and A.N. Devochkin, *J. Adv. Mater.,* 1994, **5**, 471.
7. A.N. Pestryakov, A.A. Fyodorov, V.A. Shurov, M.S. Gaisinovich and I.V. Fyodorova, *React. Kinet. Catal. Lett.,* 1994, **53**, 347.
8. A.N. Pestryakov, A.A. Fyodorov, M.S. Gaisinovich, V.P. Shurov, I.V. Fyodorova and T.A. Gubaydulina, *React. Kinet. Catal. Lett.,* 1995, **54**, 167.
9. A.N. Pestryakov, E.N. Yurchenko, A.E. Feofilov, *Catal. Today,* 1996, **29**, 67.
10. A.N. Pestryakov, V.V. Lunin, A.N. Devochkin, L.A. Petrov, N.E. Bogdanchikova, V.P. Petranovskii, *Appl. Catal. A.,* 2002, **227**, 127.
11. A.N. Pestryakov, A.N. Devochkin, *React. Kinet. Catal. Lett.,* 1994, **52**, 429.
12. A.N. Devochkin, A.N. Pestryakov, L.N. Kurina, *React. Kinet. Catal. Lett.,* 1992, **47**, 13.
13. Z.R. Ismagilov, O.Yu. Podyacheva, O.P. Solonenko, V.V. Pushkarev, V.I. Kuz'min, V.A. Ushakov and N.A. Rudina, *Catal. Today,* 1999, **51**, 411
14. O.Y. Podyacheva, A.A. Ketov, Z.R. Ismagilov, V.A. Ushakov, A. Bos, H.J. Veringa, *React. Kinet. Catal. Lett.,* 1997, **60**, 243
15. D.V. Saulin, I.S. Puzanov, A.A. Ketov, S.V. Ostrovskii, *Russ.J.Appl.Chem.,* 1998, **71**, 288

HIGHLY ACTIVE SILICA SUPPORTED PHOSPHOTUNGSTIC ACID CATALYST FOR ACYLATION REACTIONS

J.A. Gardner* G. Bond and R.W. McCabe

Centre for Materials Science, University of Central Lancashire, Preston, PR1 2HE

1 INTRODUCTION

Acylation of aromatic compounds is a widely used reaction for the production of fine chemicals.[1] The classic methodology for acylation is Friedel-Crafts, this uses stoichiometric amounts of Lewis acids, such as metal halides, but some Brönsted acids, such as polyphosphoric and sulphuric acid, can also be used as catalysts.[2] Moreover, the reaction is carried out homogeneously and the catalyst must be destroyed to obtain the final products, generating a great amount of corrosive and toxic waste.

Unfortunately, most heterogeneous catalysts investigated for this type of reaction have demonstrated only limited success, showing only high levels of activity for the acylation of activated rings with carboxylic acids.[3,4] Rare earth exchanged zeolites have been shown to have some activity for acylation of toluene with carboxylic acids.[5]

2 EXPERIMENTAL

2.1 Catalyst Characterisation

2.1.1 Silica Supported Heteropoly Acids. The catalysts were prepared using an incipient wetness method, whereby the appropriate amount of phospho-12-tungstic acid (Aldrich) was dissolved in water and slurried with sufficient silica (aerosil 380, Degussa), to give supported catalysts with loading of 20, 30 and 40% by weight. These will be referred to as HPW20, HPW30 and HPW40 respectively.

2.1.2 Cerium Exchanged Zeolite-Y. The ion-exchanged zeolites were prepared using the method described by Carvajal et al.[6] Powdered type Y molecular sieves in the sodium form as supplied by Strem Chemicals were used throughout.

Initially the sodium ion was exchanged for ammonium using the following procedure. Zeolite Y (30 g) was slurried in 3 litres of 1 M ammonium nitrate solution, at 70°C for 24 hours, with stirring. The slurry was filtered, under vacuum, and washed with deionized water (1 litre); washing was repeated five times. The zeolite in the ammonium form was subsequently exchanged with Ce^{3+}. This was achieved by stirring 10 g of NH_4Y with 1 M cerium nitrate, at 70°C for 24 hours. The ion exchanged zeolite was filtered under vacuum, washed with 1 litre of deionised water, and then dried at 120°C for 24 hours, prior to being calcined at 300°C for 24 hours in air.

2.2 Catalyst Characterisation

2.2.1 Total Surface Area Determination. The total surface area of the catalysts was determined using nitrogen adorption at 77 K, using the BET method. T he experiments were performed using a Micromeritics 2010 gas adsorption analyser.

2.2.2 Surface Acidity. The total numbers of acid sites were probed via pyridine adsorption using a purpose built vacuum microbalance. Subsequent analysis using infrared spectroscopy provided a measure of the fraction of Brönsted and Lewis sites from the absorption bands at 1545 and 1445 cm^{-1} respectively, according to the method described by Emeis et al.[7]

2.3 Catalyst Testing

Catalyst testing has been performed in a stirred batch reactor operating under reflux conditions. Catalyst evaluation was performed using 1.0 g of catalyst with 0.57 moles of aromatic hydrocarbon and 1.8 $\times 10^{-3}$ moles of carboxylic acid. Samples were taken periodically and analysed using gas liquid chromatography. The gas chromatograph (Varian CP-3380) was fitted with a flame ionisation detector. Separation of the products was achieved using a 25 m BP5 (5% phenylpolysiloxane) column. The isomer selectivity was confirmed by isolation of the product and subsequent analysis using nmr spectroscopy.

3   RESULTS

3.1 Catalyst Characterisation

The surface areas determined from nitrogen adsorption at 77 K are contained in Table 1.

Table 1   *Total surface area determined using Nitrogen adsorption at 77 K*

Catalyst	Surface Area / m^2g^{-1}
HPW20	214
HPW30	176
HPW40	132
CeY	431

The number of acid sites as measured by pyridine adsorption using a vacuum microbalance is expressed as milligrams of pyridine adsorbed per gram of catalyst. The fraction of Brönsted and Lewis sites were determined from spectroscopic measurements.

Table 2 *Surface acidity as determined by pyridine adsorption*

Catalyst	Pyridine adsorption / mg g$_{cat}^{-1}$	Brönsted sites / %	Lewis sites / %
HPW20	119	53	47
HPW30	139	58	42
HPW40	125	45	55
CeY	486	13	87

3.2 Catalyst Testing

3.2.1 Effect of Phosphotungstic Acid Loading. The effect on catalytic activity of varying the loading of phosphotungstic acid on the silica support was investigated using the acylation of toluene with decanoic acid. The reactions were performed at 110°C; the conversion as a function of time for the three catalysts is shown in Figure 1.

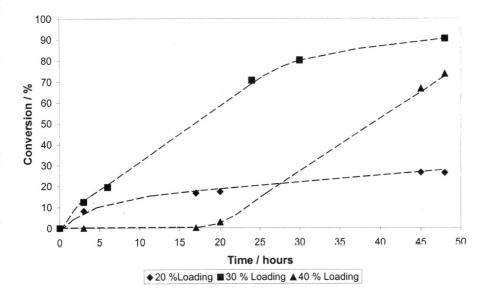

Figure 1 *Effect of phosphotungstic acid loading on catalyst activity for the acylation of toluene using decanoic acid.*

3.2.2 Nature of the Aromatic Hydrocarbon. Catalytic activity has been assessed using both anisole and toluene. Conversion, as a function of time, for the HPW30 and the CeY catalysts are shown in Figures 2 and 3.

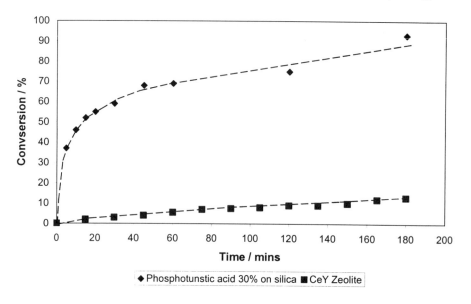

Figure 2   *Comparison of the catalytic activity for HPW30 and CeY catalyst for the acylation of anisole with decanoic acid.*

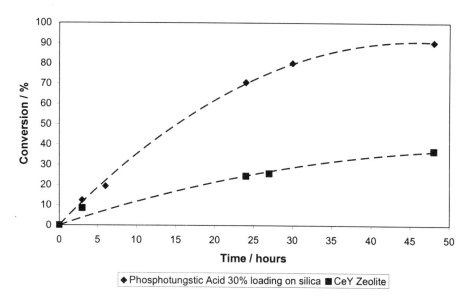

Figure 3   *Comparison of the catalytic activity for HPW30 and CeY catalyst for the acylation of toluene with decanoic acid.*

3.2.3 Effect of Chain Length of the Acylating Agent on Catalyst Activity. The effect of varying the chain length of the acylating agent was studied using a range of carboxylic acids, ranging from propanoic to hexadecanoic acid. The percentage conversion and selectivity to the *para* isomer are contained in Table 3.

Table 3 *Effect of varying the chain length of the acylating agent on catalytic activity.*

Chain length of	CeY		HPW30	
Acylating agent	Conversion / %	Selectivity / %	Conversion / %	Selectivity / %
3	1.5	97	98	80
6	4	98	87	88
8	16	94	87	87
10	24	94	90	79
12	19	86	45	78
16	0.4	55	45	82

4   DISCUSSION

Supported heteropoly acids have been shown to be active catalysts for the acylation of aromatic compounds using organic acids as the acylating agents. Their activity being far superior to Ce and La exchanged zeolites, that have previously been reported to be active for these reactions.[5]

In order to obtain maximum catalytic activity there is an optimum loading of heteropoly acid on the surface of the silica support. This is likely to result from excess heteropoly acid restricting access of the reactants acid sites within the pores of the catalyst. This is borne out by the reduced surface areas observed as the heteropoly acid loading is increased. The behaviour of the HPW40 catalyst as a function of time is quite remarkable. Initially, the catalyst exhibits virtually no activity however, after a period of time the catalyst starts to exhibit activity almost identical to that of the 30% loaded material (Figure 1). This is likely to be the result of excess surface heteroploy acid, which is weakly bound to the catalyst surface, being leached into solution. In the optimised catalyst leaching, and hence deactivation of the catalyst, does not appear to be problematic. This has been proved by recycling and reusing the catalyst. After recycling five times the catalysts exhibited no significant decrease in activity.

Gauthier et al.[8] infer that it is the Brönsted sites rather than Lewis which are responsible for catalysing acylation reactions involving carboxylic acids. This study is in agreement with their findings. While the CeY zeolite adsorbs considerably more pyridine, hence has more acid sites per gram of catalyst, the infrared study indicates that only 13% of these are Brönsted sites. This gives the HPW20 and the CeY catalyst very similar numbers of Brönsted sites per gram of catalyst. These catalysts exhibit remarkably similar activities for the acylation of toluene with decanoic acid giving support to the theory that it is the Brönsted sites that are responsible for imparting the catalytic activity. The HPW40 catalyst does not follow this trend; however, this is most probably due to there being an increase in accessible Brönsted sites due to the leaching of excess heteropoly acid as described previously.

The ion exchanged zeolite catalyst exhibits a maximum in activity for the acylation of toluene with decanoic acid. This trend has been previously observed by Chiche et al.[9] when using this type of catalyst. The HPW30 catalyst appears to be much more active for reactions involving shorter chain acids however, there is a significant reduction in rate when the chain length of the acylating agent exceeds ten carbon atoms. The selectivity, as expected, predominately favours the formation of the *para*-isomer. The zeolite being slightly more selective, due to the steric constraints imposed by the cage structure of the zeolite.

5   CONCLUSION

The supported heteropoly acid catalysts offer a significant improvement in terms of catalytic activity compared to ion exchanged zeolites. However, there is the need for further improvements before they could be considered commercially viable for acylation of non activated aromatic rings.

References

1   G.A. Olah, *Friedel-Crafts and Related Reactions,* Vols. 1-4, Wiley, New York, 1963-1964.
2   M. Spagnol, L. Gilbert, D. Alby, in: J.R. Desmurs, S. Rattoy (Eds.), *The Roots of Organic Development*, Elsevier, Amsterdam, 1996, p. 29.
3   A. Corma, M.J. Climent, H. Garcia, J. Primo, *Appl. Catal.*, 1989, **49,** 109.
4   H. van Bekkum, A.J. Hoefnagel, M.A. van Koten, E.A. Gunnewegh, A.H.G. Vogt, H.W. Kouwenhoven, *Stud. Surf. Sci. Catal,* 1994, **83,** 374.
5   B. Chiche, A. Finiels, C. Gauthier, P. Geneste, J. Graille, D. Pioch, *J. Org. Chem.,* 1986, **51,** 2128.
6   R. Carvajal, P.J Chu, J.H Lunsford, *J. Catal,.* 1990, **125,** 123-131.
7   C.A. Emeis, *J. Catal.,* 1993, **141,** 347-354.
8   B. Chiche, A. Finiels,C. Gauthier, P. Geneste, *Appl. Catal.* 1987, **30,** 365-369.
9   B. Chiche, A. Finiels, C. Gauthier, P. Geneste, J. Graille, D. Pioch, *J. Org. Chem.* 1986, *51,* 2128-2130.

THE EFFECT OF PREPARATION VARIABLES ON Pt AND Rh/Ce$_x$Zr$_{1-x}$O$_2$ WATER GAS SHIFT CATALYSTS

J.P. Breen, R. Burch and D. Tibiletti

School of Chemistry, Queen's University Belfast, David Keir Building, Stranmillis Road, Belfast BT9 5AG, N. Ireland

1 INTRODUCTION

The Water Gas Shift (WGS) reaction has been widely used in the chemical and petrochemical industry for many years. Typical commercial catalysts for this reaction have contained Fe, Cu and Zn. Until recently WGS was regarded as a mature technology, however, the current interest in PEM fuel cell systems for mobile applications has resulted in renewed interest in the WGS reaction as a means of decreasing levels of CO (a fuel cell poison) from hydrogen rich streams. The traditional Cu/Zn catalysts are not suitable for these new mobile applications. As a result, research has focused on new catalyst formulations which will give the required activity at high space velocities, are stable over time, are not pyrophoric and do not require a reductive pre-treatment. Pt loaded CeO$_2$ catalysts are one such group of catalysts that are close to meeting these requirements and have been the subject of recent research. The advantages of CeO$_2$ as a catalytic support are its ability to: stabilise well-dispersed noble metals on its surface [1]; promote the water gas shift and hydrocarbon reforming reactions [2-5] and store oxygen [4, 6]. Ceria's ability to store oxygen can be further improved by the addition of zirconia [6, 7]. Finally, Ce/Zr mixed oxides give improved resistance to sintering when submitted to high temperatures under oxidising or reducing environments. The work presented here focuses on these Ce$_x$Zr$_{1-x}$O$_2$ materials with respect to their properties for the WGS reaction.

2 EXPERIMENTAL

Various catalysts have been prepared, characterized and tested for their activities for the WGS reaction. Ce$_x$Zr$_{1-x}$O$_2$ supports were prepared by a citric acid method (CA) and a co-precipitation method (CP) with varying Ce/Zr ratios to investigate the effect of the preparation method on the properties of the support. The details of the preparation methods are given as follows:

Citric Acid Method (CA): Ce and Zr nitrate (Ce(NO$_3$)$_3$•6H$_2$O and ZrO(NO$_3$)$_2$•6H$_2$O) precursors were dissolved in methanol (ratio of methanol/nitrate solution = 2:1 (v/v)) and excess citric acid was added (molar ratio of NO$_3^-$/acid = 6:1.15). The resulting solution was stirred at room temperature for at least 12 hours. The solvent and the reaction products were eliminated using a rotary evaporator at room temperature. The most important step in the preparation of the amorphous precursor is the rapid dehydration of the solution at low temperature [8]. Dehydration was completed in an oven at 110°C overnight and finally the support precursor was calcined at 500°C for 5 hours to obtain a yellow powder. This temperature is sufficient for the pyrolysis of the precursor.

Co-precipitation Method (CP): the cerium and zirconium nitrate precursors, $Ce(NO_3).6H_2O$ and $ZrO(NO_3)_2.6H_2O$ with the desired Ce/Zr ratio were dissolved in water (ratio of water/nitrate solution = 10:1 (v/v)). A solution of NH_3 (14.7 M) was prepared and added under vigorous stirring into the $Ce(NO_3)_3$ and $ZrO(NO_3)_2$ solution until a precipitate was formed [3]. The solution was left stirring for 20 hours, filtered and then redispersed in a 0.25 M solution of NH_3. The resulting precipitate was then filtered, dried in the oven at 60°C for 1 hour, then at 110°C for 20 hours and finally calcined at 500°C for 5 hours [4].

The supports were loaded with Pt and Rh by an impregnation method using different precursors (H_2PtCl_6, $RhCl_3$ and Pt-DNDA).

X-Ray powder diffraction experiments were carried out using a Siemens D5000 X-Ray Diffractometer equipped with a CuK_α X-Ray source. The interplanar distances (d) and the particle sizes (l) were calculated by the Bragg and the Debye-Scherrer equations, respectively. Activity tests for the WGSR were carried out in a tubular quartz reactor loaded with 100 mg of catalyst. The feed gases (5% CO, 15% H_2O, He to balance, total flow rate 100 cm^3 min^{-1}) were introduced to the catalyst bed at a space velocity of 60,000 h^{-1} at several temperatures between 200 and 500°C.

3   RESULTS AND DISCUSSION

3.1   Catalyst Characterisation

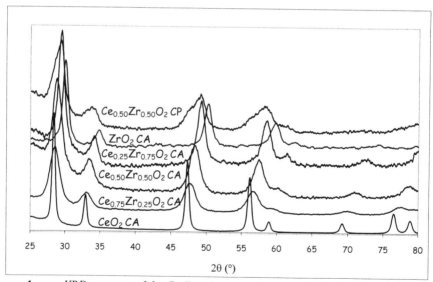

Figure 1   *XRD patterns of the $Ce_xZr_{1-x}O_2$ supports prepared by the CA and CP methods.*

Fig. 1 shows the XRD patterns for the catalyst supports prepared by the citric acid (CA) co-precipitation (CP) method. Table 1 reports the diffraction angle, θ, of the most intense peak, the particle sizes, the interplanar distances and the reference value (PCPDFWIN software) of the interplanar distance for each support.

Table 1   *XRD parameters for the support prepared by the CA and CP method*

Support	$2\theta_1$ (deg)	1 (Å)	d (Å)	d_{ref} (Å)
CeO_2 CA	28.36	158	3.1443	3.1234
$Ce_{0.75}Zr_{0.25}O_2$ CA	28.56	50.97	3.1277	3.0900
$Ce_{0.50}Zr_{0.50}O_2$ CA	28.82	63.84	3.0951	3.0800
$Ce_{0.25}Zr_{0.75}O_2$ CA	29.42	91.26	3.0334	-
ZrO_2 CA	29.95	107.36	2.9809	2.9950

The diffraction angle shifts to a lower value with increasing cerium content and the interplanar distance rises because Ce has a bigger atomic radius than Zr. The interplanar distances are close to the reference values specified in the literature, indicating that the CA method can be used to give well-mixed $Ce_xZr_{1-x}O_2$ materials with well-defined crystal structures. The $Ce_{0.50}Zr_{0.50}O_2$ support made by the co-precipitation method contains some free CeO_2 in the lattice, evident from the asymmetric peaks in the XRD pattern.

3.2   Reaction Kinetics

The effect of the support on the WGS reaction was probed by testing the activity of a series of catalysts containing Pt supported on either Al_2O_3 (CK300), SiO_2, CeO_2 or ZrO_2. In this case, the Pt was impregnated onto the support using chloroplatinic acid. The activities of these catalysts were compared to those of two commercial Fe/Cr high temperature shift catalysts. Fig. 2 shows the average CO conversions versus temperature for these experiments. The 2% Pt/ZrO_2 catalyst was the most active of the series (equilibrium CO conversion was achieved at 375 °C), the order of reactivity at 350 °C was 2% Pt/ZrO_2 > 2% Pt/CeO_2 > 2% Pt Al_2O_3 > 2% Pt/SiO_2 $\not\subset$ HT Com. Cat. 1 $\not\subset$ HT Com. Cat. 2. These results confirm that zirconia and ceria are two of the most active supports for the WGS reaction.

The choice of precursor salt is important and can have an effect on the characteristics of the final catalyst. Fig. 3 and 4 compare the activities of 2% Pt/CeO_2 and 2% Pt/ZrO_2 catalysts prepared using chloroplatinic acid or Pt-DNDA as precursors. The figures also show the activities of 2% Rh/CeO_2 and 2% Rh/ZrO_2 catalysts, the Rh being impregnated onto the supports using $RhCl_3$ as a precursor. The results clearly show that for the Pt catalysts the choice of Pt precursor is very important. The DNDA precursor produces a much more active catalyst than the chloroplatinic acid for the ceria supported Pt. The reverse is true for the Pt/ZrO_2 materials: in this case the chloride precursor produces a more active catalyst. The most active catalyst is the Pt/CeO_2 material prepared using Pt-DNDA as a precursor. Rh is a much less active metal than Pt regardless of the support used.

Given that CeO_2 and ZrO_2 were the most active of the single oxide supports for the WGS reaction, it was of interest to determine how mixed Ce/Zr oxides performed. Fig. 5 shows the average CO conversions versus temperature for a series of 2% Pt/$Ce_xZr_{1-x}O_2$ catalysts: in this case the Pt was impregnated onto the support using Pt-DNDA. The Pt-DNDA/$Ce_{0.50}Zr_{0.50}O_2$ catalyst was the most active of the series and it reached equilibrium CO conversion at 300 °C. The order of activity at 250 °C was Pt/$Ce_{0.50}Zr_{0.50}O_2$ >

$Pt/Ce_{0.75}Zr_{0.25}O_2 \not\subset Pt/Ce_{0.25}Zr_{0.75}O_2 > Pt/CeO_2 > Pt/ZrO_2$. Thus, for the Pt-DNDA, the solid solution supported catalysts were more active for the WGS reaction than those supported on the pure Ce and Zr oxides.

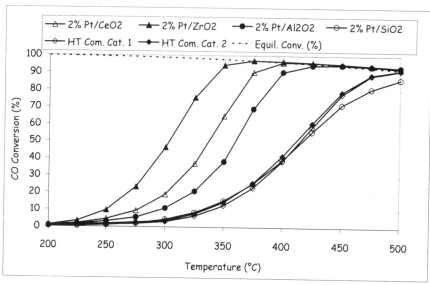

Figure 2 *Activity test for the water gas shift reaction on various supported catalysts: comparison of the average CO conversion. (CeO$_2$ and ZrO$_2$ were prepared using the CA method and Pt impregnated using a chloroplatinic acid precursor)*

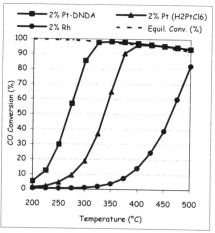

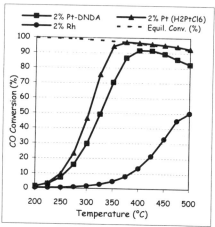

Figure 3 *The effect of precursor salt and metal on the activity of CeO$_2$ based catalysts for the WGS reaction. (CeO$_2$ was prepared using the CA method).*

Figure 4 *The effect of precursor salt and metal on the activity of ZrO$_2$ based catalysts for the WGS reaction.(ZrO$_2$ was prepared using the CA method).*

Rh/Ce$_x$Zr$_{1-x}$O$_2$ catalysts were also prepared and tested for their activities for the WGS reaction (results not shown here). These tests showed that Rh was considerably less active for the reaction than Pt. Thus for a given Ce$_x$Zr$_{1-x}$O$_2$ support the order of reaction for the metal salt used in the preparation of the catalyst was Pt-DNDA > H$_2$PtCl$_6$ > RhCl$_3$.

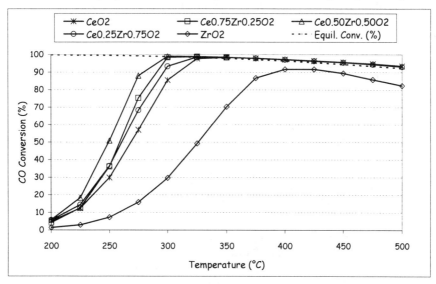

Figure 5   *Effect of Ce/Zr ratio on the activity of Pt-DNDA/Ce$_x$Zr$_{1-x}$O$_2$ catalysts for the WGS reaction. (Ce$_x$Zr$_{1-x}$O$_2$ was prepared using the CA method)*

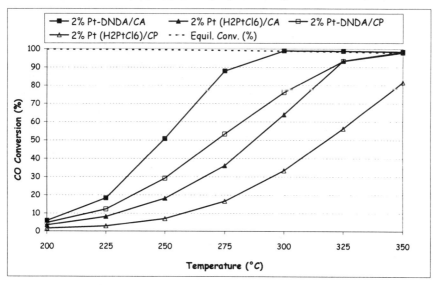

Figure 6   *Effect of support preparation method and Pt precursor on the activity of 2% Pt/Ce$_{0.5}$Zr$_{0.5}$O$_2$ catalysts for the WGS reaction.*

The $Ce_xZr_{1-x}O_2$ supports were prepared using two methods; the citric acid and co-precipitation methods. The effect of the support preparation method and metal precursor on the activity of a $Pt/Ce_xZr_{1-x}O_2$ catalyst was investigated by preparing four different 2% $Pt/Ce_{0.5}Zr_{0.5}O_2$ catalysts using either the CA or CP methods of preparing the support and loading the Pt using either chloroplatinic acid or Pt-DNDA. These catalysts were then tested for their activities, the results of which are shown in Fig. 6. It is evident, that for the mixed oxide, the CA method of support preparation produces considerably more active catalysts than the CP method. In addition, impregnation using Pt-DNDA gives more active catalysts than those prepared using chloroplatinic acid regardless of the method of preparation of the support.

2 CONCLUSIONS

XRD analysis confirmed that a series of well-defined $Ce_xZr_{1-x}O_2$ materials could be prepared using the citric acid (CA) method of preparation. No evidence was found for the presence of single CeO_2 or ZrO_2 phases in these materials. This was in contrast to the $Ce_{0.5}Zr_{0.5}O_2$ material prepared using the co-precipitation (CP) method, which contained some 'free' CeO_2 in the mixture.

A series of Pt catalysts loaded (using a chloroplatinic precursor) on different single oxide supports were prepared and their activities compared to those of two commercial high temperature WGS catalysts (HT Com. Cat.). Their activities at 350 °C were in the order 2% Pt/ZrO_2 > 2% Pt/CeO_2 > 2% Pt Al_2O_3 > 2% Pt/SiO_2 $\not\subset$ HT Com. Cat. 1 $\not\subset$ HT Com. Cat. 2. However, it was found that the order of activity between Pt/ZrO_2 and Pt/CeO_2 could be reversed by using Pt-DNDA instead of chloroplatinic acid as a precursor. Rh supported on CeO_2 and ZrO_2 catalysts were compared to the Pt catalysts and were found to be less active than the Pt catalysts.

A series of $Ce_xZr_{1-x}O_2$ supported catalysts were prepared using either a citric acid (CA) or a co-precipitation method (CP) to prepare the support. Pt was impregnated onto the support using either H_2PtCl_6 or Pt-DNDA as precursor solutions. The results of the catalytic testing showed that the mixed $Ce_xZr_{1-x}O_2$ oxides were more active than any of the single oxide supports. The CA method of preparation of $Ce_xZr_{1-x}O_2$ produced more active catalysts than the CP method and the Pt-DNDA precursor produced more active catalysts than chloroplatinic acid. The increased activity of the catalysts prepared using the CA method may be related to the fact that the support is a single phase $Ce_{0.5}Zr_{0.5}O_2$ oxide, whereas the CP support also contains a CeO_2 phase, which is less active for the reaction than the mixed oxide.

References

[1] J.R. Gonzáles-Velasco, Appl. Catal. B: Environmental 3, 1994, 191
[2] B.I. Whittington, Catal. Today 26, 1995, 41
[3] T. Shido, J. Catal. 141, 1993, 71
[4] R.J. Gorte, Appl. Catal. B: Environmental 15, 1998, 107
[5] J.R. Gonzáles-Velasco, Appl. Catal. B: Environmental 12, 1997, 61
[6] P. Farnasiero, J. Catal. 164, 1996, 173
[7] A. Trovarelli, Catal. Rev. Sci. Eng. 38, 1996, 439
[8] A. Yee et al. Catal. Today 63, 2000, 327-335
[9] A. I. Kozlov, Journal of Catalysis 209, 417-426, 2002

INVESTIGATION OF THE ACID-BASE PROPERTIES OF AN MCM-SUPPORTED RUTHENIUM OXIDE CATALYST BY INVERSE GAS CHROMATOGRAPHY AND DYNAMIC GRAVIMETRIC VAPOUR SORPTION

Frank Thielmann, Majid Naderi, Dan Burnett, Helen Jervis

Surface Measurement Systems Ltd., 3 Warple Mews, Warple Way, London W3 ORF, United Kingdom

1 INTRODUCTION

Thermodynamic parameters of catalysts are highly relevant for practical applications. In particular, surface energies and acid-base properties of catalysts are of high interest since they reflect properties of active sites involved in the catalytic process and the initial adsorption step.

Dynamic vapour phase techniques are interesting tools for the determination of these properties. When compared to standard wettability experiments, they provide two main benefits. They can easily and reproducibly be applied to powders and a wide variety of probe molecules can be selected. In the current study dynamic gravimetric vapour sorption (DVS) and inverse gas chromatography (IGC) have been used to characterise the energetic and acid-base properties of a calcined ruthenium oxide / MCM41 catalyst as well as the corresponding MCM41 support.

2 EXPERIMENTS AND METHODS

2.1 Methods

Dynamic gravimetric vapour sorption involves the use of an ultra-sensitive microbalance. The solid sample is placed in the sample pan at the end of the hang-down wire while the probe molecule is vaporised in an inert carrier gas stream. Adsorption of vapour is measured as a gain of weight, desorption as a loss of weight. From these changes of weight at different vapour concentrations sorption isotherms can be determined.

In order to calculate energetic properties it is possible to combine Young's equation with Fowkes theory[1]:

$$W_A = 2\gamma_L + \pi_e = 2\, (\gamma_L^d \cdot \gamma_S^d)^{1/2} \tag{1}$$

In this equation γ_L and γ_S are the surface tensions of the liquid and the solid, respectively and W_A the work of adhesion.

The spreading pressure π_e can be calculated from the total amount adsorbed (Θ) as a function of partial pressure (p) and the specific surface area (σ), as shown in equation (2).

$$\pi_e = \frac{RT}{\sigma} \int \Theta d \ln p \tag{2}$$

To solve this equation the adsorption isotherm needs to be measured over a wide range of partial pressures and integrated. As a result the work of adhesion W_A, the free energy ΔG and the dispersive and polar surface energy, γ_S^d and γ_S^p can be obtained.

Inverse gas chromatography involves the sorption of a known probe molecule (adsorbate, vapour) and an unknown adsorbent stationary phase (solid sample). IGC may be experimentally configured for finite or infinite dilution concentrations of the adsorbate. The latter method is excellent for the determination of thermodynamic properties such as surface energies and Lewis acid-base parameters. Measurements in this range are extremely sensitive due to the low concentration regime where the highest energy sites of the surface interact with the probe molecules.

For an IGC experiment, different probe molecules are injected and retention times can be determined. The corresponding net retention volumes V_N are computed using equation 3.

$$V_N = j/m \cdot F \cdot (t_R - t_0) \cdot \frac{T}{273.15} \tag{3},$$

where T is the column temperature, m the sample mass, F is the exit flow rate at 1 atm and 273.15K, t_R is the retention time for the adsorbing probe and t_0 is the mobile phase hold-up time (dead time). "j" is the James-Martin correction, which corrects the retention time for the pressure drop in the column bed.

The surface energy can be obtained from the equation of Schultz et al[2], who derived it from Fowkes equation.

$$RT \ln V_R^0 = 2N_A \left(\gamma_S^D \right)^{1/2} a \left(\gamma_L^D \right)^{1/2} + const. \tag{4}.$$

If $RT\ln V$ is plotted versus $a(\gamma_L^D)^{1/2}$ for a series of alkanes a straight line results and the dispersive contribution of the surface energy can be calculated from the slope. If polar probe molecules are injected, specific interactions can be determined. In the above-mentioned plot, points representing a polar probe are located above the straight line. The distance is equal to the specific component of the free energy ΔG_{SP} (equation 5).

$$\Delta G_{SP} = RT \ln V_N - RT \ln V_N^{ref} \tag{5}$$

To calculate acid-base numbers the Gutman or van Oss concept can be applied[3,4].

The Gutman approach relates the enthalpy to the acid-base numbers according to equation (6).

$$\Delta H = K_{a*} DN + K_{b*} AN^*$$ (6)

The enthalpy term in this equation can be replaced by the free energy if the entropy term in the Gibbs-Helmholtz equation is assumed to be negligible. In the experience of the authors this is generally a reasonable assumption. If the acceptor and donor number for the probe molecules are known then K_a and K_b, the acid and base number of the surface can be computed.

In the theory of van Oss a similar expression (equation (7)) can be used and electron acceptor and donor (γ_S^+ and γ_S^-) values for the surface are obtained, if the electron acceptor and donor parameters (γ_L^+ and γ_L^-) of the probe molecule are known.

$$\Delta G = N_{A*} a *2 *((\gamma_L^+ * \gamma_S^-)^{1/2} + (\gamma_L^- * \gamma_S^+)^{1/2})$$ (7)

2.2 Experimental

For DVS experiments, the sample was placed into a glass pan and the vapour concentration was stepped up by 2.5% increments until 40%, 5% until 70% and 10% until 90%. Hexane and Dichloromethane have been used as probe molecules. The measurements were carried out at 25 C and two adsorption and desorption cycles have been recorded to check for hysteresis and reproducibility of data. Nitrogen was used as a carrier gas. Samples were equilibrated in-situ for 30 min in a pure gas stream prior to the experiment.

For IGC experiments, samples have been packed into silanised glass columns (400 mm length and 3 mm ID). Pulse experiments have been carried out at 100 C with a wide range of probe molecules (Hexane, Heptane, Octane, Nonane, Dichloromethane, Ethyl acetate, Ethanol, THF, Acetone and Acetonitrile). The injection concentration was 3% and helium was used as the carrier gas at a flow rate of 10 ml/min. Prior to the experiment the samples were preconditioned at 350 C in a pure helium stream for 2 hours.

3 RESULTS AND DISCUSSION

3.1 DVS Results

Figure 1 shows two adsorption / desorption cycles for hexane on MCM41 and the catalyst. All isotherms are highly reproducible and show no sign of irreversible adsorption. The catalyst has a significantly reduced uptake compared to the pure support material. This indicates that the active compound is blocking some of the surface centres which were involved in the interaction with hexane and that the surface of the metal oxide has a much lower uptake.

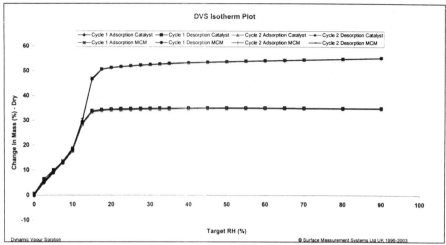

Figure 1   *Two complete Hexane adsorption/desorption cycles for MCM41 and the catalyst, measured by DVS at 25 C.*

BET surface areas have been calculated form these measurements. The MCM41 yielded to 1250.3 m^2g^{-1} while the catalyst gave an area of 792.0 m^2g^{-1}. This is in good agreement with nitrogen measurements (1301.5 m^2g^{-1} and 823.8 m^2g^{-1}) given the uncertainty introduced by the difference in cross sectional area between nitrogen (assumed to be 1.62 nm) and Hexane (assumed to be 5.15 nm).

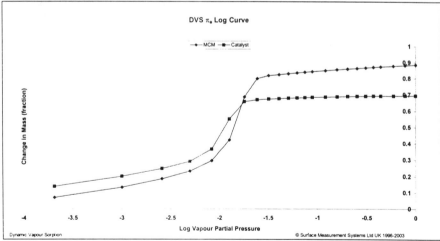

Figure 2   DVS spreading pressure curves for the 2 samples (obtained from the adsorption branch of the first cycle of a Dichloromethane experiment).

Figure 2 shows the DVS spreading pressure curves for both samples (obtained with Dichloromethane).

After the determination of the dispersive work of adhesion from the Hexane experiment the total work of adhesion can be determined for Dichloromethane. This probe molecule represents a monopolar acid according to the van Oss concept and can be therefore used to determine the base number (γ_S^-) of the surface. For the corresponding determination of the acid number the measurement with a monopolar basic probe molecule is required. Typical candidates for this would be either Toluene or Ethyl acetate. These measurements haven't been completed yet. For this reason the γ_S^+-values in the result table are missing.

The results for the DVS measurements are summarised in Table 1.

Table 1   *Thermodynamic parameters obtained from a DVS experiment.*

Sample	W_A^d (mJ/m^2)	γ_S^d (mJ/m^2)	γ_S^- (mJ/m^2)	γ_S^+ (mJ/m^2)
MCM 41	58.54	47.89	79.68	-
Catalyst	69.30	67.11	119.14	-

It can clearly be seen that the catalyst shows higher values for all determined thermodynamic parameters compared to the MCM support. From the base numbers it can be concluded that the surface of the MCM41 seems to be less basic (or more acidic) than the catalyst surface. The determination of the acidic contribution should give a more detailed picture, allowing an evaluation of the relative acidity / basisity of the materials.

3.2 IGC Results

Experiments were again fully reproducible suggesting reversible adsorption / desorption processes for all probe molecules.

Dispersive surface energies and specific free energies are shown in Table 2.

Table 2   *Thermodynamic parameters obtained from a DVS experiment.*

Sample	γ_S^d (mJ/m^2)	γ_S^- (mJ/m^2)	γ_S^+ (mJ/m^2)	Ka	Kb
MCM 41	41.35	64.25	101.8	0.22	0.08
Catalyst	95.63	140.79	155.67	0.27	0.12

Due to the high number of tested probe molecule it was possible to apply both, the van Oss concept using monopolar probe molecules and the Gutman approach using bipolar probes. γ_S^+ and γ_S^- numbers as well as K_a and K_b are summarised in Table 3 together with the dispersive surface energies.

The Gutman acid-base plot is shown in Figure 3.

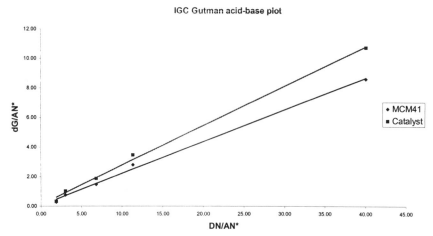

IGC Gutman acid-base plot

Figure 3   *Gutman acid-base plot (from an IGC experiment with Ethyl acetate, Acetone, Acetonitrile, THF and Ethanol)*

The dispersive surface energies a nd the specific free e nergies as well a s the acid-base numbers are significantly higher for the catalyst in comparison to the pure support.
Both the van Oss and the Gutman numbers indicate that the MCM surface is predominantly acidic. This was anticipated by other research groups based on the observation in hydrogenation experiments[5]. When the catalyst is considered the ratio of the acid to the base numbers becomes smaller suggesting a relative increase in basisty, although the catalyst appears to be predominately acidic.
  If the results are compared to those obtained from DVS experiments, it becomes obvious that the relative trends for the dispersive surface energy as well as the γ_S^- numbers are similar.
  For the MCM41 there is a remarkable agreement between the results from DVS and IGC given the different theoretical and experimental approach. For the catalyst however, absolute numbers appear to be significantly higher in the IGC results. One explanation would be the difference in the pre-treatment and measurement temperature. IGC pre-conditioning and experiments take place at 350 C and 100 C respectively while both procedures are carried out at ambient temperature in the case of the DVS. For this reason some of the more active sites might be still blocked by pre-adsorbed water in the DVS experiments. Another reason for this effect is the enhanced sensitivity of the IGC experiment under the chosen infinite dilution conditions, where only the highest energy sites are involved in the interaction while in the DVS experiment all energy sites are considered, so that values represent an average of all energy sites.

4 CONCLUSIONS

A Ruthenium oxide / MCM41 catalyst as well as the pure MCM41 support were studied by inverse gas chromatography and dynamic gravimetric vapour sorption. The obtained energy parameters were all higher for the catalyst, despite the vapour uptake and the surface area being higher for the pure MCM41. Both techniques suggest a more acidic nature of the two materials studied. The relative basisity of the catalyst seemed to be higher.

Acknowledgements

The authors would like to thank Prof. Duncan Bruce, Department of Inorganic Chemistry, Exeter University, UK for supplying the materials used in this study.

References

1   K. Mital, Contact Angle, *Wettability and Adhesion*, VSP, Utrecht 1993
2   P. Mukhopadhyay and H. Schreiber, *Colloid and Surfaces A* 1995, **100**, 47
3   V. Gutman, V., *Coordination Chemistry Reviews* 1967, **2**, 239
4   C. van Oss, R. Good and M. Chaudhury, *Langmuir* 1988, **4**, 884
5   Y. Chen and C. Li, *J. Chin. Inst. Ch*

DEVELOPMENT OF NOVEL SUPPORTED Mo₂C CATALYSTS: CARBURIZATION KINETICS AND OPTIMAL CONDITIONS

T.H. Nguyen[1], Y.J. Lee [1] E.M.T. Yue[1], A. Khodakov [2], M.P. Brungs[1] and A.A. Adesina[1*]

[1] School of Chemical Engineering and Industrial Chemistry, University of New South Wales, Sydney, Australia 2052
[2] CNRS-Laboratiore de Catalyse Homogene et Hetrogene, Universite des Sciences et Technologies, Lille, France

1 INTRODUCTION

There is a global interest in the development of transition metal carbides for various applications because of their unique physical and chemical properties which combine the distinctive attributes of covalent solids, ionic crystals, and transition metals [1]. Levy and Boudart [2] first demonstrated that tungsten carbide has comparable activity to noble metals for hydrogenolysis, hydrogenation and dehydrogenation reactions. This was ascribed to similarity in the electronic structure of the early transition metal carbide to those in the noble metal group. Since then, applications to other industrially important petrochemical reactions have been reported [2-5]. Metals carbides, however, require high temperature synthesis, which is often accompanied by poor surface areas and low porosities – a disincentive for catalysis. Oyama [7] has reviewed the preparation of high surface area metal carbides and recent studies at Oxford [8-9] have also shown that physiochemical property and ultimate catalytic activity are determined by the type of carburising (hydrocarbon) agent. Typically, these metal carbides are obtained from metal oxide precursors.

Studies in our laboratory have also shown that high surface area MoS₂ catalyst for catalytic H₂S decomposition may be obtained via precipitation from homogeneous solution (PFHS) of the sulphide compound on the support [10-12]. These catalysts are thermally stable even at 873-1073K. Thus it seems logical to employ PFHS – synthesized metal sulphides as starting material in place of low surface area oxide. In particular such a catalyst has the potential to be sulphur-tolerant, a definite advantage when utilized as Fischer-Tropsch catalyst. Additionally methane/H₂ mixture is commonly used as the carburising gas, however, higher hydrocarbons are more readily activated and thus, require low temperature, which further reduces the risk of loss in surface area [9]. Even so, excessive carbon deposition associated with hydrocarbon dehydrogenation may reduce catalytic activity via site blockage. Thus, a balance must be struck between competing demands and expected performance. In this study H₂/propane mixture was used as the carburising gas, in accordance with;

$$MoS_2 + (\frac{3-x}{3})C_3H_8 + 5H_2 \Rightarrow MoC_{1-x} + 2H_2S + C_2H_6 \qquad (1)$$

where 0<x<1 allows the formation of different molybdenum carbide species [13] Clearly, x=0.5 refers to the carbide Mo₂C. Although catalyst design is a multivariable optimization problem, the synthesis and characterization approach involving one-factor-at-a-time method is both labour-intensive and time consuming when optimum preparations are being sought. Thus, a statistical strategy is employed in this study. Moreover, temperature-programmed carburization would be useful in the identification of different phases obtainable and the influence of heating rate (if any) on the appearance and relative magnitude of these phases, especially as different phases may contain different catalytic active centres. Analysis of the carburization kinetics would also be beneficial in preparing tailored catalyst for future use.

2 EXPERIMENTAL

2.1 Wet Chemistry

Catalyst with 12% Mo loading was deposited on silica support via precipitation of metal sulphide from a homogenous solution containing calculated amount of ammonium molybdate ($(NH_4)Mo_7O_{24}.4H_2O$), urea, thioacetamide and nitric acid. Reaction was carried out in a water bath shaker kept at 90°C for 3 hours. The slurry was then filtered and the precipitate was dried in oven at 120°C for 14 hours. All chemicals were obtained from Aldrich as AnalaR grade and ultra-pure water was used to prepare the solutions

2.2 Carburization

Catalyst carburization was carried out in a quartz tube fixed bed reactor (I.D=6mm) centrally placed in a temperature controlled tubular furnace. A two level factorial design for 3 factors, temperature (674 and 973K), $H_2:C_3H_8$ ratio (1 and 5) and carburization holding time (1 and 4 hours) was employed. With flowrate for each experiment kept at 100mlmin⁻¹. These limits were selected based on information from the literature and stoichiometric considerations. BET surface area was measured on a Micromeritics TriStar 3000 with N₂ adsorption at 77K. FT-IR spectra were obtained from a Nicolet Nexus FT-IR spectrophotometer using Omnic Software. The effect of heating rate during carburization was studied on a ThermoCahn TGA 2121.

3 RESULTS

3.1 Surface Area Analysis

Table 1 shows the range of BET surface areas (from 2 replicates of each catalyst) for all 8 specimens. Yates' analysis was used to obtain the variance estimates associated with the relevant effects given in the 'effect identification' column. The sign of the variance is a reflection of the direction of the effect. It is seen that the interaction between time and the $H_2:C_3H_8$ ratio has the smallest variance estimate. F-values for other effects were calculated using the smallest variance as the denominator as shown in the last column. From standard tables, $F_{4,4}$ at 95% confidence level is 6.39, this immediately suggest that the statistically-

Table 1 *Surface area of Mo$_2$C prepared under various conditions*

Sample Number	Temp (°C) T	Ratio (H$_2$:C$_3$H$_8$) C	Time (hrs) t	BET (m^2/g)	Degrees of Freedom	Variance estimates	Effect Identification	F Values
1	400	1	1	73.35	8	84.35	Average	
2	700	1	1	85.23	4	10.2125	T	20.02
3	400	5	1	78.86	4	8.965	C	17.58
4	700	5	1	98.67	4	1.112	TC	2.18
5	400	1	4	77.28	4	0.64	t	1.25
6	700	1	4	83.6	4	-5.6325	Tt	11.04
7	400	5	4	87.45	4	-0.51	Ct	1
8	700	5	4	90.34	4	-2.8525	TCt	5.59

significant variables of temperature, H$_2$:C$_3$H$_8$ ratio and the time-interaction. The positive sign for H$_2$:C$_3$H$_8$ ratio indicates that within the range investigated, BET surface area increased with increase in H$_2$:C$_3$H$_8$ value. This is expected since a strongly reducing environment opens up the porous structure during the conversion of the solid precursor. Similar effect of composition has been reported by Claridge et. al. [14] albeit with H$_2$/CH$_4$ as the carburising gas. Increasing temperature also seemed to promote higher BET surface areas. This may also be due to thermally induced fine pore creation with the decomposition of organic species in the solution matrix. Interestingly the holding time, temperature-composition interaction and the 3-factor interaction did not have substantial influence. This would suggest that the lower limit of 1 hour was adequate. On the other hand, the statistical significance of the temperature-time interaction also indicates that heating rate may be an important factor. This was later explored for the carburization kinetics.

On the strength of this finding a multivariable regression for predicting surface are is

$$y_{BET} = a_0 + a_1 x_1 + a_2 x_2 + a_3 x_1 x_3 \tag{2}$$

where,

$$x_1 = \frac{T - 400}{300}$$

$$x_2 = \frac{C - 1}{4}$$

$$x_3 = \frac{t - 1}{3}$$

and a$_0$ is the BET surface area at the lowest level of all the factors. Parameters estimate (a$_0$, a$_1$, a$_2$, and a$_3$) from a fit of Eq. 2 of the whole data gave:

a$_0$= 75.594   a$_1$=12.154   a$_2$= 8.404   a$_3$= -4.98

Constrained optimization using equation (2) as objective function yield the conditions for 'best' BET surface area as T = 700°C, C = 5 and t = 2.44hr

3.2 Carburization Kinetics

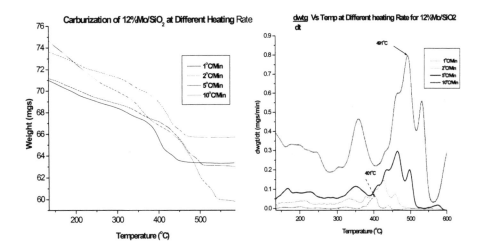

Figure 1 *Carburization of 12%MoS₂/SiO₂ at different heating rate*

Figure 2 *dwtg/dt with respect to temp at various heating rate*

Carburization kinetics was studied using heating rate of 1, 2, 5 and 10 °C/min in a ThermoCahn TGA with carburising gas of $H_2:C_3H_8=5$ and up to 700°C. Fig 1 shows the weight-temperature profiles at different heating rates. It is apparent that % weight drop increased with increasing heating rate. Moreover, the temperature at which the weight became constant increased with heating rate. These values correspond to different times (temperature/heating rate) for complete conversion. Fig 2 shows the plot of the rate of weight change with respect to temperature. A peak suggests the formation of a MoC_{1-x} phase. Only peaks at T>300°C correspond to a MoC_{1-x} peak. There are 4 peaks identifiable. There is a temperature shift in the location of the peak with increase in heating rate. Green et. al. [10] have also observed three phases of molybdenum carbide in a study with butane as the carburising gas. In our case peaks correspond to non-stoichiometric forms of molybdenum carbide.

Although qualitative identification of the molybdenum carbide phase corresponding to each phase is not available, it seems that the increase in peak height (intensity) with increased heating rate is an indication that the amount of each phase increases with time. In particular, the temperature at which each peak is form shifted to higher values at higher heating rate further confirming the increased amount of each phase. Indeed, a plot of the heating rate verses peak temperature, Figure 3, for each peak is essentially linear over the 4 heating rates used. The slope of this line is a characteristic of that phase for the gas-solid conversion at the average peak temperature

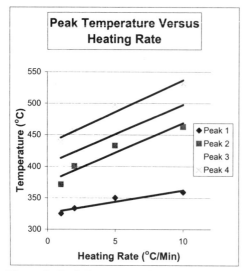

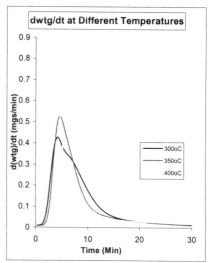

Figure 3. *Peak Temp versus heating rate*   **Figure 4.** *Change in weight w.r.t time*

It is apparent that 3 of the phases are structurally-similar, however, the lower temperature peak may be a sub-stoichiometric molybdenum carbide.

Further to this, isothermal runs were conducted at 300, 350, and 400°C by heating MoS_2 sample in helium to the required temperature and held constant at this level for the next 2 hours. However, after 30 minutes in helium flow (to remove possible volatiles), the gas was switched to $H_2:C_3H_8$ (=5) mixture. The resulting weight drop with time was tracked continuously. Figure 4 shows the weight drop profile.

Since a constant $H_2:C_3H_8$ (=5) was used, the MoS_2 carburization rate may be written in power law form as

$$- r_{MoS_2} = k C^n_{MoS_2} \tag{3}$$

where k is a function of temperature and $H_2:C_3H_8$
Thus

$$- \frac{d W_{MoS_2}}{d t} = k \left[W_{MoS_2} (1 - x_{MoS_2}) \right]^n \tag{4}$$

$$\therefore - \frac{d W_{MoS_2}}{d t} = k_{rxn} (1 - x_{MoS_2})^n \tag{5}$$

where

$$k_{rxn} = kW_{MoS_2,ini}^n (MW_{MoS_2})^{1-n}$$

and $kW_{MoS_2,ini}^n$ =initial weight of MoS₂ in the TGA boat.

Since $\ln(-\dfrac{dW_{MoS_2}}{dt})$ can be obtained from the TGA data, a plot of

$\ln(\dfrac{dW_{MoS_2}}{dt})$ Versus $\ln(1-x_{MoS_2})$ was used to calculate k_{rxn} and n for the data sets.
Although n values varied between 1.87 to 2.02, Arrhenius expression for k_{rxn} was obtained as

$$k_{rxn} = 1.1483x10^{11}e^{\frac{-18853}{t}}$$

which gives an activation energy, E_A, of 156.74 kJmol⁻¹ for the carburization of MoS₂. The second order (n=2) dependency of MoS₂ content may be a reflection of the fact that two MoS₂ entities are required for the formation of a Mo₂C species.

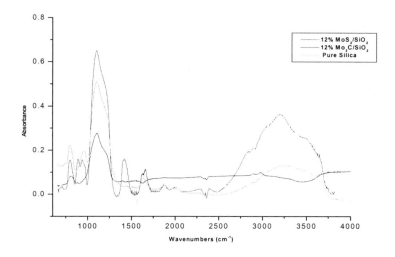

Figure 5. *FT-IR Spectra*

Figure 5 shows the FT-IR spectra of the 12%Mo₂C along with the uncarburized MoS₂ and the pure silica support. Broad IR absorption peak in the region between 3000 to 3600 cm⁻¹ may be assigned to the OH stretching mode due to the presence of water in the silica and MoS₂ specimens. The cluster of peaks in the 780 to 1100 cm⁻¹ region corresponds to different molybdenum species. For example, the sharp peak at 960 cm^{-1} in the uncarburized catalyst finger points a terminal Mo=O double bond, while peaks appearing between 1400 to 1660 cm^{-1} are various organosulphur bonds

4 CONCLUSION

Supported Mo$_2$C catalyst have been obtained from the carburization of MoS$_2$ using H$_2$:C$_3$H$_8$ mixtures over 673-973K for different periods. Analysis of the 2^3 experimental design shows that only H$_2$:C$_3$H$_8$, temperature and temperature time interaction has significant effect on BET surface area at 95% confidence level. The resulting regression model was used to obtain the optimal carburizing conditions. Subsequent temperature programmed carburization indicates that 4 molybdenum carbide phases were obtained at different temperature approximately 598, 648, 677 and 708K. The kinetics analysis of the carburization resulted a 2nd order dependency on initial MoS2 and an activation energy of 157kJmol^{-1}.

References

1. Oyama S. T. *The Chemistry of Transition Metal Carbides and Nitride*, Blackle Academic and Professional, Glasgow 1996, p5
2. Levy, R. L., Boudart, M., *Science* 1973, **181**, 547
3. Leclercq, L.; Almazouari, A.; Dufour, M.; Leclercq, G. *The Chemistry of Transition Metal Carbides and Nitrides*; Oyama, S. T., Ed.; Blackie Academic & Professional: Glasgow, 1996; pp 345- 361
4. l. Toth, L. E. *Transition Metal Carbides and Nitride,* Academic Press, New York 1971
5. Sherif, F.; Vreugdenhil, W. *The Chemistry of Transition Metal Carbides and Nitrides*; Oyama, S. T., Ed.; Blackie Academic & Professional: Glasgow, 1996; pp 414-425.
6. Boudart M.; Volpe L. *J. Solid State Chem.* 1985, **59**, 348
7. Oyama S. T. *Catal. Today* 1992, 15, 179
8. Green, M. L. H.; Xiao, T.; York, A. P. E.; Williams V. C. *Chem. Mater.* 2000, **12**, 3896
9. Xiao T., York A. P. A., Coleman, K. S., Claridge J. B., Sloan J. Charmock J., Green M.L.H., *J. Mater. Chem.* 2001, **11**, 3093-3098
10. Adesina A. A.; Meeyoo V.; Foulds G. *React. Kinet. Lett.* 1995, **56**, 231
11. Adesina A. A.; Moffat S. C. *Catal. Lett.* 1996, **37**, 167
12. Adesina A. A.; Gwuana M. *React. Kinet. Lett.* 1997, **55**, 62
13. N. Arul Dhas, A. Gendanken, Chem. Mater. 1997, **9**, 3144-3154
14. Claridge J.B., York A. P. E., Brungs A. J., Green M. L. H., Chem. Mat, 2000,**12**, 132

KETO-ENOL ISOMERISM ON TRANSITION METAL SURFACES, A DENSITY FUNCTIONAL THEORY STUDY

Rajinder Mann[1], Graham J. Hutchings[1], Werner van Rensburg[2] and David J. Willock[1].

[1]Department of Chemistry, Cardiff University, Cardiff CF10 3TB.
[2]Current Address: Sasol Technology Ltd, 1 Klasie Havenga Road Sasolburg, PO Box 1, Sasolburg, 1947 South Africa.

1 INTRODUCTION

The hydrogenation of ketones over metal surfaces is an important step in many applications in heterogeneous catalysis, such as the production of unsaturated alcohols from α-β unsaturated ketones. Another important example is the enantioselective hydrogenation of pyruvate esters to lactates over modified supported Pt and Pd catalysts which has been widely studied due to the high *e.e.*s that can be achieved[1]. In the molecular level models of these reactions it has been assumed that the pyruvate ester adsorbs in a π-bonded mode with the molecular plane parallel to the surface[2,3,4]. Recently, however, XPS and UPS results have suggested that the adsorption of ethyl pyruvate on Pt(111) involves the interaction of the *keto* oxygen atom with the surface. This would suggest that the molecular plane is not parallel to the surface and so will have important implications in mechanistic models of the reaction. The lone pair bonding mode has also been observed for simpler ketones such as acetone. Some time ago Avery, using a combination of EELS and temperature programmed desorption experiments, suggested that at high coverage acetone adsorbs on Pt(111) through the lone pair of the oxygen taking up an "endon" orientation [η^1(O)-acetone] with the C=O bond toward the surface[5]. At low coverage a more strongly adsorbed species with a lower C=O stretch frequency was observed and assigned to a η^2(C,O)-acetone molecule with a di-σ adsorption geometry. In this adsorption mode the C=O bond is parallel to the surface but the bond order is reduced due to interaction of both the C and O atoms with Pt atoms. More recently high resolution RAIRS experiments have suggested that acetone on Ni(111) adsorbs in the enolate form[6].

This range of possible adsorption modes for acetone prompted us to study the alternative adsorption possibilities for formaldehyde, as an example where the *enol* form is not possible, and acetone on Pt and Pd surfaces using periodic density functional theory. A previous theoretical study in this area has been published by Dumesic *et al.*[7] however they have concentrated on the reaction pathway for hydrogenation of acetone over Pt(111) assuming the η^1(O)-acetone is the only adsorbed state present. In this contribution we compare adsorption of acetone in the *enol* and *keto* isomers on both Pt(111) and Pd(111).

We have previously studied the adsorption of ethene to the (111) surface of Pt[8] and Cu[9] using the VASP simulation package. However more recently we have tested the alternative CASTEP code and a part of this work is to report a comparison of these two programs for calculations involving these types of molecule. Accordingly, in the methodology section we compare CASTEP calculations on ethene adsorption to Pt(111) with our earlier work. In the results section we then cover the calculations of adsorption modes for formaldehyde and acetone using only the CASTEP program.

2 METHODOLOGY

2.1 Parameters for periodic DFT calculations

Periodic density functional theory as implemented in CASTEP[10] was used to study formaldehyde and acetone adsorption. Input files were generated using Cerius[2] and carried out using the parallel academic version of the code on national supercomputer facilities[11]. CASTEP uses a planewave basis set to describe the electron density of a three dimensional periodic system. Exchange and correlation energies were included via the generalized gradient approximation with the Perdew-Wang 91 (GGA-PW91) functional[12]. Core states of each atom were described by ultra-soft pseudopotentials[13] allowing a relatively small planewave basis to be employed which was set with a cutoff of 380 eV. Reciprocal space sampling used the Monkhurst-Pack[14] scheme with a k-point grid density of $3\square\ 3\square\ 3$ (14 unique k-points). For the face centred cubic unit cell of Pt and Pd the lattice constants optimised using CASTEP were 3.9755Å and 3.9158Å respectively, which are in good agreement with the experimental values of 3.9891Å and 3.9239Å.

For surface calculations we use a slab consisting of 3 atomic layers separated from its periodic image perpendicular to the surface by a vacuum gap of at least 10Å. Optimization of the adsorbate structures included the metal atom positions as degrees of freedom and so the binding energy values quoted for adsorbates are referenced to a fully optimised clean surface. To minimize the interaction of adsorbates with their periodic images we use a 3×3 surface unit cell (p(3×3)) so that 27 metal atoms are included. The larger real space unit cell reduces the number of k-points required and a 3×3×1 k-point grid (equivalent to 5 unique k-points) was used.

The adsorption energy, ΔE_{ads}, of a molecule was calculated from the energies of the adsorbed state E_{as}, the optimised free surface, E_s, and the gas-phase molecule, E_g, using:

$$\Delta E_{ads} = -(E_{as} - (E_s + E_g))\qquad\qquad\qquad (1)$$

Each energy is calculated based on the same periodic simulation box and with the same planewave cutoff and k-point sampling grid. The adsorption energy, then, is simply the energy evolved during adsorption and will be positive for exothermic adsorption.

2.2 Comparision of VASP and CASTEP programs

In earlier work[8] we considered the adsorption of ethene on Pt(111) using the VASP program[15] and so we initially carried out duplicate calculations with CASTEP for comparison. Both programs implement periodic DFT using a planewave basis set and allow the core electrons to be replaced by ultrasoft pseudopotentials. For this comparison a p(2×2) surface was used with a 9.2 Å vacuum gap and the parameters from reference 8 (cutoff (300 eV), 9×9×1 k-point sampling). This gave adsorption energies converged to within approximately 3 kJmol^{-1} in the VASP calculations. The main difference in the approaches used in these programs is the treatment of the pseudopotentials. In CASTEP the pseudopotentials that were available to us were calculated at the local density level but were used in gradient corrected calculations. In VASP both the pseudopotential derivation and the calculations employed gradient corrections.

A comparison of the calculated adsorption energies using the two programs is given in table 1. Both CASTEP and VASP give the same ordering: Cross bridge has the molecule adsorbed equidistant from two surface Pt atoms with the C-C bond perpendicular to the Pt-Pt direction, giving an extremely weak interaction with a binding energy of around 10 kJ mol^{-1}. In atop adsorption the ethene molecule is co-ordinated to a single Pt atom in a π-bonded mode. The two alternatives differ in the orientation of the C-C bond relative to the neighbouring atoms, being aligned with bridge sites (atop-bridge) or hollow sites (atop-hollow). These alternatives are almost iso-energetic and so the orientation of the ethene in an atop site would not be fixed. The most favourable adsorption site is the bridge where the two carbon atoms interact with separate Pt atoms in a di-σ bonded species.

However, the CASTEP estimated adsorption energies are consistently higher than those obtained with VASP. We have previously shown[8] that the VASP calculations gave extremely good agreement with the available thermochemical data. This suggests that CASTEP is overestimating the binding energy of ethene by between 10 and 20%.

Table 1 *VASP and CASTEP energies for ethene on Pt(111).*

	Adsorption Energy (kJmol^{-1})	
Type of molecular geometry	VASP[a]	CASTEP
Atop – bridge	85.8	105.7
Atop – hollow	84.8	103.9
Bridge	127.3	143.3
Cross bridge	9.2	11.0

[a]data from reference 8.

The calculated geometry of ethene adsorbed in the bridge site is summarized in table 2. The di-σ bonding mode gives a considerable distortion of the molecule away from the planar geometry of the gas phase, as can be seen from the C-CH$_2$ angle. This is accompanied by a longer C-C bond, suggesting that the bond order for C-C has reduced as the carbon hybridization changes from sp^2 to sp^3. The geometries from the two programs show general agreement and certainly lead to the same conclusions regarding the change of molecular geometry compared to the corresponding gas phase data. However the stronger interaction from the CASTEP results leads to shorter Pt-C distances and a slightly more pronounced movement of the co-ordinated Pt atoms out of the surface plane.

Table 2 *Geometry of adsorbed ethene on Pt(111) from VASP and CASTEP.*

	Simulation Method			
	VASP		CASTEP	
Property	Gas-phase	Bridge	Gas-phase	Bridge
C-C (Å)	1.334	1.485	1.318	1.472
C-H (Å)	1.099	1.096	1.084	1.090
C-CH$_2$ angle[a]	180.0	138.1	180.0	141.9
C-C-H angle	122.2	112.1	122.0	114.6
Pt-C (Å)		2.117		2.099
Δz (Pt)[b] (Å)		0.235 (×2)		0.252/0.250

[a]angle of C-C bond to CH$_2$ plane. [b]displacement of co-ordinated Pt atoms from surface plane.

3   RESULTS AND DISCUSSION

3.1 Formaldehyde adsorption on Pt(111) and Pd(111)

Table 3 gives the adsorption energies and geometric data for formaldehyde adsorption on Pt and Pd (111) surfaces. In both cases we find two distinct adsorption modes. In the weaker interaction only the oxygen atom is closely co-ordinated to the surface, interacting with a single M atom in a $\eta^1(O)$ configuration. The molecule remains almost planar with very little affect on the C=O bond length or the O-CH$_2$ angle. The adsorption energy on Pt is 28 kJ mol^{-1} greater t han for Pd and the M-O distance is correspondingly shorter b y 0.171Å. In the case of Pd the molecule is orientated almost perpendicular to the surface, as evidenced by the large value of the M-O-C angle, whereas for Pt the C=O bond makes a much smaller angle to the surface. In the $\eta^1(O)$ adsorption mode the perturbation of the surface by the presence of the molecule is small but the stronger interaction with Pt results in a slightly greater movement of the co-ordinated metal out of the surface plane.

If we start the calculation with the molecular plane parallel to the surface so that both C and O atoms are close to surface m etal a toms optimization results in structures reminiscent of the di-σ bonded ethene used as a comparison between CASTEP and VASP in the methodology section. In this case, however, the more electronegative oxygen atom forms a shorter bond than carbon to the surface. The extension of the C=O bond for adsorption to Pd is greater than that for Pt even though the adsorption energy to the latter is greater.

Table 3   *Geometry of adsorbed formaldehyde on Pt and Pd(111) surfaces*

M = Pd	ΔE_{abs} (kJ mol^{-1})	C=O (Å)	O-CH$_2$[a] ($^\circ$)	M-C (Å)	M-O (Å)	M-O-C ($^\circ$)	M Δz (Å)
$\eta^1(O)$	13.2	1.228	180	3.527	2.320	172	0.058
$\eta^2(C,O)$	60.3	1.347	148	2.117	2.040	111	0.120(C) /0.083(O)
M = Pt							
$\eta^1(O)$	41.2	1.215	180	3.007	2.149	123	0.068
$\eta^2(C,O)$	77.6	1.304	137	2.107	2.039	111	0.155(C) /0.119(O)

[a]Angle between CO bond and CH$_2$ plane.

3.2   Acetone adsorption on Pt(111) and Pd(111)

For a cetone adsorption we considered both the *keto* and *enol* isomers of the a dsorbate. These have a calculated gas phase energy difference of 51 kJ mol^{-1} with the *keto* form favoured. Table 4 summarizes the calculated adsorption data for the *keto* isomer. As in the case of formaldehyde two initial geometries were considered, one with the molecular plane parallel to the surface and one with it perpendicular. For acetone both of these alternatives lead to a $\eta^1(O)$ or end-on optimized final geometry and the di-σ bonding mode is not observed. For both metals the final geometries differ in the orientation of the molecular plane with respect to the surface with the parallel starting point leading to a smaller O-Me$_2$/Pt angle. The calculated value of the adsorption energy on Pt(111) in this mode is in very good agreement with the estimate made from TPD results[5] (45 kJ mol^{-1}). For the Pd case the two adsorption modes have very similar energies but in the case of Pt the perpendicular orientation of the molecule is preferred by 30 kJ mol^{-1}. In these *keto-*

isomer adsorption modes the molecule remains planar on adsorption and the bonding can be thought of as a weak interaction formed by the oxygen lone pair with a single surface M atom.

Table 5 summarises the results for adsorption of the *enol* isomer. This shows a higher adsorption energy on both Pd and Pt. The adsorption energies quoted in table 5 are referenced to the gas phase *enol*-isomer with the energy referenced to normal gas phase acetone in brackets. The adsorption energies for the *enol*-isomer are significantly higher than for the ketone calculations (table 4) and are sufficient to overcome the gas phase energy difference between the two isomers.

Table 4   *Geometry of adsorbed keto isomer of acetone on Pt and Pd(111) surfaces*

	ΔE_{abs} (kJ mol^{-1})	C=O (Å)	O-Me$_2$/Pt^c (°)	M-C (Å)	M-O (Å)	M-O-C (°)	Δz^d (Å)
M=Pd							
$\eta^1(O)^a$	18.8	1.237	29	3.126	2.353	134	0.114
$\eta^1(O)^b$	20.8	1.226	87	3.367	2.154	169	0.052
M=Pt							
$\eta^1(O)^a$	12.9	1.233	29	3.302	2.655	141	0.042
$\eta^1(O)^b$	42.9	1.241	89	3.257	2.189	142	0.154

^{a}Structure located starting from molecular plane parallel to surface. ^{b}Structure located starting from molecular plane perpendicular to surface. ^{c}Angle defined between molecular plane (O and methyl carbons taken as reference points) and surface plane. ^{d}Displacement of co-ordinated M atom from surface plane, ^{f}Angle formed by plane of 3 atoms and the bond indicated by dash.

Table 5   *Geometry of adsorbed enol isomer of acetone on Pt and Pd(111) surfaces*

	ΔE_{abs} (kJ mol^{-1})	C-OH (Å)	C-CH$_2$^{a} (°)	C-CCO^a (°)	M-C (Å)	M-C-C (°)	Δz^b (Å)
M=Pd							
$\eta^2(C,C)^c$	71.1 {21.1}^{d}	1.362	156	163	2.180 /2.312	77	0.286
M=Pt							
$\eta^1(C)$	117.5 {66.5}^{d}	1.304	139	176	2.130 /2.804	103	0.190

^{a}Angle formed by plane of 3 atoms and the bond indicated by dash. ^{b}Displacement of co-ordinated M atom from surface plane. ^{c}Both atoms coordinate to a single Pd atom. ^{d}Adsorption energy relative to gas phase *keto*-isomer.

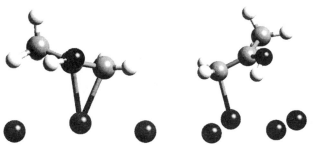

Figure 1   *Adsorption geometry of the enol-isomer of acetone on a) Pd(111) and b) Pt(111). Most of the metal atoms have been removed for clarity.*

3   CONCLUSIONS

For formaldehyde adsorption we have found both a $\eta^1(O)$ and a $\eta^2(C,O)$ adsorption mode on both Pt(111) and Pd(111), with the calculated energies suggesting the latter form is more strongly bound.

For acetone the $\eta^2(C,O)$ adsorption mode did not form even if the starting geometry was biased toward it, suggesting there is a barrier to the formation of the di-σ bonded molecule. The *enol* isomer was found to adsorb in a $\eta^2(C,C)$ geometry on Pd(111) and in a $\eta^1(C)$ alkyl alcohol mode on Pt(111) as shown in figure 1. In both cases the adsorption energy in the *enol* form is sufficient to compensate for the gas phase energy difference between *keto* and *enol* isomers. This suggests that thermodynamically the *enol* form will be favoured on both metals. However, we have not considered the barrier to the formation of the *enol* or the possiblity of an enolate ion but these are the subject of current work.

Acknowledgements

We acknowledge funding for a Compaq SC cluster at the RAL purchased and supported with funding from the JREI (JR99BAPAEQ) and Compaq. Part funding for this project came from an EPSRC IMI project involving Johnson Matthey, BP-Amoco, Robinson Brothers, Grace Davidson and Accelrys.

References

1   G.Webb and P.B.Wells, *Catal.Today,* 1992, **12**, 319.
2   K.E.Simons, P.A.Meheux, S.P.Griffiths, I.M.Sutherland, P.Johnston, P.B.Wells, A.F.Carley, M.K.Rajumon, M.W.Roberts and A.Ibbotson, *Recl. Trav. Chim. Pays-Bas*, 1994, **113**, 465.
3   O.Schwalm, B.Minder, J.Weber and A.Baiker, *Catal. Lett.*, 1994, **23**, 271.
4   A.Baiker, *J.Mol.Catal.A*, 1997, **115**, 473.
5   N.R.Avery, *Surf. Sci.*, 1983, **125**, 771.
6   W-S. Sim, T-C.Li, P-X.Yang and B-S Yeo, *J.Am.Chem.Soc.*, 2002, **124**, 4970.
7   R.Alcala, J.Greeley, M.Mavrikakis and J.A.Dumesic, *J.Chem.Phys.*, 2002, **116**, 8973.
8   G.W.Watson, R.P.K.Wells, D. J. Willock and G. J. Hutchings, *J.Phys.Chem. B,* 2000, **104**, 6439.
9   G.W.Watson, R.P.K.Wells, D. J. Willock and G. J. Hutchings, *Surface Science,* 2000, **459**, 93.
10   CASTEP 4.2 Academic Version, Lincensed under the UKCP-MSI Agreement 1999, M.C.Payne, M.P.Teter, D.C.Allan, T.A.Arias and J.D.Joannopoulos, *Rev. Mod. Phys.*, 1992, **64**, 1045.
11   Calculations carried out using Mott (see acknowledgement) and the T3E facility, Turing, through the Materials Chemistry consortium.
12   J.P.Perdew, *Phys. Rev. B*, 1986, **33**, 8822.
13   D.Vanderbilt, *Phys. Rev. B*, 1990, **41**, 7892.
14   H.J.Monkhurst and J.D.Pack, *Phys. Rev. B,* 1976, **13**, 5188.
15   G.Kresse, J.Furthmüller, *J. Phys. Rev B.*, 1996, **54**, 11169, Kresse, G., Furthmüller, *Comput. Mater. Sci.*, 1996, **6**, 15, G.Kresse, J.Hafner, *J. Phys: Condens. Matter.*, 1994, **6**, 8245.

DIRECT TRANSFORMATION OF METHANE TO HIGHER HYDROCARBONS IN PRESENCE OR ABSENCE OF CARBON MONOXIDE

J.L. Rico[1], J.S.J. Hargreaves[2] and E.G. Derouane[3]

[1]Laboratorio de Catálisis, FIQ e IIM, Universidad Michoacana, Morelia Mich., Mexico.
[2]Department of Chemistry, Joseph Black Building, University of Glasgow, Glasgow, G12 8QQ, U.K.
[3]Leverhulme Centre for Innovative Catalysis, Department of Chemistry, University of Liverpool, PO Box 147, Liverpool, L69 3BX, U.K.

1 INTRODUCTION

In recent years, a lot of research effort has been directed towards dehydroaromatisation of methane in which methane is converted to aromatic products such as benzene and naphthalene in addition to hydrogen.[1] Perhaps the most well studied system has been that employing Mo/ZSM-5 based catalysts, where the bifunctional interaction between the zeolite Bronsted acidity and molybdenum species is well recognised. Under reaction conditions, the active molybdenum species are known to be in the form of carbides [2,3] or oxycarbides, and recently it has been proposed that the α-MoC_{1-x} phase is the most active form.[4] Deactivation, primarily due to coke formation, is well precedented in this reaction and represents a major obstacle to be overcome in the successful application of these catalysts. In this respect, it is interesting to note that Ichikawa and co-workers have published studies indicating that the inclusion of low levels of CO or CO_2 in the feed can promote the reaction via the suppression of coke formation in the case of both Mo/HZSM-5 [5,6] and Re/HZSM-5 [7] catalysts. Other approaches adopted towards this aim have been the inclusion of second metal components [8] and a reduction of the acid strength of the HZSM-5 support.[9]

In this study, we report our results aimed investigating the potential promotion of catalyst activity/lifetime by inclusion of a low level of CO in the feed. Attention has been directed towards a study of the nature of the coke species formed both in the presence and absence of CO, over ZSM-5 (Si/Al= 47) based catalysts which have been deliberately prepared so as to contain the α-MoC_{1-x} phase.

2 METHOD AND RESULTS

2.1 Catalyst preparation and activity testing

10wt% Mo/HZSM-5 catalysts were prepared by impregnation of dried ZSM-5 with $(NH_4)_6[Mo_7O_{24}].4H_2O$, dissolved in deionised water. The resultant slurry was slowly evaporated at 80 C, and the solid was then dried at 120 C for 2 h, before being calcined in air at 400 C for 1 h. In order to obtain the supported α-MoC_{1-x} phase, 0.73 g of catalyst was activated in a continuous laboratory quartz tubular reactor, by heating from room temperature to 330 C at 16 C/min under a flow of ethane/H_2 (8.33% of ethane in H_2), for 24 h. The sample was then carburised from 330 C to 700 C under a flow of methane/H_2 (10% methane in H_2). Further details about the activation and carburisation procedure can be found elsewhere.[10] Once at 700 C, the sample was exposed to pure methane or a mixture of CH_4-CO (2% CO), using a space velocity of 1002 ml/h g cat. The gas products were monitored by gas chromatography using FID and TCD detectors and Porapak Q and Mol Sieve columns. To avoid condensation of the high boiling point compounds, the line from the reactor outlet to the sampling valve was heated up to 200 C. The samples were characterised b y n itrogen B ET, a nd T G-DSC. F or t he l atter p urpose, a portion o f s pent sample was treated under air from room temperature to 600 C at a heating rate of 0.5 C/min.

2.2 Results

The main gas products detected in the transformation of methane are hydrogen, benzene, toluene, n aphthalene, 1 - a nd 2 - m ethylnaphthalene a nd h eavier h ydrocarbons. F igures 1 and 2 show that greater B/M (mol of benzene per inlet mol of methane) ratios are obtained when CO is included in the feedstream, as reported by Liu et al.[5] The ratio is *ca.* 25 % higher in presence of CO during the early stage of reaction, and the effect slowly decreases at longer times on-stream. No conversion of CO itself could be detected. After 50 hours on stream, the ratios B/M are practically the same for both catalysts, showing no advantageous effect of CO. Figures 1 and 2 also show that the mol percentage of benzene among the hydrocarbons produced for both catalysts are practically the same, showing no effect of CO, see also Figures 3 and 4 for benzene and toluene selectivity. The increase in benzene and toluene selectivity with time on-stream for both catalysts is the result of the slow disappearance of the heavier hydrocarbons from the gas product. Benzene and hydrogen are the only g as products detected at the last stage of the e xperiments, where catalysts are expected to be heavily coked. Post-reaction characterisation was therefore undertaken in an attempt to elucidate the origin of the effect of CO.

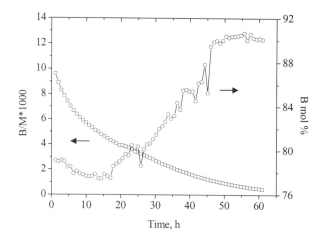

Figure 1   *Catalytic activity of Mo/ZSM-5 in the reaction of methane without CO*

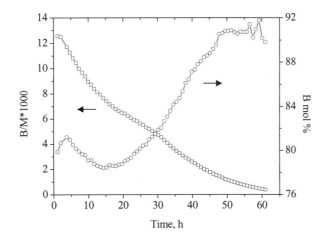

Figure 2   *Catalytic activity of Mo/ZSM-5 in the reaction of methane with CO*

The specific surface areas of the fresh and spent catalysts are shown in Table 1. The latter samples show a significant decrease of surface area/porosity, indicating that substantial pore blockage by the carbonaceous species occurred during the methane reaction. In addition, it is notable that there is a difference in the surface areas and average pore diameters of the two post-reactor samples, which may be indicative of a difference in the amount/nature of the coke formed during reaction. On this basis, TG-DSC studies were undertaken to determine the quantity and nature of carbonaceous species in the catalysts after reaction.

Table 1   *Specific surface areas of the fresh and spent samples*

Sample	Areas, m^2/g	Micropore volume, cm^3/g	BET average pore diameter, A
ZSM-5	389	0.107	27
Mo/ZSM-5	339	0.093	25
Spent without CO	24	0.003	71
Spent with CO	20	0.002	96

In line with previous observations reported in the literature, where both two[6] and three[11,12] species have been reported, our TG-DSC studies shown in Figures 5 and 6 indicate the existence of at least two different carbonaceous species on both catalysts after reaction. We have assigned these species as low and high temperature species, LTS and HTS respectively. In the literature, low temperature species have been assigned to coke which is associated with Mo species[6], and higher temperature species have been related to coke associated with Bronsted acidic sites.[13]

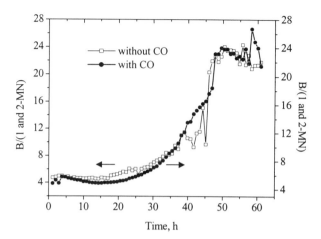

Figure 3   *Benzene selectivity related to 1 and 2-methylnaphthalene*

In the present study, although the carbonaceous species formed on the spent catalysts appear to be generally similar in both the presence and absence of CO on first inspection, careful comparison indicates that there are differences. As shown in Table 2, a greater total amount of carbonaceous species (11.5% more) is evident on the catalyst exposed to CO during reaction. Therefore, the beneficial role of CO inclusion in the feedstream, cannot simply be taken as due to a reduction in the total amount of coke formed, *per se.* Furthermore, whilst the amount of HTS is generally similar in both the presence and absence of CO, the major difference is in the amount of LTS, with CO enhancing its formation.

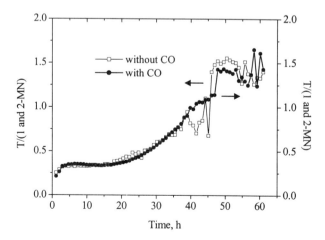

Figure 4   *Toluene selectivity related to 1 and 2 methylnaphthalene*

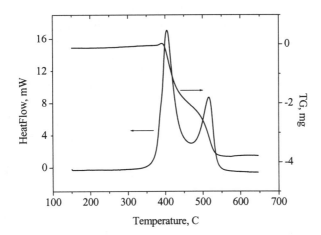

Figure 5   *TG-DSC graph of spent α-MoC$_{1-x}$/ZSM-5 in absence of CO*

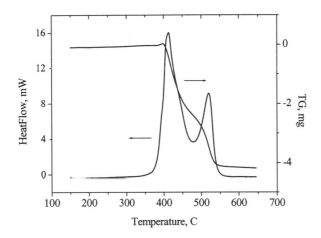

Figure 6   *TG-DSC graph of spent Mo/ZSM-5 with CO*

The analogous oxidation of pure Mo$_2$C using comparable experimental conditions as the TG-DSC shown above, shows a single exothermic signal with a maximum at *ca.* 457 C and a total gain in weight of 40.1% resulting from the formation of molybdenum oxide. Similar experiments performed on the α-MoC$_{1-x}$/ZSM-5 phase prior to reaction show a complex oxidation pattern with a number of temperature maxima all occurring below 350 C, which may be a reflection that the value of x in the molybdenum carbide phase can vary between 0 and 1 and/or of differences in dispersion. These results will be reported in detail elsewhere. Therefore, based on these observations, it can be concluded that the initial properties of the molybdenum carbide are changed throughout the methane reaction due to coke formation, and that the heat release together with the weight gain due to the formation of molybdenum oxide during combustion, are probably overshadowed by the oxidation of coke. An additional observation in our post reactor characterisation studies, Table 3, is that there is a notable shift in the temperature maxima for both HTS and, in particular, LTS which indicates that the inclusion of CO in the feed has modified the reactivity of the resultant carbonaceous species. It is therefore concluded that one of the effects of CO is related, directly or indirectly, to the increase of the amount and/or reactivity of LTS and finally, the total amount of coke. On the basis of these observations, it may be argued that the greater total percentage of carbonaceous deposits observed after 72 h on-stream with the use of CO, might be related to higher benzene production in the first 50 h of reaction.

Table 2   *Coke content on the spent Mo/ZSM-5 after 72 h on stream, analysed by TG-DSC*

Sample	Total % of coke	Coke, %	
		LTS	HTS
Without CO	17.4	9.0	8.4
with CO	19.4	10.9	8.5

LTS and HTS stand for low and high temperature species, respectively.

Liu et al.[5] have suggested that CO is effective at suppressing coke formation and is involved in the reaction by supplying an active form of carbon which reacts to form products, leaving a surface oxygen residue which regenerates CO via reaction with less active carbonaceous species. It is clear that whilst our studies do not support the proposal that the beneficial effects of CO inclusion in the feed relate to the reduction of coke content *per se*, the role of CO could be to modify the reactivity of carbonaceous species by such a mechanism. In this context, it is notable that we have observed the greatest effect of the inclusion of CO to be on the formation of the quantity and the reactivity of LTS, which others[6,13] have assigned to carbonaceous species associated with the molybdenum component of the catalyst.

Table 3   *TG-DSC heat analysis of coke burned from the spent α-MoC$_{1-x}$/ZSM-5*

Sample	LTS		HTS	
	Temperature C	Heat released, J/mg	Temperature C	Heat released, J/mg
without CO	404.6	47.3	516.1	25.4
with CO	413.2	45.2	519.8	24.6

Whilst no net conversion of CO was observed in our studies, it is interesting to note that Lunsford and co-wokers[12] have demonstrated that CO pre-treatment at the temperatures employed in the current study results in the formation of a coke associated with the molybdenum carbide phase in Mo_2C/HZSM-5 catalysts. This observation lends support to the observation that the inclusion of CO in the feedstream modifies the LTS content in our studies, possibly via the mechanism proposed by Liu et al.[5]. In making comparisons between the present study and those in the literature, however, it should be borne in mind that comparison between the various studies is not straightforward, because the forms of the molybdenum carbide phases are different. To our knowledge, our extended experiments are the first to be performed directly with the α-MoC$_{1-x}$ phase. Given the assignment of HTS to coke associated with the ZSM-5 framework[13], the general similarity in the concentration and reactivity of such species in both our spent catalysts is hardly surprising. The difference in both the quantity and behaviour of the LTS carbonaceous species observed in the TG-DSC experiments is a significant observation, as there is ample evidence for the role of carbonaceous species facilitating catalytic reactions involving hydrogen transfer in the literature.[14,15]. At present, it is not clear whether the differences in the temperature of reaction of LTS observed in the TG-DSC studies are a consequence of differing properties/reactivity of the coke formed, eg C/H ratios, aromaticity etc, or whether some effect directly on the α-MoC$_{1-x}$ component is operative, eg modification of morphology or dispersion. Further studies are required to elucidate this and perhaps should relate to times on stream where the effects of added CO is observed to be greatest.

3 CONCLUSION

Based on previous observations reported in the literature, we have undertaken a study of the effects of CO addition to the feedstream of α-MoC$_{1-x}$/ZSM-5 catalysts applied to the aromatisation of methane. It is apparent that CO enhances the formation of benzene in the first 50 hours on-stream, with the effect gradually dying off at prolonged run times. Post-reaction characterisation after 72 h on stream indicates that catalysts are heavily coked, with a large drop in surface area and accessible micropore volume being evident. TG-DSC analysis indicates differences in the content and reactivity of carbonaceous species between catalysts run in the presence and in the absence of co-fed CO. The inclusion of CO slightly enhances the production of coke, possibly as a consequence of the enhancement of the formation of benzene in the first 50 h on-stream. The largest effect of CO addition appears to be on the quantity and reactivity of low temperature species which may be associated with the Mo containing phase. This study has shown that the benefits of CO inclusion in the feedstream for methane aromatisation in the presence of α-MoC$_{1-x}$/ZSM-5 are not simply due to suppressed coke formation, as suggested by some other workers for other related systems.

References

1 Y. Xu, X. Bao and L. Lin, *J. Catal.* 2003, in press.
2 F. Solymosi, J. Cserenyi, A. Szoke, T. Bansagi and A. Dszko, *J. Catal.* 1997, **165**, 150.
3 D. Wang, J. H. Lunsford and M. P. Rosynek, *Topics in Catal.* 1996, **3**, 289.
4 C. Bouchy, I. Schmidt, J. R. Anderson, C. J. H. Jacobsen, E. G. Derouane and S. D. Derouane Abd. Hamid, *J. Mol. Catal. A* 2000, **163**, 203.
5 S. Liu, Q. Dong, R. Ohnishi and M. Ichikawa, *Chem. Commun.*,1998 1217.
6 Y. Shu, R. Ohnishi and M. Ichikawa, *J. Catal.* 2002, **206**, 134.
7 L. Wang, R. Ohnishi and M. Ichikawa, *J. Catal.* 2000, **190**, 276.
8 S. Liu, Q. Dong, R. Ohnishi and M. Ichikawa, *Chem. Commun.*, 1997, 1455.
9 Y. Lu, Z. Xu, Z. Tian, T. Zhang and L. Lin, *Catal. Lett.* 1999, **62**, 215.
10 J. R. Anderson, *PhD Thesis*, 2001, University of Liverpool, Liverpool,UK.
11 W. Ding, S. Li, G. D. Meitzner and E. Iglesia, *J. Phys. Chem. B* 2001, **105**, 506.
12 B. M. Weckhuysen, M. P. Rosynek, J. H. Lunsford, *Catal. Lett.* 1998, **52**, 31.
13 H. Jiang, L. Wang, W. Ciu and Y. Xu, *Catal. Lett.* 1999, **57**, 95.
14 S. J. Thomson and G. Webb, *J. Chem. Soc., Chem. Commun.*, 1976, 526.
15 L. E. Cadus, O. F. Gorriz and J. B. Rivarola, *Ind. Eng. Chem. Res.* 1990, **29**, 7, 1143.

CATALYTIC PROPERTIES OF DAWSON-TYPE HETEROPOLYACIDS FOR ALCOHOL DEHYDRATION AND ALKENE ISOMERISATION.

Federica Donati and Paul McMorn.

Cardiff University, Chemistry Department, Main Building, Park Place, CF10 3TB, Cardiff, UK. E-mail: McMornP1@cf.ac.uk.

1 INTRODUCTION

Heteropolyanions (HPA) may be represented by the general formula $[X_xM_mO_y]^{q-}$ ($6 \leq m/x \leq 12$) where M is usually Mo or W, or less frequently V, Nb, Ta, in their higher oxidation state. There is no restriction on heteroatom X, thus resulting in an incredibly high number of possible complexes[1].

HPA have become interesting materials for catalysis because of their tunable redox and acid-base properties and their stability in the solid state. In fact it is not only possible to change the heteroatom and the framework metal atom, drastically affecting the chemistry of the compound, but it is also possible to exchange the counter cations, thus modifying the overall acid site strength of the catalyst and ultimately the activity and selectivity of the catalyst towards a particular reaction.

Keggin-type structures $[XM_{12}O_{40}]^{q-}$ have been widely studied in homogeneous and heterogeneous catalysed reactions[2,3] and significant attention has been paid to elucidating the catalytic behaviour and "pseudo-liquid" active state. The use of Dawson-type $[X_2M_{18}O_{62}]^{q-}$ HPA has been limited to liquid-phase applications[4] and few gas-state reactions such as isobutane dehydrogenation[5] and methyl-*tert*-butyl ether (MTBE) synthesis from methanol and isobutene[6]. Valuable information on the solid state properties of the Dawson-type heteropolyacid $H_6P_2W_{18}O_{62}$, detailing its acidity-activity correlation using probe reactions, such as alcohols dehydration and butenes isomerization, has been obtained and compared to the well-known Keggin-type heteropolyacid $H_3PW_{12}O_{40}$. This study is part of a larger project aimed at the design of a catalyst that can be used for the first one–step gas-phase synthesis of MTBE, using *t*-butanol or 1-butene as starting reagents.

2 EXPERIMENTAL

2.1 Catalysts preparation

$H_6P_2W_{18}O_{62} \cdot nH_2O$ (Dawson-type heteropolyacid) was synthesized according to the literature[7]. The structure of the heteropolyacid was confirmed by ^{31}P NMR and IR spectroscopy[8]. $H_3PW_{12}O_{40}$ was purchased from Aldrich. Silica supported HPA were prepared according to the incipient wetness impregnation method. Different amounts of the acid, corresponding to loadings of 2, 5, 10, 20, 50 and 75% (wt.), were dissolved in distilled water, added dropwise to the silica (fumed silica CAB-O-Sil M5, specific area $\cong$ 200m^2/g) and oven dried at 80°C for 24hrs.

2.2 Reactor studies

Catalytic reactions were performed in a fixed-bed quartz tube reactor (d= 0.12mm). For the alcohol dehydration reaction the amount of catalyst was kept constant (0.5g), while for butene isomerization the volume of catalyst was kept constant (0.3ml). Reaction temperature ranged from 50 to 200 °C. The total flow to the reactor varied according to the nature of the experiment involved. For alcohol dehydration the total flow was 50ml/min. Helium w as flowed t o a s aturator k ept a t constant t emperature i n o rder to o btain a 9% concentration of 2-butanol and *t*-butanol, and 4.5% concentration of 1-butanol. For 1-butene and isobutene experiments, the total flow was 10ml/min and the concentration around 20% for the alkene. For the isobutene reaction the micro-reactor was modified in order to collect the liquid products at the outlet of the reactor (ice trap at T = 0°C), while the eventual gaseous products and the reagent were analysed on line by GC. All the catalysts were pretreated in situ at T=150°C under He (40ml/min) for 1 h. The data were collected after 2.5hrs of reaction at each temperature. Product analysis was performed by on line PE 8700 GC equipped with FID detector. Alltech Silanized Glass column (2m x 2mm ID, 0.19% Picric Acid on 80/100 Carbograph) was used for 1-butene and isobutene isomerization reactions. Chrompack DB5 was used for the dehydration of butanols.

3 RESULTS AND DISCUSSION

3.1 *t*-Butanol, 2-Butanol, 1-Butanol dehydration

Alcohol dehydration generally occurs by heating the alcohol in the presence of a strongly acidic compound (H_2SO_4, $KHSO_4$ or H_3PO_4 in the liquid phase, SO_4^{2-}/ZrO_2 or SiO_2/Al_2O_3 in gas-phase reactions). Product selectivity is closely linked to the nature of the acid sites: butyl-ether and 1-butene formation requires weak acidity while isomerization of 1-butene to 2-butene (*cis* and *trans* isomers) is favored by strong Brønsted acidity. Keggin-type and Dawson-type heteropolyanion salts have been proved to be inactive for acid-base catalyzed reactions, so demonstrating that only Brønsted acidity plays a fundamental role for the catalyst activity. Acid-base catalyzed reactions have been investigated mainly with Keggin-type heteropolyacids, whose acidity is stronger than that of Dawson-type (acid strength order determined by NH_3 heat absorption: $H_3PW_{12}O_{40}$ > $H_4SiW_{12}O_{40}$ > $H_6P_2W_{18}O_{62}$)[9].

In our studies, the catalytic activity of Dawson-type $H_6P_2W_{18}O_{62}$ for the dehydration of 1-, 2- and *tert*-butanol has proved to be greater than the catalytic activity of Keggin-type $H_3PW_{12}O_{40}$ (Fig.1). Characterisation of the acid sites, using standard techniques like ammonia and pyridine Temperature Programmed Desorption[8], have indicated Dawson-type heteropolyacids possess lower acidity that their Keggin counterparts, factors other than acidity must be considered. In particular, it was found that lower activation temperatures a re r equired f or t he D awson-type heteropolyacid i n o rder to d ehydrate t he alcohols when compared to the Keggin. Substantial difference in the butenes yield between Dawson and Keggin-type heteropolacids was observed at low temperatures for *t*-butanol and 2-butanol, while they were similar for the dehydration of 1-butanol. This suggests that there is a strong dependence of the catalytic performance on the nature of the carbocation, as intermediate of reaction.

The type of by-products formed depends on the alcohol and the catalyst. 1-Butanol and 2-butanol dehydration over all the catalysts produces 1-butene, 2-butene, isobutene, and dibutylether. The ether formation is favoured at the lowest temperatures, thus the selectivity towards butenes decreases with decreasing temperature. An exception is found in the dehydration of *t*-butanol, as selectivity is higher at higher temperatures. This is due to the great numbers of by-products formed during the first stages of the reaction at temperatures as low as 55°C , this being especially true for the Dawson-type acid.

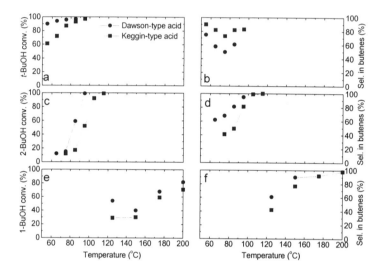

Figure 1   *a,c,e: conversion of t-butanol, 2-butanol and 1-butanol respectively.*
b,d,f: selectivity in butenes for the dehydration of t-butanol, 2-butanol and 1-butanol respectively. t-Butanol, 2-butanol 9%, 1-butanol 4.5% in Helium, flow rate 50ml min⁻¹, 0.5g catalyst.

3.2 The effect of supporting Dawson-type and Keggin-type HPA on SiO₂ in the isomerisation of 1-butene.

Supporting HPA is convenient when looking at surface-type reactions as the bulk catalyst has a very low surface area and thus most of the protons are not accessible for the reaction. The choice of the support material must take into account the possible interaction and structural rearrangement of the heteropolyacid on the support surface. Keggin-type heteropolyacids supported on SiO_2 or C has been proved to retain the structure, although ^{31}P MAS NMR studies suggested that the silanol groups lower the acid strength of the catalyst, resulting in deprotonation of the Keggin unit[10].

Surface area measurments and XRD patterns indicate that the dispersion of the catalyst over the silica is efficient only at low loading (5-10wt.%), while formation of catalyst bulk is possible at higher loadings. IR spectra show that the primary structure of the heteropolyacids is retained after impregnation. TGA data indicates that release of water from t he s upported K eggin c atalysts t akes p lace a t t emperatures l ower t han t hat f or t he pure acid: this suggests that the interaction between the acid and the silica is weaking the capacity of the catalyst to form strong hydrogen bonds.

1-Butene isomerisation over the pure Dawson and pure Keggin-type heteropolyacids has shown that the Keggin-type is more active and selective towards *trans* 2-butene at low temperatures (Fig.2). Traces of isobutene were detected at the highest temperatures, for the pure Keggin acid but not at all for the pure Dawson acid. Some authors have associated higher selectivities towards the *trans* isomer to be due to higher a cid strength[11]. On the basis of conversion and selectivity, it is possible to conclude that the pure Keggin acid has the higher acid strength, as expected from the heat of adsorption of ammonia. Measurements of the catalytic activity over time (not shown) have indicated that the pure Keggin-type heteropolyacids undergoes abrupt changes in conversion and selectivity, which may be due to rearrangements in the secondary structure by release of water or by proton migration from the catalyst bulk to the surface. The Dawson-type acid instead did not show any sign of bulk activity. By increasing the reaction temperature the difference in activity between the two type of heteropolyacids decreases, as also confirmed by studies of catalytic activity over time (not shown). The difference in activity also decreases by supporting the two heteropolyacids on fumed silica.

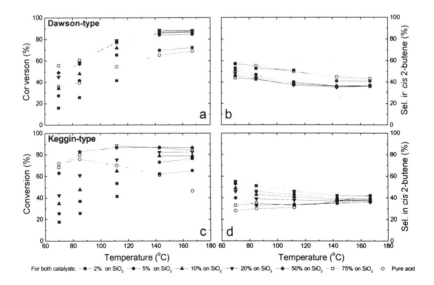

Figure 2   *a,b: 1-butene conversion and selectivity towards cis 2-butene for the Dawson-type heteropolyacid. c,d: 1-butene conversion and selectivity towards cis 2-butene for the Keggin-type heteropolyacid. Total flow 10ml min^{-1}, 20% 1-butene in Helium, 0.3ml catalyst.*

By supporting the Keggin-type heteropolyacid on silica (Fig.2) the conversion and selectivity towards the *trans* isomer decreases to lower values, being minimum for the 2wt.% supported sample. By supporting the Dawson-type heteropolyacid on silica (Fig.2), the conversion values increase for the samples with loading $\geq$ 20wt.%. Traces of C8 products and isobutene were also found for the highest loading (75wt.% and 50wt.%); selectivity towards the *trans* isomer continuously increases with increasing loadings.

In c onclusion, b y supporting t he K eggin-type h eteropolyacid o n s ilica t he r esulting catalysts appear to be more stable against deactivation, this resulting in higher conversion values at higher temperatures. Neverthless, the acid strength of the catalysts seems to decrease by impregnation of the Keggin acid on silica. Supporting the Dawson-type heteropolyacid on silica is convenient not only because the catalysts are stable against deactivation, but also because the acidity of the catalyst seems to increase.

3.3 Isobutene polymerisation

Isobutene isomerisation over Keggin and Dawson-type heteropolyacids has been investigated in order to assess whether isobutene, which is the desired reacion product from 1-butene isomerisation, is stable under the reaction conditions used.

As shown in Fig.3, all the catalysts are extremely active for the polymerisation of isobutene. Activity drops drastically for 2wt.% Keggin on SiO_2, 2wt.% Dawson on SiO_2 and pure Keggin acid after only 10 min of reaction. The pure Dawson acid, although very active, deactivates much more slowly. When increasing the temperature the difference in activity between the supported and unsupported catalysts, and between the Dawson and the Keggin acids reduces.

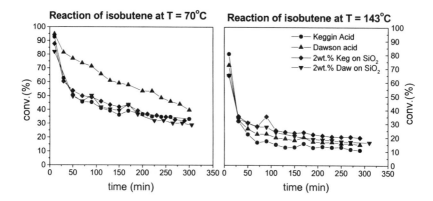

Figure 3   *Isobutene conversion over 2wt.% Keggin on SiO_2, 2wt.% Dawson on SiO_2, pure Keggin acid and pure Dawson acid at 70°C and at 143°C. Total flow 10ml min⁻¹, 20% isobutene in helium, 0.3ml catalyst.*

4 CONCLUSION

The behaviour of Dawson and Keggin-type heteropolycids has been studied over surface-type and bulk-type reactions. Both heteropolyacids have shown similar activity for all the reactions investigated, with the biggest differences found for *t*-BuOH dehydration and isobutene polymerisation. It is possible for *t*-BuOH dehydration that the pseudo-liquid phase is responsible for the higher peformance of the Dawson, as suggested by Misono for MTBE synthesis. However this does not explain why the Dawson, which is the most selective catalyst at low temperature for 1-BuOH and 2-BuOH dehydration, suddenly becomes the less selective for *t*-BuOH dehydration.

When assessing the acid strength of the Keggin and the Dawson heteropolyacids on surface-type reactions, the Keggin is the stronger acid at the lowest temperatures. By increasing the reaction temperature the activity of the two heteropolyacids is comparable and a contribution of the catalyst bulk to the reaction over the Keggin at low temperatures has been suggested.

Finally, the effect of supporting the heteropolyacids on silica depends the type of acid being supported. By supporting the Keggin-type the stability of the catalyst against deactivation is higher, although the acidity might be lower, while by supporting the Dawson-type acid on silica both stability and conversion are higher at higher loadings.

Acknowledgements

We would like to thank the EPSRC for funding.

References

1   M.T Pope, *Heteropoly and Isopoly Oxometalates*, Springer-Verlag, Berlin, 1983.
2   M. Misono, Catal. Rev. Sci. Eng. 1987, **29**, 269.
3   Y. Ono, in *Perspectives on Catalysis,* ed. J.M. Thomas and K.I. Zamaraev, Blackwell Sci., London, 1992, p.341
4   C.L. Hill, *Coord. Chem. Rev.*, 1995, **143**, 407.
5   C. Comuzzi, A. Primavera, A. Trovarelli, G. Bini, F. Cavani, *Catalysis Letters*, 1996, **36**, 75.
6   S. Shikata, T. Okuhara, M. Misono, *J. of Mol. Cat.*, 1995, **100**, 49.
7   H.Wu, *J. Biol. Chem.,* 1920, **43**, 189.
8   T. Okuhara, N. Mizuno, M. Misono, *Adv. Catal.*, 1996, **41**, 113.
9   M. Misono, *J.Chem.Soc. Chem. Comm.*, 2001, **13**, 1141.
10 I.V. Kozhevnikov, *Chem. Rev.*, 1998, **98**, 171.
11 T. Matsuda, M. Sato, T. Kanno, H. Miura, K. Sgiyama, *J. Chem. Soc. Faraday Trans 1*, 1981, **77**, 3107.

CATALYTIC AIR OXIDATION OF TOLUENE IN SUPERCRITICAL CO$_2$ USING SOLID SUPPORTED SURFACTANTS CONTAINING Co(II) SPECIES

Jie Zhu, Alan Robertson and Shik Chi Tsang*

Surface and Catalysis Research Centre, Department of Chemistry, The University of Reading, Whiteknights, Reading, RG6 6AD, UK. E-mail: s.c.e.tsang@reading.ac.uk

1 INTRODUCTION

Partial oxidation of alkylaromatic hydrocarbons with air or molecular oxygen is of a great importance in industry.[1] Cobalt salts are widely used as homogeneous catalysts in the liquid phase oxidation. Reactions are usually carried out in acetic acid in the presence of a promoter(s) (Mn ions and/or bromide).[1-3] The acetic acid offers as a unique solvent to bring together the hydrophobic substrates, sufficient dissolved oxygen and the ionic catalyst/promoter species for reaction but is itself kinetically stable against oxidation. Much effort has been devoted to replace acetic acid with environmentally more benign solvents,[4] but has met with little success. Heterogenisation of the soluble Co catalysts is another subject of considerable interest since the loss of solvents and catalysts during separation leads to unacceptable levels of waste. One common approach of catalyst immobilization is the establishment of a covalent linkage(s) between the catalyst species and the surface functional groups on a support. It is unfortunate that during oxidation, polar products and oxidants could favourably chelate the metallic catalyst species resulting in severe leaching of metal components.[5] The second approach is to employ a biphasic catalysis involving two immiscible liquid phases. Catalyst species will retain in a one phase and the substrates/products in the other. Catalytic reaction takes place at the interface of the two immiscible liquids. Rates of such reactions are often limited by a slow mass transfer of species across the interface.[6]

Supercritical carbon dioxide (scCO$_2$) has many advantages as a solvent medium for catalytic oxidations including non-toxicity, chemical stability, enhanced diffusivity (mass transfer) and pressure-dependent solvation properties that facilitate simple separation. In addition, it is the least expensive solvent after water. ScCO$_2$ is recently receiving considerable and even increasing interest as a versatile new medium for catalytic oxidation reactions.[7,8] Despite the fact that a higher activity and selectivity are commonly reported in these processes than in the conventional solvents, the inability of scCO$_2$ to disperse ionic/polar species limits its wider applications. However, there have been significant advances toward the formation of aqueous emulsions in scCO$_2$ by a variety of different fluorinated surfactant molecules, which can carry water-soluble species into supercritical CO$_2$ phase[9]. We have recently demonstrated an excellent way of carrying out biphasic catalysis in scCO$_2$ without facing the slow interfacial transfer problems. That is the use of free-form fluorinated carboxylic surfactant stabilised *micelles* to bring the ionic catalyst species into contact with the alkylaromatic hydrocarbons in scCO$_2$. As a result, air oxidation of toluene can occur in water-scCO$_2$ mixture without the use of acetic acid[8]

To further advance the alkylaromatic oxidations to more greener practises here we report our initial attempt on the development of new *solid* catalysts containing Co(II) species for catalysis in supercritical CO$_2$. It is noted that the slow mass transfers in biphasic catalysis reactions involving two immiscible liquid phases is facilitated if one phase is carried by a modified high surface area porous solid [10-12]. Thus, this new work involves covalent linkage of fluorinated surfactant molecules with the silanol surface groups on ultra-high-pore-volume (UHPV) silica. The porous composite solid is then allowed to carry tiny water droplets that in turn carry the ionic catalyst species. It is believed that the fluorinated anchored surfactant molecules could form similar micellar aggregates on the inner surfaces of silica (fluorinated moieties strongly interact with scCO$_2$ and the carboxylic groups with the Co(II)/NaBr). Our preliminary results clearly suggest that the solid supported assemblies improve the rate of interface reactions between the hydrophilic catalyst (Co(II) acetate and NaBr) and the hydrophobic substrate (toluene).

2 EXPERIMENTAL

2.1 Preparation of silica tethered fluorinated species

UHPV silica (Aldrich, 150Å) was pre-treated with 2M HNO$_3$, rinsed and dried prior to use. 1.723 g resulting silanol surface groups activated silica was allowed to react with 1.8 mmol aminopropyltriethoxysilane (APTES, Aldrich, 99%) in 25 ml pre-dried toluene at 120°C for 12 hours under nitrogen atmosphere. After reaction, the solid was collected by filtration and exhaustively wash with toluene and acetone to remove excess of APTES. A light yellow powder (**A**) was obtained after dried at 80°C under high vacuum (Scheme 1).
1.7 mmol hexadecafluorosebacic acid (FLUOROCHEM, 96%) was allowed to react with 1.7 mmol thionylchloride (Aldrich, 99+%) in pre-dried diethyl ether at 40°C for 12 hours. Thereafter, 1.7 g powder (**A**) was added into the mixture and the resulting mixture was kept refluxing for another 12 hours. Then, a mixture of 1.7 mmol pyridine and 5ml diethyl ether was added drop-wisely into the mixture (remove the HCl formed) and kept refluxing for 12 hours. The solid was filtered and exhaustively washed with diethyl ether. A yellow powder (**B**) was obtained after dried at 80°C.
1.503 g yellow powder (**B**) was allowed to react with 2.0 mmol cobalt(II) acetate tetrahydrate in 25 ml methanol at 60°C for 3 hours, followed by addition of a mixture of 1.5 mmol pyridine and 10 ml methanol dropwise. The mixture was allowed to reflux for another 12 hours before it was recovered by filtration and followed by washing with methanol and diethyl ether. A offwhite powder (**C**) was obtained after it was dried at room temperature. When 1.516 g yellow powder (**B**) was allowed to react with 5ml of 2M ammonia/ methanol solution in 25 ml methanol for 4 hours. A yellow powder (**D**) was obtained after solvent was removed at 40°C.

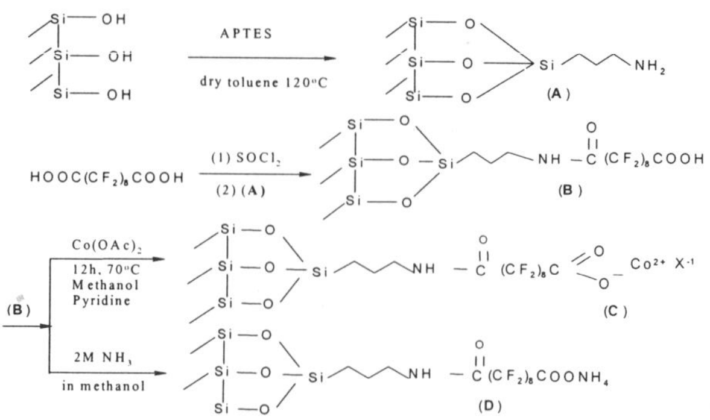

Scheme 1   *Synthesis of immobilised fluorinated species*

2.2 Catalytic tests

All the catalytic tests were carried out in a valid volume of 111 ml Parr batched autoclave with a Teflon cup insertion. All the components were added into the autoclave directly, followed by adding a trace quantity of internal standard (methane). 10 bar dioxygen was charged from a cylinder directly followed by charging carbon dioxide using a booster (Haskel) to the desired pressure. The reaction was stirred at the set temperature for the desired reaction time. Stirring was achieved by means of an overhead magnetic stirrer; the motor was set at ½ maximum speed, giving approximately 360 rpm. Quantities of toluene and the internal standard were monitored with an online GC. High boiling products were collected by releasing the supercritical fluid into a liquid trap containing 50 ml CH_3CN cooled by dry ice. The products remained in autoclave were extracted by another 50 ml CH_3CN. The two solutions were combined and analysed quantitatively using HPLC equipped with a UV-visible detector.

3 RESULTS AND DISCUSSIONS

3.1 Characterization of silica immobilised fluorinated acid (surfactant)

The aim of this study was to investigate the possibility of heterogenisation of similar micellar species containing cobalt (II) catalyst on silica surface.[8] The methodology of APTES immobilisation onto silica surface was achieved according to a modified method reported in our previous work.[12] FTIR spectra (Figure 1a and 1b) of silica before and after modification (powder **A**) clearly shows that the modified silica displays a much weaker O-H stretch absorption peak at approximately 3448 cm^{-1} than the unmodified silica. Undoubtedly, the difference on adsorption can be attributed to formation of Si-O-Si bonds and elimination of EtOH from the surface of silica (this eliminates most of the O-H groups). After the attachment of fluorinated acid (powder **B**), FTIR spectra (Figure 1c) shows a new adsorption peak at approximately 1685.5 cm^{-1}, which can be attributed to the C=O stretch adsorption from fluorinated acid. Separate experiments show that no adsorption change was detected from FTIR spectra before and after silica was immersed initially with APTES and fluorinated acid mixture at room temperature, then filtered and Soxhlet washed under identical treatments. These results suggested the fluorinated acid had been chemically tethered onto the surface of silica in the case of powder **B** rather than physically retained inside the porous structure.

Figure 1   *FTIR spectra:* unmodified silica (a); APTES-modified silica of powder **A** (b); fluorinated surfactant acid-modified silica of powder **B** (c).

A thermogravimetry study (TG) was performed on samples of the silica materials at various preparative stages. TG curve (a ramp rate of 10°C min⁻¹ in air) of dried pure silica in Fig. 2a indicated weight loss (~0.9%) at approximately 800°C for pure silica was partly attributed to the removal of a small amount of physisorbed water (at ~ 100°C) and to surface silanol dehydroxylation reaction at elevated temperatures. In the case of APTES modified silica (Fig. 2b), the weight loss (~2%) below 100°C was attributed to removal of the physisorbed water. Notice that the significant weight loss (~4.8%) between 300-700°C was assigned to the combustion of organic moieties attached upon the silica. This value corresponded to 0.8 mmol. -CH₂CH₂CH₂NH₂ per gram silica (nearly a monolayer coverage for this high surface area silica). Figure 2c shows that, after immobilisation of the fluorinated acid, there was in total 10% weight loss after the removal of physisorbed water. The higher weight loss (compare curves b and c in Fig. 2) was attributed to the combustion of new immobilised organic species after the fluorinated acid attachment. This value corresponded to the surface coverage of the fluorinated carboxylic acid of approximately 0.1mmol/g silica (*ca*.8 times lower than the stoichiometric value). We believe that the bulk fluorinated molecules may have encountered some accessibility problems, perhaps to the inner internal surfaces deep inside the small pores. This accounts for the low attachment value of the fluorinated molecules.

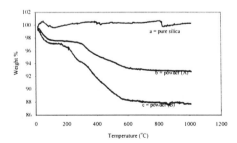

Figure 2   *TG curves:* UHPV silica (a); APTES-modified silica of powder **A** (b); fluorinated surfactant acid-modified silica of powder **B** (c).

3.2 Catalytic oxidation of toluene

3.2.1 New Solids immobilised cobalt (II) Loading cobalt species to the surface tethered acid groups was achieved according to a similar method reported by Clark *et al* .[13] While the precise structure of the surface-bound cobalt complex is yet unknown the elemental analysis (atomic absorption) shows a cobalt loading of 0.08 mmol g^{-1}, a similar value as the surface attached fluorinated acid groups (80% loading). The deviation from stoichiometric value may suggest that two surface carboxylic groups compete for a single Co(II) site. The supported cobalt reagent was tested as a catalyst in aerial oxidation of toluene in the presence of sodium bromide and water in scCO$_2$. It is noted that our previous study showed that cobalt acetate/NaBr in scCO$_2$-water mixture gave almost no conversion after a prolonged period of time indicating the poor interface does not allow efficient catalysis if two bulk solvent systems (H$_2$O and scCO$_2$) were directly used[8]. Here, it is shown from Figure 3a that high conversion (98.6%) and high selectivity (98.9% to benzoic acid) was achieved with an estimated average turnover frequency of 1.5 x 10^{-3} s^{-1} over the new silica catalyst containing the Co/NaBr species (powder **C** + NaBr + H$_2$O). Comparing this result with the inactivity and poor selectivity of other high surface solid cobalt catalyst (i.e. Co/Al$_2$O$_3$) [14] this is an enormous enhancement in the catalytic activity and selectivity. Two points deserve our attentions, namely the good interface between Co species with the substrate and the ease of converting Co(II) to and from Co(III) in the catalysts should be ensured for efficient catalysis. It is believed that Co species firmly bound within solid lattices in other Co catalysts show far less activity than our weakly carboxylic groups bound Co. Although the reaction proceeds smoothly by using fresh catalyst, our attempts to reuse the catalyst encountered some problems. An obvious attenuation in catalytic activity was observed (Figure 3b) when the supported cobalt catalyst was recovered by filtration and washing with acetonitrile/water to remove the product(s). After this post-treatment 20% of Co species (AA) was indeed lost, which could account for the loss of activity. It is interesting to observe that the deactivated catalyst could partly re-gain its initial activity by addition of 0.1mmol cobalt acetate (Figure 3c).

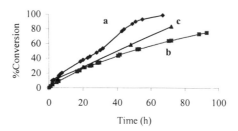

Figure 3   *Catalytic oxidation of toluene, testing conditions: 0.5g immobilised fluorous cobalt (powder C), 0.2 mmol NaBr, 14.1 mmol toluene, 500 µl H$_2$O, trace methane as internal standard, 10 bar O$_2$ and 150 bar CO$_2$ were blended at 120°C:* performance of the fresh catalyst (a); a repeated test of recovered catalysts (b); performance of a recovered catalyst from b with addition of 0.1mmol cobalt acetate (c).

Similarly, we show that using silica immobilised fluorinated ammonium surfactant species, powder **D** (more simple to prepare) can directly carry aqueous cobalt acetate by ion pair interactions (Scheme 1). The rate of reaction was slightly slower than that (1.5 x

10^{-3} s^{-1}) of powder **C**. Again the preliminary results show that the catalyst system can be fully recovered to its initial activity when the used catalyst was washed and re-loaded with the same amount of cobalt acetate, NaBr and water under identical conditions. Higher alkylaromatic molecules are rather soluble in scCO₂ as well as the corresponding acids[16]. It is envisaged that such new solid catalyst developments for air-oxidation of alkylaromatic molecules in supercritical CO₂ could lead to a new continuous process which facilitates products separation and catalysts regeneration in a more greener manner.

4 CONCLUSIONS

This work has demonstrated for the first time that chemically tethered fluorinated surfactant molecules on the surface of silica can act as a new carrier for water-soluble catalytic species in scCO₂, which facilities excellent mass transfers for biphasic catalysis. Thus, a cleaner process for the oxidation of alkylaromatic molecules without the use of corrosive acetic acid is reported.

Acknowledgements

SCT acknowledged the initial support of this work from the EPSRC ROPA schemes and to the University of Reading for the studentship to JZ.

References

1   A. K. Suresh, M. M. Sharma and T. Sridhar, *Ind. Eng. Chem. Res.*, 2000, **39**, 3958.
2   M. P. Czytko and G. K. Bub, *Ind. Eng. Chem. Prod. Rev. Dev.*, 1981, **20**, 481.
3   W. Partenheimer, Chem. Ind., 1990, **40**, 321.
4   M. Hronec, M. Haruštiak and J. Ilavský, *React. Kinet. Catal. Lett.*, 1985, **27**, 231.
5   A. J. Butterworth, J. H. Clark, P. H. Walton, S. J. Barlow, *Chem. Commun.*, 1996, 1859.
6   R. V. Chaudhari, B. M. Bhanage, R. M. Deshpande, H. Delmas, *Nature*, 1995, **373**, 501.
7   A. M. Steele, J. Zhu and S. C. Tsang, *Catal. Lett.*, 2001, **73**, 9.
8   J. Zhu, A. Robertson and S. C. Tsang, *Chem. Commun.*, 2002, 18, 2044.
9   K. P. Johnston, K. L. Harrison, M. J. Clarke, S. M. Howdle, M. P. Heitz, F. V. Bright, C. Carlier and T. W. Randolph, *Science*, 1996, **271**, 624.
10   R. Neumann and M. Cohen, *Angew. Chem. Int. Ed. Engl.* 1997, 36, 1738.
11   J. DeSimone, M. Selva, and P. Tundo, *J. Org. Chem.* 2001, **66**, 4047.
12   S. C. Tsang, N. Zhang, L. Fellas and A. M. Steele, *Catal. Today*, 2000, **61**, 29.
13   B. K. Das and J. H. Clark, *Chem. Commun.* 2000, 7, 605.
14   K. M. Dooley and F. C. Knopf, *Ind. Eng. Chem. Res.*, 1987, **26**, 1910.
15   J. Zhu and S. C. Tsang, *Catal. Today*, in press.
16   K. D. Bartle, A. A. Clifford, S. A. Jafar, G. F. Shilstone, *J. Phys. Chem. Ref. Data*, 1991, **20**, 713.

SELECTIVE HYDROGENATION REACTIONS IN IONIC LIQUIDS

Kris Anderson,[a,b] Peter Goodrich,[a] Christopher Hardacre[a,c] and David W. Rooney[a,b]

[a] The QUILL centre, [b] School of Chemical Engineering, [c] School of Chemistry, The Queen's University of Belfast, Belfast BT9 5AG, Northern Ireland

1 INTRODUCTION

Since their humble beginnings as alternative battery electrolytes, ionic liquids have emerged as one of a small number of new and exciting solvents capable of meeting the needs of tomorrow's chemical industry. In less than thirty years they have evolved from these electrolytes into tuneable, environmentally friendly, solvent/catalyst systems. In general their power stems from their unique ability to dissolve and therefore bring together a wide range of molecules (inorganic, organic and catalytic) to perform otherwise difficult homogeneous reactions, although there are considerable engineering benefits as well. Of these, the lack of any significant vapour pressure, firstly eliminates any hazard associated with solvent volatilisation and secondly, allows for the direct recovery of low boiling products. In addition, it is recognised that these solvents can be combined with, or retrofitted to, traditional methods of production leading to improvements in selectivity, catalyst recycle and an overall reduction in the huge quantities of organic solvents used worldwide.

The specific and selectable properties of these solvents allow a diverse range of reactions to be carried out including alkylations[1], C-C bond coupling reactions[2], polymerisations[3] , hydrogenations[4], hydroformylation[5], and alkoxycarbonylations[6] These reactions have been extensively reviewed recently[7]. Of all the reactions studies only one, to our knowledge has been performed using a heterogeneously supported catalyst suspended in the ionic liquid. In this paper Hagiwara et al.[8] investigated the Heck reaction using a carbon supported palladium in [bmim][PF$_6$]. Here they presented results indicating both good recyclability and activity of the catalyst. In another study, Carlin et al[9] also immobilised palladium on carbon in an ionic liquid co-polymer membrane for gas phase (propene) hydrogenation. Both of these studies suggest that palladium catalysts are effective in ionic liquids.

In this paper we will report on the selective hydrogenation of α,β unsaturated aldehydes using a supported palladium catalyst in a range of ionic liquids. We will also compare their reactivity and selectivity to conventionally used organic solvents. Two systems have been investigated, cinnamaldehyde to form hydrocinnamaldehyde and citral to form citronellal. The two reaction schemes with the range of products expected are shown in scheme 1.

(a)

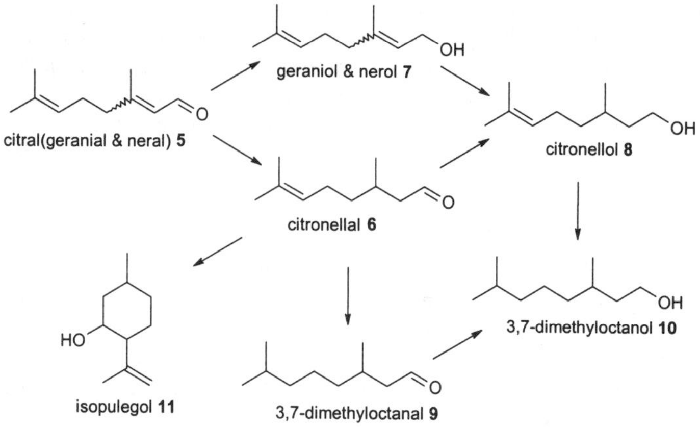

(b)

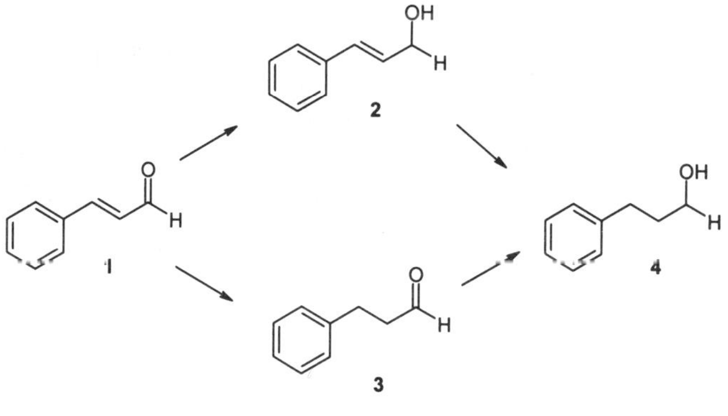

Scheme 1 *Schematic of (a) citral and (b) cinnamaldehyde reduction pathways*

2 EXPERIMENTAL

(Lancaster, 99%) and HPLC grade organic solvents were used as received. The ionic liquids tested were prepared in house following standard preparative procedures, the details may be found elsewhere.[10] In each case the ionic liquid was treated under vacuum at 60 °C in order to minimise water and were then washed extensively to ensure that halide contamination from the preparation process was < 10 ppm. 10% Pd on activated carbon was obtained from Aldrich and 5% Pt on graphite, pre-treated in flowing hydrogen at 350 °C for 1 hour, from Johnson Matthey. Research grade gases were supplied by BOC.

All reactions were carried out in a Baskerville mini autoclave under hydrogen pressure. In a typical reaction ionic liquid or organic solvent (2 cm^3), catalyst (5.5 mg) and substrate (4 mmol) were degassed in the autoclave, pressurised with hydrogen and heated to the desired temperature. Samples were analysed by GC-FID. Ionic liquid samples were extracted using diethyl ether and for the recycle tests, after extraction, the ionic liquid was charged with fresh substrate and hydrogenation repeated. In the case of the platinum catalysts, the IL/catalyst system was pre-treated at the reaction temperature under hydrogen for 30 mins prior to addition of the reactants and recycled.

3 RESULTS AND DISCUSSION

Table 1 compares the results from the selective hydrogenation of citral, at a variety of pressures and temperatures in both ionic liquids and organic solvents using palladium supported on carbon (Pd/C) and platinum supported on graphite (Pt/G). It is clear that the ionic liquid systems are highly selective solvents for the reduction of C=C without any reduction of the carbonyl group over palladium and similar selectivities can be achieved for the carbonyl group over platinum.

Table 1 *Percentage distribution with solvent type for citral(1), citronellal (2), geraniol+nerol (3), citronellol (4), 3,7-dimethyloctanal (5), 3,7 dimethyloctanol (6), isopulegol (7) and selectivity towards citronellal using Pd/C and towards geraniol+nerol using Pt/G at various temperatures and hydrogen pressures for 4 hours unless otherwise stated. (Abbreviations used- [C$_8$Py]: N-octylpyridinium, [emim]: 1-ethyl-3-methylimidazolium, [bmim]: 1-butyl-3-methylimidazolium, and [NTf$_2$]: bis(trifluoromethylsulfonyl)amide)*

Catalyst-Solvent[§]	Temp./°C	Press./bar	%1	%2	%3	%4	%5	%6	%7	%selectivity
Pd/C-[bmim][BF$_4$]	60	40	24	75	-	-	-	-	<1	99
	30[a]	10	48	52	-	-	-	-	-	100
	30[*]	10	80	19					<1	95
Pd/C-[C$_8$Py][BF$_4$]	60	40	44	56		-	-	-	-	100
	30	10	68	32	-	-	-	-	-	100
Pd/C-[emim][NTf$_2$]	60	40	-	81	-	17	-	<1	<1	81
	30	5	-	98	-	<1			<1	98
	30	5 (16h)	-	98	-	<1	-	-	<1	98
Pd/C-cyclohexane	30	3 (1h)	-	62	-	-	5	27	6	62
Pd/C-toluene	30	3 (1h)	31	53	3	<1	<1	5	6	77
Pd/C-dioxane	30	3 (1h)	44	43	<1	2	<1	4	5	77
Pt/G-[emim][NTf$_2$]	60[b]	40	18	1	81	-	-	-	-	99
	60[**]	40	64	2	34	-	-	-	-	94
Pt/G-toluene	30	3 (1h)	42	<1	28	29	-	-	-	48
Pt/G-cyclohexane	30	3 (1h)	46	<1	15	38	-	-	-	28
Pt/G-dioxane	30	3 (1h)	53	1	18	28	-	-	-	38

The ionic liquid results may be compared with those performed in conventional organic solvents such as toluene, dioxane and cyclohexane. At all conversions, the selectivity is found to be much higher in the ionic liquid than for the organic solvents. For example, the maximum selectivity found for citral to citronellal using Pd/C was 90 % in toluene but this was only achieved at 22 % conversion compared with 98% selectivity at 100% conversion in [emim][NTf$_2$]. Similarly, in the case of Pt/G, a maximum of 48% selectivity towards geraniol+neriol was found but again only at 58% conversion in toluene compared with 99% selectivity at 81 % conversion, again in [emim][NTf$_2$].

In both Pd/C- and Pt/G-ionic liquid systems, higher hydrogen pressures and temperatures increased the rate of hydrogenation, as expected, whereas, only changes in temperature lead to a significant change in the selectivity observed. In all ionic liquid systems studied, increases in temperature resulted in a decrease in selectivity. For example, above 60 °C, the selectivity towards citronellal in [bmim][BF₄] using Pd/C, decreases rapidly with the formation of citronellol. However, the selectivity-temperature variation for each ionic liquid was dependent on the anion used with the high selectivity found over a wider temperature window in [BF₄]⁻ based ionic liquids than in the equivalent [NTf₂]⁻ based systems. The temperature variation on selectivity and the invariance in the selectivity with pressure found in the ionic liquids is consistent with results reported previously in organic solvents.

Whilst the selectivity of the ionic liquid systems are generally much higher than with the organic solvents, the conversions tend to be lower. In Table 1, the results from the organic solvents were taken after one hour reaction whilst those in the ionic liquids were after four hours. This reduction in rate is most likely to be due to mass transfer effects. The viscosity of ionic liquids is much higher than found in commonly used organic solvents and it has been reported that the solubility of hydrogen in ionic liquids is small. Interfacial shear normally developed through mixing of the slurry is considerably reduced when using ionic liquids due the higher viscosity and lower relative density between the catalyst and the solvent. Overall, the mass transfer of the reactants to the surface of the catalyst will be reduced when compared with the organic solvents and this in turn will lead to a lower rate.

As well as increased selectivity, the other main advantage of the ionic liquid systems is the ability to recycle the reaction medium without loss of solvent. To facilitate the recycle, the reaction mixture was simply extracted with diethyl ether resulting in a mass recovery > 99 %. The extraction was performed with the catalyst *in-situ* and no catalyst was found to traverse into the extractant phase either as particulate matter or as the result of leaching of the metal from the support. As shown in Table 1, a drop in activity for the platinum catalyst, from a conversion of 82% to 36 % after 4 hours reaction, was observed on the first recycle when using [emim][NTf₂].

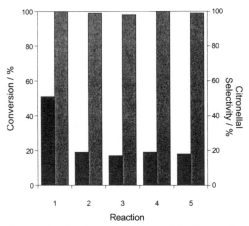

Figure 1 *Variation of citral conversion (black) and selectivity towards citronellal(grey) over Pd/C on recycle in [bmim][BF₄] at 30 °C and under 10 bar hydrogen*

Figure 2 shows a series of recycles using Pd/C in [bmim][BF₄]. Again a drop in activity is shown but this reduced activity is maintained on subsequent recycles and with extended reaction times complete conversion was achieved. It should be noted that the decrease in activity is not due to a dissolved platinum group metal species in the ionic liquid. Reactions performed on filtered ionic liquid showed no activity and ICP results indicated that the metal concentration was below the detection limit. During all recycles, the selectivity remained higher than in the organic solvents although a small decrease was observed. Comparison with direct vacuum distillation from the ionic liquid, again with the catalyst *in-situ* showed a similar decrease in the conversion and selectivity.

Figure 2 shows the conversion of cinnamaldehyde with respect to time and the selectivity towards hydrocinnamaldehyde in 1-butyl-3-methyl imidazolium tetrafluoroborate ([bmim][BF₄]) and 1-octyl pyridinium tetrafluoroborate ([C₈Py][BF₄]) ionic liquids at 40 bar and 60 °C as well as in 1-ethyl-3-methyl imidazolium bis(trifluoromethylsulfonyl)amide ([emim][NTf₂]) at 5 bar and 30 °C. The reactions in all the ionic liquids proceed slowly when compared with organic solvents. For example, at 5 bar and 30 °C in toluene, dioxane and cyclohexane, 100 % conversion of cinnamaldehyde is obtained in 45 minutes compared with 240 minutes in the best ionic liquid used, [emim][NTf₂]. However, the selectivity towards hydrocinnamaldehyde is constantly above 97%, even at 100 % conversion in these ionic liquids compared with a maximum of 85 % (toluene), 89 % (dioxane) and 79 % (cyclohexane) over a range of temperatures (20-120 °C) and pressures (3-43 bar).

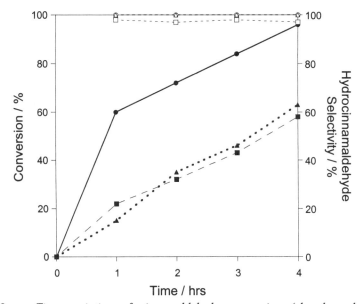

Figure 2   *Time variation of cinnamaldehyde conversion (closed symbols) and selectivity (open symbols) towards hydrocinnamaldehyde in [bmim][BF₄] (circles) and [C₈Py][BF₄] (squares) at 60 °C and 40 bar and in [emim][NTf₂] (triangles) at 30 °C and 5 bar*

4 CONCLUSIONS

In this study we have shown that ionic liquids are highly selective solvents for the C=C double bond hydrogenation of cinnamaldehyde and citral forming hydrocinnamaldehyde and citronellal, respectively. We have shown that by changing the anion associated with the ionic liquid it is possible to vary the selectivity and that selectivities close to 100 % are possible when using a heterogeneous 10% Pd/C catalyst. It has also been observed that the rate of reaction is slower than it would be in conventional organic media. However the ionic liquid system facilitates facile organic solvent extraction or vacuum distillation of the product phase without the requirement of removing the catalyst. Rapid deactivation of the catalyst on the first recycle is currently under investigation, however further deactivation on subsequent recycles appears to be negligible and therefore the system shows good recyclability.

References

1   C. E. Song, W. H. Shim, E. J. Roh, J. H. Chio, *Chem. Commun.*, 2000, 1695.
2   C. J. Mathews, P. J. Smith, T. Welton, *Chem. Commun.*, 2000, 1249
3   A. J. Carmichael, D. M. Haddleton, S. A. F. Bon, K. R. Seddon, *Chem. Commun.*, 2000, 1237
4   Y. Chauvin, L. Mussman, H. Olivier, *Angew. Chem. Int. Ed. Engl.*, 1995, **34**, 2698
5   P. W. N. M. vanLeeuwen, P. C. J. Kamer, J. N. H. Reek, P. Dierkes, *Chem. Rev*, 2000, 100
6   D. Zim, R. F. deSouza, J. Dupont, A. L. Monteiro, *Tetrahedron Lett.*, 1998, **39**, 7071
7   For example, T. Welton, *Chem. Rev.*, 1999, **99**, 2071, C.M. Gordon, Appl. Catal. A **222**, 101 (2002), P. Wasserscheid and W. Keim, *Angew. Chem. Int. Ed.*, 2000, **39**, 3772
8   H. Hagiwara, Y. Shimizu, T. Hoshi, T. Suzuki, M. Ando, K. Ohkuboc and C. Yokoyamac, Tet. Lett. 42 (2001) 4349
9   RT Carlin and J Fuller, J. Chem. Soc. Chem Comm. ,1997, 1345.
10   K.R. Seddon, A. Stark and M.J. Torres, Pure Appl. Chem., 2000, 72, 2275

ENANTIOSELECTIVE HYDROGENATION OF METHYL PYRUVATE IN THE GAS PHASE OVER CINCHONIDINE-MODIFIED PLATINUM

N.F. Dummer, R.P.K.Wells, S.H.Taylor, P.B.Wells and G.J. Hutchings

Department of Chemistry, Cardiff University, P.O. Box 912, Cardiff, CF10 3TB, UK

1 INTRODUCTION

Liquid-phase enantioselective hydrogenations of α-ketoesters by alkaloid-modified Pt has been investigated intensively and reviewed.[1-3] The first reported reactions of this type were the hydrogenations of methyl pyruvate, MeCOCOOMe, and of methyl benzoyl formate, PhCOCOOMe, over Pt modified by the cinchona alkaloids.[4]

The simplest cinchona alkaloids, cinchonidine and its 10,11-dihydro-derivative, have been shown by D-tracer studies and by NEXAFS and ATR-IR spectroscopy to adsorb by interaction of the quinoline moiety with the platinum surface.[5-7] Mechanistic studies have established that a site exists adjacent to the open-3 conformation of adsorbed cinchonidine at which pyruvate ester can undergo selective enantioface adsorption.[8] Hydrogenation of the preferred enantioface results in preferential formation of one enantiomer of the product, methyl lactate, MeC*H(OH)COOMe. Pt modified by cinchonidine provides R-lactate preferentially, whereas the near enantiomer cinchonine provides S-lactate in excess. Values of the enantiomeric excess of 75% can be obtained without optimisation, and of 98% under special conditions.[9] In solution, conditions that achieve enantioselectivity normally promote the reaction rate by a factor of 2 to 100 depending on conditions.[1,2]

Recently, we have obtained reproducible gas phase enantioselective reaction using 5% Pt/alumina and 5% Pt/silica.[10] In this paper, we extend this initial study and, in particular, show the results of the gas phase reaction of methyl pyruvate over a non-porous 1% Pt/α-alumina, a 2.5% Pt/silica, and the standard reference catalyst EUROPT-1 (6.3% Pt/silica) each modified by cinchonidine.

2 EXPERIMENTAL

2.1 Materials and procedures

Methyl pyruvate (Fluka), cinchonidine (CD) (Fluka) and dichloromethane (Fischer) were used as received. 2.5% Pt/silica (Johnson Matthey) and 6.3% Pt/silica (EUROPT-1) were re-reduced in flowing 5% H_2/Ar at 473 K for 2 h before use.

1% Pt/α-alumina was prepared as follows. γ-Alumina (Sasol Puralox SBA 200) of surface area 202 m^2 g^{-1} was heated in air at 1473 K for 5 h. XRD confirmed that the product was α-alumina, and the surface area was 5 m^2 g^{-1}. A sample of this α-alumina was immersed for 5 h in an aqueous solution of hydrogen hexachloroplatinate(IV) hydrate (the precursor having been dissolved in the minimum amount of distilled water). The impregnated material was filtered, dried by heating in air at 333 K for 3 h, and reduced in flowing 5% H$_2$/Ar at 473 K for 2 h.

Catalysts were pre-modified by immersion in solutions of cinchonidine in dichloromethane. 200 mg samples of catalyst were stirred for 10 min in 25 ml of 0.34 or 3.4mM solutions of alkaloid. Modified catalyst was filtered and dried under vacuum before use. 25 mg (Pt/silica) or 50 mg (Pt/α-alumina) samples were used in the reactor.

Hydrogenations were conducted in a flow microreactor (i.d. 3 mm) constructed in glass and located in a tube furnace. The catalyst bed was supported by glass wool. The reactor temperature of 313 K was monitored by a thermocouple located inside the microreactor. Catalyst samples were pre-conditioned in a He flow (80 ml min^{-1}) for 10 min followed by a 3:1 He:H$_2$ flow (80 ml min^{-1}) for a further 10 min. This He:H$_2$ mixture was then diverted through the saturator at 293 K to deliver methyl pyruvate vapour to the catalyst. The exit gas was analysed by chiral gas chromatography (25 m β-Chirasil-dex).

$$ee/\%(R) = 100([R\text{-}] - [S\text{-}])/([R\text{-}] + [S\text{-}])$$

3 RESULTS

1% Pt/α-alumina was active in the unmodified state for racemic hydrogenation of methyl pyruvate and in the cinchonidine-modified state for enantioselective hydrogenation of the ester. Conversions observed in reactions over unmodified and modified catalysts over the first hour of time on line are shown in Figure 1. Activity decayed with time in each case and modification by alkaloid reduced catalytic activity.

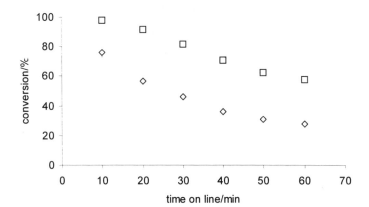

Figure 1   *Variation of conversion with time on line for methyl pyruvate hydrogenation at 313 K over Pt/α-alumina. Squares: unmodified catalyst; diamonds: catalyst pre-modified in 0.34mM cinchonidine*

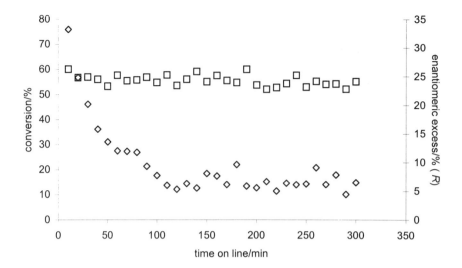

Figure 2   *Variation with time on line of conversion (diamonds) and of enantiomeric excess (squares) for methyl pyruvate hydrogenation at 313 K over 1% Pt/α-alumina pre-modified in 0.34mM cinchonidine*

Figure 2 shows that the activity of modified Pt/α-alumina fell to a constant value after 2 h and that the enantiomeric excess was constant at 25 ± 2% throughout a period of 5 h.

Pt/silica catalysts were used to determine the effects of modifier concentration and temperature on enantiomeric excess. 2.5% Pt/silica, pre-modified in 0.34mM cinchonidine and used at 313 K gave an initial enantiomeric excess of 40%(R) but this value decayed to zero over 3.5 h (Figure 3). By contrast, pre-modification in 3.4mM cinchonidine gave a catalyst which improved in performance over the first hour of use and thereafter gave ee = 34 ± 2%. Temperature was varied over the range 298 to 313 K; Figure 4 shows that 313 K provided optimum performance.

The enantiomeric excess provided by the standard reference catalyst EUROPT-1 (6.3% Pt/silica) operated under the same conditions as the 2.5% Pt/silica, was 41 ± 2% (Figure 3); the value was constant within these limits over 5 h on line.

4 DISCUSSION

4.1 Activity, and the effect on activity of adsorbed alkaloid

Alkaloid-free Pt/α-alumina showed high activity for pyruvate hydrogenation, 100% conversion being initially achieved (Figure 1). This is a notable result, because reaction in the liquid phase in a stirred batch reactor is normally slow, the rate reduces with time in a manner indicating a self-poisoned reaction, and 100% conversion is seldom achieved.

Liquid phase reaction gives higher molecular weight products.[13,14] In a gas phase reaction these will tend to accumulate at the Pt surface which may interpret the decay of activity with time in both the modified and unmodified reaction.

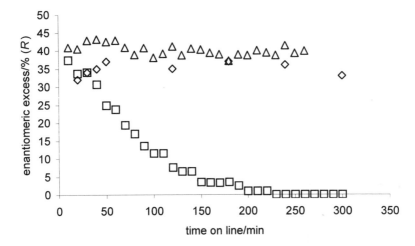

Figure 3 *Effect of pre-modification conditions on enantiomeric excess for Pt/silica catalysts at 313 K. Squares: 2.5% Pt/silica pre-modified in 0.34mM CD solution; diamonds: 2.5% Pt/silica pre-modified in 3.4mM CD solution; triangles: 6.3% Pt/silica (EUROPT-1) pre-modified in 3.4mM CD solution. Conversion = 100%*

Cinchonidine adsorption on Pt/α-alumina depressed the rate by a factor of 1.3 – 2.1. A similar reduction has been observed for the gas-phase reaction over Pt/alumina.[12] This contrasts with the situation in liquid-phase reactions where the action of cinchonidine in inducing enantioselectivity is accompanied by a substantial rate enhancement, the origin of which has been much debated.[1,15] The decoupling of rate and enantioselectivity has been observed in another context[13] but for different reasons. The present case merits further investigation.

4.2 Enantioselectivity

Pt/α-alumina modified in 0.34mM alkaloid was enantioselective, giving ee = 25%(R) independent of time on stream (Figure 2). This provides a clear demonstration that enantioselectivity can indeed be achieved in a gas-phase reaction; better performance may be obtained when the system has been optimised.

Pt supported at 2.5 wt % on porous silica, modified in 3.4mM alkaloid, gave a higher enantiomeric excess of 34%(R) (Figure 3). When modified by 0.34mM alkaloid, ee = 40%(R) was achieved initially, but enantioselectivity decayed to zero with time on stream. Remarkably, high activity (100% conversion) was retained throughout. This represents a second circumstance in which fast racemic reaction has been achieved.

Enantiomeric excess varied with temperature. At 50 minutes time-on-stream the values at 100% conversion were 31% at 298 K, 34% at 306 K, and 36% at 313 K. At 323 K, the enantiomeric excess was 32% for a reaction which did not attain 100% conversion. This temperature variation follows the behaviour of EUROPT-1 in the liquid phase[11] where a decrease in performance at above 318 K has been attributed to changes in the adsorbed state of cinchonidine[6,11] or to hydrogenation of the quinoline ring; either process would lead to less effective adsorption of the modifier and thereby to a reduction in the number of enantioselective sites.

Enantiomeric excess diminished with time on stream when the modification concentration was 0.34mM, but remained constant when the concentration was 3.4mM (Figure 3). It is known for Pt/silica that alkaloid adsorption can occur on both the metal and the support[12] and in our catalysts there may have been diffusion of alkaloid molecules between the support and the active phase during hydrogenations. In such a case, modification at a higher concentration would have provided more extensive adsorption on the support, and hence a larger reservoir of cinchonidine would have been available to sustain a steady state concentration on the Pt surface or to replace any modifier rendered inactive by partial or complete hydrogenation.

4.3 Performance of the standard catalyst EUROPT-1.

EUROPT-1 modified in 3.4mM alkaloid gave ee = 41%(R) which was retained over the 5 h test period (Figure 3). This catalyst provides ee = 70%(R) in dichloromethane solution at 30 bar, and 65%(R) in ethanolic solution at 10 – 110 bar.[15] That gas phase reaction may be less selective is to be expected from the absence of a rate enhancement. In the liquid phase, enantioselective reaction is so much faster than racemic reaction (factor ~ 40) that the contribution to the enantiomeric excess from reaction at unmodified sites is small by comparison with that at enantioselective sites. By contrast, in the gas phase reaction, the rate of reaction at the modified surface is less than that at the unmodified surface, so the effect of reaction at racemic sites at a modified surface makes a larger contribution to the overall reaction.

2 CONCLUSIONS

Gas phase enantioselective hydrogenation of methyl pyruvate over cinchonidine modified Pt has been achieved. Enantioselectivity was accompanied by a reduction in reaction rate. Modifier concentration during catalyst preparation determined catalyst performance.

References

1   (a) P.B. Wells and R.P.K. Wells, *Chiral Catalyst Immobilisation and Recycling*, eds., D.E. De Vos, I.F.J. Vankelecom, P.A. Jacobs, Wiley-VCH, Weinheim, 2000, Chapter 6, pp. 123-154. (b) A. Baiker, loc. cit., Chapter 7, pp. 155-171.
2   M. von Arx, T. Mallat and A. Baiker, *Topics in Catal.*, 2002, **19**, 75.
3   H.U. Blaser, *J. Chem. Soc. Chem. Commun.* 2003, p.293.
4   Y. Orito, S. Imai and S. Niwa, *Nippon Kagaku Kaishi*, 1979, p. 1118.
5   G. Bond and P.B. Wells, *J. Catal.*, 1994, **150**, 329-334.
6   T. Evans, A.P. Woodhead, A. Gutierrez-Sosa, G. Thornton, T.J. Hall, A.A. Davis, N.A. Young, P.B. Wells, R.J. Oldman, O. Plashkevych, O. Vahtras, H. Agren and V. Carravetta, *Surf. Sci.*, 1999, **436**, L691 – L696.
7   D. Feri, T. Burgi and A. Baiker, *J. Catal.*, 2002, **210**, 160.
8   K.E. Simons, P.A. Meheux, S.P. Griffiths, I.M. Sutherland, P. Johnston, P.B. Wells, A.F. Carley, M.K. Rajumon, M.W. Roberts and A. Ibbotson, *Recueil Trav. Chim. Pays-Bas*, 1994, **113**, 465-474.
9   B. Torok, K. Balazsik, G. Szollosi, K. Felfoldi and M. Bartok, *Chirality*, 1999, **11**, 470.

10 M. von Arx, N. F. Dummer, D.J. Willock, R.P.K. Wells, P.B. Wells and G.J. Hutchings, submitted for publication.
11 P.A. Meheux, A. Ibbotson and P.B. Wells, *J.Catal.*, 1991, **128**, 387.
12 I.M. Sutherland, A. Ibbotson, R.B. Moyes and P.B. Wells, *J. Catal.*, 1990, **125**, 77.
13 J.A. Slipszenko, S.P. Griffiths, P. Johnston, K.E. Simmons, W.A.H. Vermeer and P.B. Wells, *J. Catal.*, 1998, **179**, 267.
14 G.A. Attard, D.J. Jenkins, O.A. Hazzazi, P.B. Wells, J.E. Gillies, K.G. Griffin and P. Johnston, paper presented at this Symposium.
15 A. Vargas, T. Burgi and A. Baiker, *New J. Chem.*, 2002, **26**, 807.

ENANTIOSELECTIVE HYDROGENATION OF *N*-ACETYL DEHYDROPHENYL-ALANINE METHYL ESTER (NADPME) AND SOME RELATED COMPOUNDS OVER ALKALOID-MODIFIED PALLADIUM

N.J. Caulfield[1], P. McMorn[1], P.B. Wells[1], D. Compton[2] K. Soars[2] and G.J. Hutchings[1]

[1]Chemistry Department, Cardiff University, Cardiff, CF10 3TB, UK
[2]Robinson Brothers, Phoenix Street, West Bromwich, West Midlands, B70 0AH, UK

1 INTRODUCTION

Enantioselective hydrogenation of carbon-carbon double bonds using cinchona-modified Pd has received much attention.[1-5] The highest enantioselectivities have been achieved using reactants which possess an acidic moiety, such as α,β-unsaturated carboxylic acids (50 -72%)[1-5] and hydroxypyrone (~86%).[6] The enantioselectivities achieved have been attributed to the crucial role that the acidic function plays in interaction with the quinuclidine-N atom of the adsorbed modifier. This interaction is not available to the corresponding α,β-unsaturated esters and hence, under the same experimental conditions, no enantioselectivity is observed.[2] Here we report the first enantioselective hydrogenation of an unsaturated ester over a cinchona-modified heterogeneous metal catalyst; this has been achieved by further functionalisation of the ester so as to restore the possibility of a reactant-modifier interaction. The target reaction in this work has been the hydrogenation of *N*-acetyl dehydrophenylalanine methyl ester (**I**) to *N*-acetyl phenylalanine methyl ester (Figure 1). The target reaction is catalysed by Rh(DIPAMP)-complexes in solution[7] and by these complexes supported on oxides.

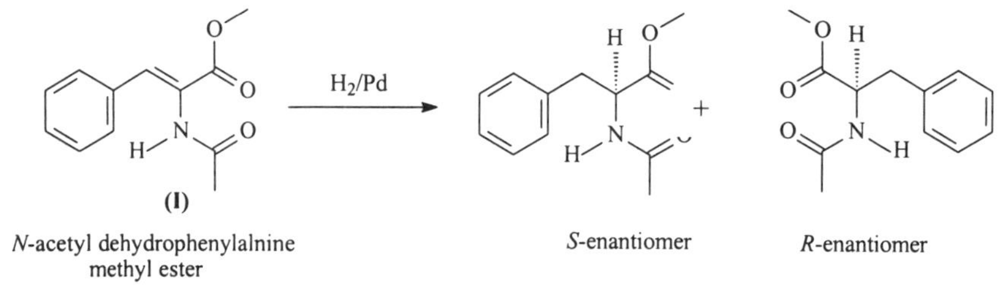

(I)
N-acetyl dehydrophenylalnine
methyl ester

S-enantiomer   *R*-enantiomer

Figure 1   *Hydrogenation of N-acetyl dehydrophenylalanine methyl ester (I)*

2 METHODS

2.1 Materials

N-acetyl dehydrophenylalanine and its methyl ester (**I**) were prepared using conventional methods. Cinchonidine, cinchonine, quinine and quinidine (Fluka) (Figure 2), methanol (Fisher Chemicals) and 5% Pd/alumina (Johnson Matthey) were used as received.

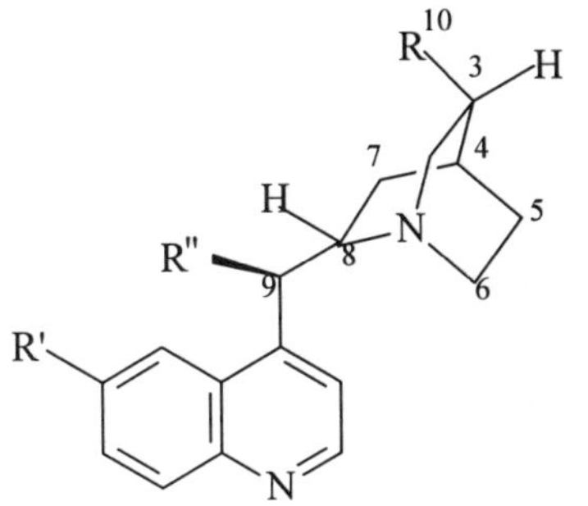

(**1**)	(**2**)
(**1**) C$_8$= *S*, C$_9$ = *R*	(**2**) C$_8$= *R*, C$_9$ = *S* **Substituents**

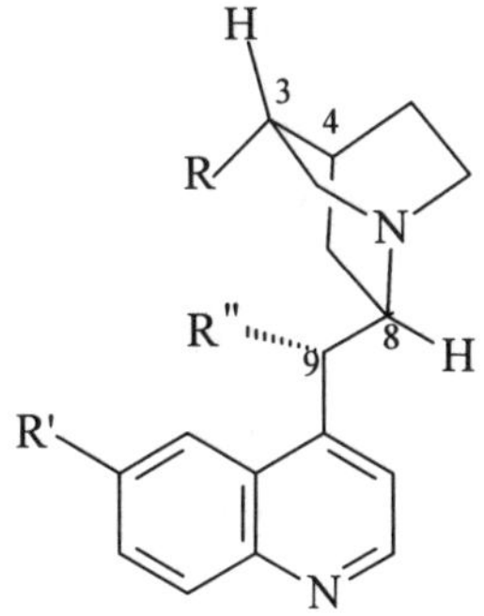

Cinchonidine (CD)	Cinchonine (CN)	R = C$_2$H$_3$ R'= H R" = OH
Quinine (QN)	Quinidine (QD)	R = C$_2$H$_3$ R'= OMe R" = OH

Figure 2   *Structures and configurations of four natural cinchona alkaloids*

2.2   Procedures

Standard reaction conditions. Reactions were carried out in a Parr high pressure steel reactor. A solution in methanol (10 ml) of reactant (100 mg) and modifier (15 mg) was added to the catalyst in the reactor, which was then closed, purged twice with 3 bar helium and twice with 3 bar hydrogen, and pressurised with hydrogen to 10 bar. Reactions occurred at ambient temperature (293 – 295 K) and the mixture was stirred at 1000 rpm. Because of the small quantities of reactant used (0.46 mmol for (**I**)) hydrogen uptake, and hence conversion, could not be measured accurately. Consequently, reactions were allowed to proceed for 3 h within which time 100% conversion was normally achieved.

   Conditions for measurement of kinetics. The kinetics of NADPME hydrogenation were investigated using larger quantities of reactant so that measurable hydrogen uptakes were observed. The procedure used was that set out above. A Buchi Pressflow unit maintained hydrogen pressure at the set value and recorded total hydrogen uptake as a function of time.

   After reaction, the catalyst was removed by filtration, and the solution analysed by chiral gas chromatography using a 25 m Chirasil-L-Val column. The enantiomeric excess, defined as e.e./%(*S*) = ([*S*-] – [*R*-])/([*S*-] + [*R*-]), was determined by integration of the peak areas for the *R*- and *S*-products. Values were reproducible to ± 0.7.

3 RESULTS

3.1 Enantioselectivity in NADPME hydrogenation

The effectiveness of representative members of three families of alkaloids were evaluated as modifiers for the hydrogenation of NADPAME (**I**) over Pd/alumina (Table 1). These alkaloids all induce enantioselectivity in α-ketoester hydrogenation.[1] The results are shown in Table 1.

Table 1   *Values of the enantiomeric excess observed in various alkaloid-modified hydrogenations of NADPME under standard conditions (100% conversion).*

Modifier	e.e./%
cinchonine	10 (*S*)
cinchonidine	4 (*R*)
dihydrocinchonidine	4.5 (*R*)
quinine	1 (*R*)
quinidine	0
brucine[a]	0
codeine[b]	2 (*R*)

[a] strychnos alkaloid, [b] morphine alkaloid

Only t he c inchona a lkaloids w ere e ffective m odifiers, o f w hich c inchonine a fforded t he highest enantiomeric excess. Enantioselectivity in the opposite sense was achieved using cinchonidine and dihydrocinchonidine, as expected,[1] but both were inferior to cinchonine under the chosen conditions. Quinine, quinidine, brucine and codeine were ineffective.

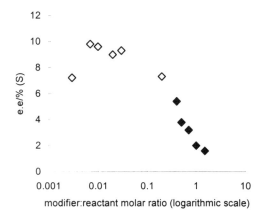

Figure 3   *Variation of enantiomeric excess with cinchonine concentration for NADPME hydrogenation under standard conditions. [NADPME] = constant throughout. Points in black indicate solutions saturated in cinchonine*

The effect of cinchonine concentration on enantiomeric excess is shown in Figure 3. Under standard conditions the highest values of the enantiomeric excess were obtained over a range of CN:NADPME molar ratios from 0.007:1 to 0.03:1 after which any further increase was detrimental. Cinchonine has a lower solubility than the other cinchona alkaloids and this resulted in reaction solutions becoming saturated when the CN:NADPME molar ratio exceeded 0.4:1. Before saturation, the decline may be attributable to an unfavourable increase in the pH of the reaction medium.[8] These solutions (black points, Figure 3) were all of equal cinchonine concentration; however, the enantiomeric excess did not plateau at this point but continued to decrease steadily so that, at a CN:NADPME molar ratio of 1.8, the activity of the catalyst had diminshed to zero. It is probable that, under these conditions, cinchonine had crystallised onto the catalyst surface rendering it inactive.

3.2 Kinetics of NADPME hydrogenation

The initial rate of the unmodified reaction (0.2 g catalyst, 3.0 g NADPME, 10 bar) was 48 mmol h^{-1} g_{cat}^{-1}. With cinchonine as modifier, the initial rate decreased dramatically such that, under the same conditions, and using the optimum CN:NADPME molar ratio determined under standard conditions (Figure 3), the initial rate was 10 mmol h^{-1} g_{cat}^{-1}. Thus, alkaloid modifier reduced the rate by a factor of 5.

The initial rate of cinchonine-modified reaction was directly proportional to the mass of catalyst used (0.2 – 1.2 g) as expected for diffusion-free reaction. Using a constant mass of catalyst (0.2 g) and of cinchonine (60 mg) the initial rate was zero order in [NADPME] (1.5 – 4.5 g) indicating strong chemisorption of the reactant, and one-half order in hydrogen (1 – 70 bar). An order in hydrogen of one-half has been observed in other Pd-catalysed enantioselective hydrogenations;[2,9] the interpretation in the present case awaits the results of D-tracer studies.

3.3 Enantioselective hydrogenation of some compounds related to NADPME

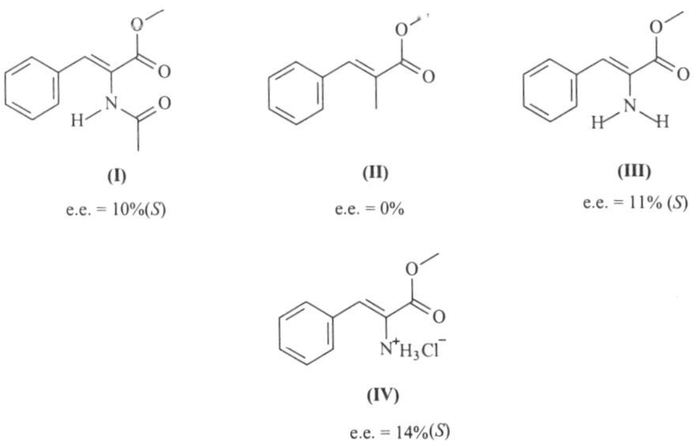

(I)

e.e. = 10%(*S*)

(II)

e.e. = 0%

(III)

e.e. = 11% (*S*)

(IV)

e.e. = 14%(*S*)

Figure 4   *Values of enantiomeric excess for NADPME and three related compounds*

Variation of reactant structure was investigated in order to probe the interaction between NADPME and cinchonine (Figure 4). In a series of cinchonine-modified hydrogenations conducted under standard conditions, Z-α-methylcinnamic acid methyl ester (**II**) gave racemic product, the free amine dehydrophenylalanine methyl ester (**III**) gave e.e. = 11%(*S*), and the hydrochloride salt of (**III**) gave e.e. = 14%(*S*).

4   DISCUSSION

Enantioselective hydrogenation catalysed by cinchona-modified Pd or Pt is proposed to occur as a result of a hydrogen bonding interaction between a functional group of the reactant and the quinuclidine-N atom of the adsorbed modifier.[1] Thus, for example, methyl-2-butenoic acid[2,10] and phenylcinnamic acid[11] are enantioselectively hydrogenated because the carboxylic acid group in each can H-bond to the modifier. Methylcinnamic acid methyl ester (**II**) has no such functional group and hence its conversion to racemic product is expected. When the α-methyl group in (**II**) is replaced by amine as in (**III**) the possibility of reactant-modifier interaction is restored, and enantioselection is again observed (Figure 4). Rendering the amine function more acidic, as in the hydrochloride of (**III**), strengthens the acid-base interaction and improves the enantiomeric excess. Protection of the amine group in (**III**) by acetylation, giving the target compound of this investigation (**I**), does not significantly diminish the enantiomeric excess, indicating that a single -N···H···N- interaction is sufficient to provide the necessary reactant-modifier complex in the adsorbed state.

Introduction of the modifier resulted in a five-fold reduction in rate. Similar rate reductions are observed in the hydrogenation of $\alpha\beta$-unsaturated acids over Pd[1,2,9] whereas enantioselective reductions of activated ketones over cinchona-modified Pt provide substantial rate enhancements.[1,12,13] In general, where the function of the reactant undergoing hydrogenation interacts directly with the quinuclidine-N atom of the modifier there is the possibility of rate enhancement; where the functional group undergoing hydrogenation is distant from the quinuclidine-N, and the modifier-reactant interaction relies on the participation of some other function in the reactant, there is no rate enhancement. In the present reactions, the amine group provides the modifier-reactant interaction and hydrogenation occurs at the remote carbon-carbon double bond; hence, no rate enhancement is expected. The rate decrease is attributed to modifier occupying a fraction of the surface and a smaller number of hydrogenation sites being available in consequence.

The enantiomeric excess favours the *S*-enantiomer in each reaction. The clear similarities between the hydrogenation of NADPME and of α, β-unsaturated acids suggests a common type of mechanism. Acid hydrogenation has been interpreted on the basis both of a monomer-modifier interaction model and a dimer-monomer interaction model.[1] In the absence of evidence for NADPAME dimerisation in solution, monomer-modifier interaction is assumed. Figure 5 is a representation of a 1:1 H-bonded interaction between NADPME adsorbed by one enantioface adjacent to cinchonine adsorbed in the open-3 conformation. Hydrogenation by addition of two H-atoms from below the plane of the adsorbed molecules would result in the formation of the *S*-enantiomer of *N*-acetyl phenylalanine methyl ester. The sense of the observed enantioelectivity is thereby interpreted. The low value of the e.e. may result from the low preference of the reactant for adsorption by the enantioface shown. Quinine and quinidine were ineffective as modifiers, which indicates that substitution of the quinoline ring by -OMe sterically hinders access to the enantioselective site. Brucine and codeine were ineffective probably because their basic-N atoms were inaccessible to the reactants. Modelling and energy calculations to assess these proposals are in progress.

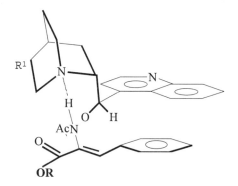

Figure 5 *The 1:1 interaction between adsorbed NADPME and adsorbed cinchonine. The model assumes that the aromatic rings are π-bonded to the Pd surface.*

5   CONCLUSIONS

The carbon-carbon double bond in dehydrophenylalanine methyl ester, its hydrochloride salt, and its *N*-acetyl derivative has been hydrogenated enantioselectively over cinchonine-modified Pd giving a low excess of the *S*-enantiomer. In each case, a H-bonding interaction of the type -N···H···N- is proposed between the amine function of the reactant and the quinuclidine-N of the modifier. On this basis, the mechanism is suggested to be analogous to that published for the hydrogenation of αβ-unsaturated acids.

References

1   P.B. Wells and R.P.K. Wells, *Chiral Catalyst Immobilisation and Recycling,* eds., D.E. De Vos, I.F.J. Vankelecom, P.A. Jacobs, Wiley-VCH, Weinheim, 2000, Chapter 6, pp. 123 – 154.
2   T.J. Hall, P. Johnston, W.A.H. Vermeer, S.R. Watson and P.B. Wells, *Stud. Surf. Sci. Catal.*, 1996, **101**, 211.
3   Y.Nitta and K. Kobiro, *Chem. Lett.,* 1994, p.1095.
4   A. Tungler, Y.Nitta, K. Fodor, G.Farkas and T.Mathe, *J. Molec. Catal. A:Chem.* 1999, **149**, 135.
5   K. Borszeky, T. Mallat and A. Baiker, *Catal Lett.*, 1999, **59**, 95.
6   W.R. Huck, T. Burgi, T. Mallet, A. Baiker, *J. Catal.*, 2001, **200**, 171.
7   C. R. Landis and J.Halpern, J. Am. Chem. Soc., 1987, 109, 1746
8   N.J. Caulfield, P. McMorn P.B. Wells, and G.J. Hutchings, unpublished work.
9   P.B. Wells and A.G. Wilkinson, *Topics in Catal.*, 1998, **5**, 39-50.
10   K. Borszeky, T. Burgi, Z. Zhaohui, T. Mallat and A. Baiker, *J. Catal.*, 1999, **187**, 160.
11   Y. Nitta, K. Kobiro and Y. Okamoto, *Proc. 70th Ann. Meeting Chem. Soc. Japan,* 1996, **1**, 573.
12   M. von Arx, T. Mallat and A. Baiker, *Topics in Catal.*, 2002, **19**, 75.
13   D. Feri, T. Burgi and A. Baiker, *J. Catal.*, 2002, **210**, 160.

ENVIRONMENTAL CATALYSTS: CATALYTIC WET OXIDATION OF DIFFERENT MODEL COMPOUNDS

I. M. Castelo-Branco, [1] S. R. Rodrigues, [1] R. Santos, [2] R. M. Quinta-Ferreira, [1]

[1] Department of Chemical Engineering, University of Coimbra, Pólo II – Pinhal de Marrocos, 3030-290 Coimbra, Portugal, E-mail: guida3@eq.uc.pt
[2] Centro Tecnológico da Cerâmica e do Vidro, Coimbra, R. Veiga Simão, Apartado 8052, 3020-053 Coimbra, Portugal

1 INTRODUCTION

Nowadays, surrounding environment has been polluted in many ways: industrial liquid and gaseous effluents, domestic wastewaters and solid residues are being released and deposed contaminated with toxic and refractory compounds that requires treatment. Industrial liquid effluents are as varied as industry itself[1] and in most cases are toxic to aquatic life causing euthrophication. On the other hand, traditional methods as biological treatment are not suitable to be used since contaminants are present at concentrations too high, yet too low to be economically feasible its recovery. Therefore, effluents have to be submitted to other treatment methods that may conduce to total degradation of the contaminants or its degradation into biodegradable compounds.

Catalytic Wet Oxidation (CWO) involves the liquid phase oxidation of organic compounds using a source of oxygen (pure oxygen, air) and a catalyst. While wet oxidation (WO) is carried out at 125-320 °C and 0,5-20 MPA [1], the operating conditions of CWO are less severe. Therefore, it is our goal to developed catalysts to be used in CWO of several organic compounds, which are toxic to the environment: formaldehyde, acrylic acid. Catalysts prepared in our laboratory were cerium based catalysts (Mn/Ce, Co/Ce and Ag/Ce) and Cu-Mn composite oxides. Cerium, manganese and copper oxides were also prepared.

2 EXPERIMENTAL PROCEDURE

All catalysts were prepared using the corresponding nitrate salts. Ag/Ce catalysts were prepared according procedure 1, while all the other catalysts preparation followed procedure 2.

Procedure 1 – NaOH(3M) was added to a mixed metal salt solution until pH of 10 is reached. The precipitate was washed several times and dried at 100°C. The solid was crushed and calcinated at 300 °C for 3 hr. [2, 5]

Procedure 2 – A mixed metal salt solution was poured into NaOH 3M solution. After filtration, the precipitate was washed and dried at 100 °C. Afterwards, the solid was crushed and calcinated at 300 °C for 3 hr. [3, 4, 5]

Catalyst notation and composition are in table 1. Catalysts were characterized by N_2 adsorption and XRD techniques.

The operating conditions used to performe activity studies are refered in table 2. The oxidation apparatus is described elsewhere.[5] 450 ml of water and 3 gr of catalyst were loaded into the reactor and heated to the desired temperature. A 50 ml of compound solution was introduced into the system followed by oxygen (time zero of the reaction). Liquid samples were withdrawn periodically and analysed for *Total Organic Carbon* (TOC) determination.

Table 1   *Catalyst notation and composition*

No.	Notation	Ag	Mn	Cu	Co	Ce
1	AgCe0,7	7	-	-	-	93
2	AgCe7	70	-	-	-	30
3	CoCe7	-	-	-	70	30
4	MnCe8	-	80	-	-	20
5	MnCe7,5	-	75	-	-	25
6	MnCe7	-	70	-	-	30
7	MnCe6	-	60	-	-	40
8	MnCe2,2	-	22	-	-	78
9	CuMn8	-	20	80	-	-
10	CuMn5	-	50	50	-	-
11	CuMn2	-	80	20	-	-
12	Ce Ox	-	-	-	-	100
13	Mn Ox	-	100	-	-	-
14	Cu Ox	-	-	100	-	-

(Catalyst / Molar composition(%))

Table 2   *Oxidation operating conditions (catalyst loading of 6 g/l and stirring speed of 350 rpm)*

Compound	TOC i (ppm)	T (°C)	P O_2 (bar)	t (min)	Catalysts
Formaldehyde	1100	200	15	180	1, 3, 4, 5, 6, 7
Acrylic Acid	500	200	15	180	2, 3, 6, 8, 9, 10-14

(Operating conditions)

3   MAIN RESULTS

3.1 Activity Studies and Leaching

The results of the activity studies of the catalysts used in CWO of formaldehyde, acrylic acid and ethylene glycol are presented in figure 1 (a and b).

3.1.1   Formaldehyde CWO.[5,7] Among all the catalysts tested on CWO of formaldehyde (figure 1a), MnCe composite oxides showed the highest activities. In fact, TOC reduction was almost 100 % (MnCe8 and MnCe7). Comparing the performance of CoCe7 and MnCe7 (same cerium content), it can be concluded that Mn has a stronger effect on this process. AgCe0,7 leads to lower TOC removal, which can be explained by the lower Ag concentration (7 %).

3.1.2   Acrylic Acid CWO. As well as in CWO of formaldehyde, MnCe7 exhibits the highest TOC removal (97%), while Mn and Ce oxides lead to 96 % and 60%, respectively (figure 1 b). For the same Ce composition, acrylic acid oxidation over AgCe7 and CoCe7 was not as efficient as over MnCe7, being therefore Mn the key component in this process.

CuMn composite oxides are also quite active and in the range 20%Mn (CuMn2) – 80% Mn (CuMn8), final TOC removal suffered almost no changes: 93 %. These catalysts can be considered promising catalysts in CWO of effluents containing acrylic acid.

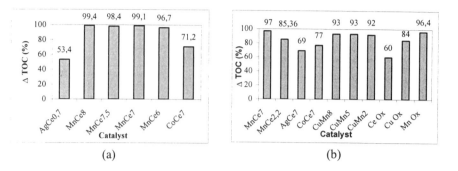

| (a) | (b) |

Figure 1 *TOC variation as a function of catalyst: a – formaldehyde; b – acrylic acid.*

3.2 Surface area analysis

The results obtained by N_2 adsorption are presented in table 3, as well as δTOC/S (TOC variation as a function of unit area) obtained for acrylic acid oxidation. MnCe7 and CoCe7 have the highest surface areas. On the other hand, Mn Ox exhibited the lowest value, only 32 m^2/g while all other surface areas were higher than 91 m^2/g. Nevertheless, the TOC variation per unit of surface area achieved with Mn Ox is the highest of all.

Table 3 *Catalyst BET surface area and TOC variation per unit surface area.*

Catalyst	S_{BET} (m^2/g)	δTOC/S (ppm/m^2)
MnCe7	102.48	1.5775
CoCe7	103.86	1.2356
CuMn8	91.91	1.6864
Mn Ox	32.19	4.7872
Ce Ox	94.38	1.0595

3.3 XRD

MnCe2,2 (fig. 2), MnCe7 (fig. 3), Ce Ox (fig. 4), Mn Ox (fig. 5), CuMn5 (fig. 6), MnCe8 and Cu Ox were submitted to XRD analysis using a Philips PW1710 diffractometer with a Cu-K_α radiation.

A "cerianite" (CeO_2) was clearly identified as the main structure of Ce Ox (JCPDS #34-394). For Cu Ox sample, monoclinic CuO is the predominant structure (JCPDS #5-661). For Mn Ox, the dominant diffraction peaks are the characteristic of Mn_5O_8 (JCPDS #39-1218). Lower intensity peaks were associated to Mn_3O_4 (JCPDS #24-734) and/or Mn_2O_3 (JCPDS #18-803). The analysis of MnCe8 identifies Mn_5O_8 and CeO_2, being Mn_3O_4 also observed and no Mn_2O_3 were detected.

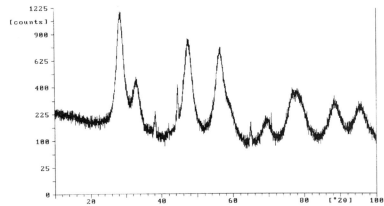

Figure 2 *X-ray diffractogram of MnCe2,2.*

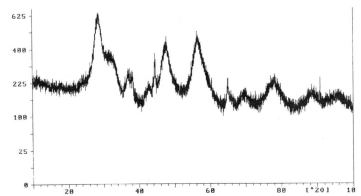

Figure 3 *X-ray diffractogram of MnCe7.*

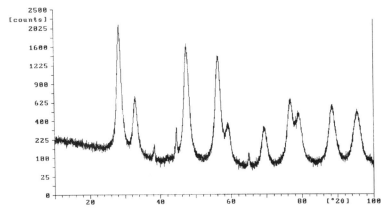

Figure 4 *X-ray diffractogram of Ce Ox.*

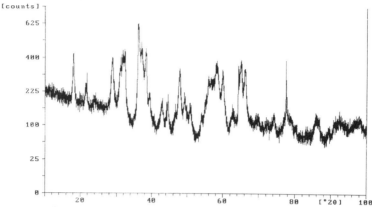

Figure 5 *X-ray diffractogram of Mn Ox.*

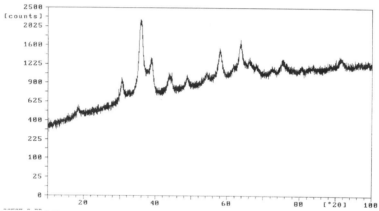

Figure 6 *X-ray diffractogram of CuMn5.*

Either in MnCe8 as MnCe7 the existence of peaks attributed to "cerianite" indicates the existence of separated solid phases of Mn and Ce oxides.[6] Although, MnCe2,2 structure is attributed to "cerianite" and no Mn oxide phases were detected, the peaks are broader than in Ce Ox (fig. 4), which can be explained by either the formation of solid-solution of Mn-Ce oxides with "cerianite" structure or to the formation of amorphous Mn oxide phases.[6] CuMn5 XRD peaks were attributed to a cubic structure of $Cu_{1.5}Mn_{1.5}O_4$ (JCPDS #35-1172) – fig. 6).

4 CONCLUSIONS

In conclusion, MnCe composite oxides were proved to be promising catalysts for CWO of formaldehyde and acrylic acid containing effluents. CuMn composite oxides were also tested in the CWO of acrylic acid and it was found out that the mixed oxides have higher performance than single oxides in what respects to TOC reduction. The N_2 adsorption characterization of some of the catalysts tested leads to the conclusion that higher BET surface areas are obtained with cerium based catalysts. Also CuMn8 and Ce Ox have a high surface area. On the other hand, Mn Ox has the lowest surface area and the highest TOC reduction as function of the area, which can be explained by the fact that Mn oxides are the active component of the catalysts. Accordingly, CoCe7 exhibits lower performance than MnCe7 (same Ce composition). Either Cu Ox, Mn Ox or CuMn composite oxides are quite active in acrylic acid CWO, being both Cu oxides and Mn oxides key components in the activity of these catalysts.

Acknowledgements

Isabel Castelo-Branco is grateful to *Fundação para a Ciência e Tecnologia*, Portugal, for her Ph.D. grant PRAXIS/BD/22298/99.

References

1   V. Mishra, V. Mahajani and J. Joshi, *Ind. Eng. Chem. Res.*, 1995, **34**, 2.
2   S. Imamura, D. Uchilori and K. Utani, *Catal. Letters*, 1994, **24**, 377.
3   S. Imamura, M. Nakamura, N. Kawabata and J.-I. Yoshida, *Ind. Eng. Chem. Res.*, 1986, **25**, 34.
4   S. Imamura, A. Doi and S. Ishida, *Ind. Eng. Chem. Prod. Res. Dev.*, 1985, **24**, 75.
5   I. Castelo-Branco, A. Silva and R. Quinta-Ferreira, "Catalysts for Wastewater Treatment" in Principles and Methods for Accelerated Catalyst Design and Testing, eds. E. Derouane, V. Parmon, F. Lemos and F. Ramôa Ribeiro, Klumer Academic Publ., Nato Science Series, Series II: Mathematics, Physics and Chemistry, Dordrecht, 2002, vol. 69, pp. 383-388.
6   H. Chen, A. Sayari, A. Adnot and F. Larachi, *Appl. Catal. B:Environ.*, 2001, **32**, 195.
7   A. M. Silva, I. Castelo-Branco and R. Quinta-Ferreira, *Chem. Eng. Sci.*, 2003, **58**, 963.
8   A. M. T. Silva, R. Neves and R. Quinta-Ferreira (submitted)

USE OF IR AND XANES SPECROSCOPIES TO STUDY NOx STORAGE AND REDUCTION CATALYSTS UNDER REACTION CONDITIONS.

J. A. Anderson[1] B. Bachiller-Baeza[1] and M. Fernández-García[2]

[1]Surface Chemistry and Catalysis Group, Division of Physical and Inorganic Chemistry, University of Dundee, DD1 4HN
[2]Instituto de Catálisis y Petroleoquímica, CSIC, Campus Cantoblanco, 28049 Madrid, Spain

1 INTRODUCTION

The ability to study catalysts under *in-situ* conditions has long been recognised. Changes during the course of reaction may not be readily detected by performing before and after type analysis. During chemical reaction for example, the presence of reactant and products at elevated temperatures and possibly elevated pressures are known to influence the surface composition and arrangement of adsorbent atoms. Sintering, poisoning, coking and surface segregation of one or more components may occur, all of which may have profound affects on the course of the reaction as a function of time. In addition to modifications occurring during chemical reaction, the ability to monitor catalysts during the preparation and regeneration stages may provide important detail regarding processes occurring during these steps which are not obvious from measurements made before and after treatment. Ideally one would like to extract maximum information from application of the minimum number of techniques. Our approach[1-5] has been to combine infrared (FTIR) and x-ray absorption spectroscopies (XAFS) to provide information concerning the surface and bulk of the supported phases, respectively.

In the application of XAFS, our attention has focussed on performing temperature-programmed type experiments, by scanning through the absorption edge of interest while treating the sample in the reactive gas flow while raising the temperature. Analysis of the edge structure using Principle Component Analysis (PCA) allows details of the number and nature of species present to be extracted. Evolution of the concentration profiles can then be followed during preparation (i.e. calcination, reduction), or regeneration treatments, or during catalytic reaction. Further benefits arise from the use of XAFS combined with simultaneous measurement of some other physical characteristic of the sample (e.g. XRD)[5-7] or catalytic behaviour (activity/selectivity measurements by GC/MS analysis of the effluent stream from the reactor).[6,7] In the present study, results will be presented from work performed on NOx storage and reduction catalysts.

2 EXPERIMENTAL

BaO/Al$_2$O$_3$ samples (10 wt% BaO) were prepared by precipitation of BaOH from a nitrate solution onto Degussa Aluminoxide C γ-alumina using an ammonia solution. The material was filtered, washed, dried at 363 K (16 h) and calcined in air (100 cm^3 min^{-1}) at 773 K (2 h). A portion of both samples was wet impregnated with 1 wt% Pt (H$_2$PtCl$_6$) and excess water removed by heating while continually stirring. The resulting powder was dried overnight at 363 K and calcined in air (100 cm^3 min^{-1}) at 773 K (2 h). DRIFT spectra (4 cm^{-1} resolution, 100 scans) were recorded using an MCT detector and Harrick environmental cell. A computer-controlled gas blender fed the IR cell, with exit gas analysis performed using a chemiluminescence detector and a quadropole mass spectrometer. Samples were calcined *in situ* in the DRIFT cell at 673 K (1 h) in a flow of dry air (50 cm^3 min^{-1}), then exposed to NO$_2$ at 673 K. This procedure has been shown to decompose bulk BaCO$_3$ as revealed by *in-situ* XRD experiments.[5] The cell was then flushed in air before a TPD of NOx was performed between 298 and 873 K at 4 K min^{-1} under a stoichiometric flow (50 cm^3 min^{-1}) containing 0.355 % propene, 7.98 % air, balance N$_2$. Combine XAFS/XRD experiments were performed at Daresbury Laboratories (station 9.3) using a double-crystal Si (220) monochromator. Samples as 13 mm diameter discs were loaded into the high temperature cell, with gases supplied *via* a computer controlled gas blender. XAFS at the Pt L$_{III}$ edge was performed in the transmission mode using 3 gas ionisation chambers with a reference foil between the 2nd and 3rd detector. X-ray diffractograms using 11.381 keV radiation were measured using a position sensitive detector collected over 139 s.

Computational details

The XANES spectra were analyzed using Principal Component Factor Analysis (PCA)[1,2] which assumes that the absorbance in a set of spectra can be mathematically modeled as a linear sum of individual components, called factors, which correspond to each of the chemical species present in a sample, plus noise.[8] To obtain the number of individual components, an F-test of the variance associated with factor k and the summed variance associated with the pool of noise factors is performed. A factor is accepted as a "pure" species (factor associated with signal and not noise) when the percentage of significance level of the F-test, %SL, is lower than a test level set by previous studies as 5%.[1-5] The ratio between the reduced eigenvalues, R(r), which approach one for noise factors, is also used to reach this decision. Once the number of individual components is determined, XANES spectra corresponding to individual species and their concentration profiles are generated by an orthogonal rotation (varimax rotation) which should align factors (as close as possible) along the unknown concentration profiles, followed by iterative transformation factor analysis (ITFA). ITFA starts with delta function representations of the concentration profiles located at temperatures predicted by the varimax rotation, which are then subjected to refinement by iteration until error in the resulting concentration profiles is lower than the statistical error extracted from the set of raw spectra.[1,2]

3   RESULTS

Figure 1 shows FTIR spectra obtained by heating a NO$_2$ pretreated sample in a stoichiometric gas mixture. Unlike Pt-free samples[9], spectra for Pt/Ba/Al$_2$O$_3$ were not dominated by bands due to NO$_2$ adsorbed on the alumina (1618, 1585, 1568 cm^{-1}), but by features at 1432 and 1359 cm^{-1}. At temperature above 411 K (Fig. 1(e)), the band at 1432 cm^{-1} was resolved to reveal two maxima at 1461 and 1439 cm^{-1}. These reached a minimum intensity in spectra recorded at 601 K (Fig. 1(j)) but were then progressively increased in intensity on raising the temperature to 873 K. The two higher frequency bands due to NO$_2$ adsorbed on the alumina support were essentially fully depleted by 727 K (Fig. 1 (m)) leaving a single component at 1545 cm^{-1} which split at higher temperatures to reveal two components at 1560 and 1543 cm^{-1} (Fig. 1 (n-p)). The loss of intensity of the 1545 cm^{-1} band between 648 and 692 K, appeared to parallel the feature at 1359 cm^{-1} suggesting that these were two vibrations of the same species.

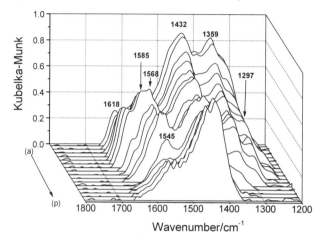

Figure 1 *FTIR spectra of Pt/Ba/Al$_2$O$_3$ after exposure to 1500 ppm NO$_2$ in N$_2$/air at 673 K and (a) cooled to 298 K. Then heated in stoichiometric propene/air mix at (b) 298, (c) 335, (d) 377, (e) 411, (f) 444, (g) 478, (h) 511, (i) 556, (j) 601, (k) 648, (l) 692, (m) 727, (n) 783, (o) 821 and (p) 873 K.*

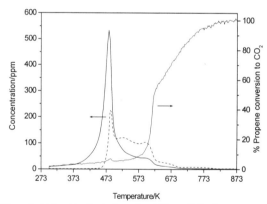

Figure 2 *TPD of NO (dashed lines), NO$_2$ (full line), and CO$_2$ released from Pt/10Ba/Al$_2$O$_3$, after exposure to NO$_2$/air/N$_2$ at 673 K, cooling to 298 K and then heating to 873 K in a stoichiometric propene/air mixture.*

The NOx profiles (chemiluminescence) and CO$_2$ signal (m/e 44) during a temperature programmed ramp are displayed in Fig. 2. NO$_2$ was released in a single, almost symmetrical peak at 481 K while NO was released in three stages: a sharp peak at 486 K and two smaller peaks at 527 and 596 K. The tail of the latter corresponded with the temperature at which an abrupt increase in CO$_2$ was observed, reaching a 50% conversion level at 613 K. Beyond this, a much more gradual increase in CO$_2$ production was observed (corresponding with a more gradual decline in C_3H_6 concentrations (m/e 39 and 41)). Under experimental conditions employed, this CO_2 released gave rise to small amounts of crystalline $BaCO_3$ as reveled by increased signal at 15.9° (2θ) between 585 and 642 K in the *in-situ* XRD patterns (Fig.3). The intensity was much less than that observed in the sample prior to calcination/673 K NO_2 treatment, suggesting that the extent of carbonate formation during reaction was limited by the rate of CO_2 bulk diffusion. This carbonate signal increased while heating between 642 and 773 K before diminishing about this temperature.

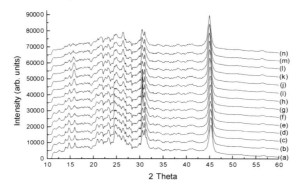

Figure 3 *XRD pattern for Pt/Ba/Al$_2$O$_3$ catalyst, calcined at 673 K and exposed to NO$_2$/air at 673 K then cooled to 298 K and heated in a stoichiometric propene/air mix at (a) 299, (b) 326, (c) 383, (d) 426, (e) 485, (f) 527, (g) 585, (h) 642, (i) 689, (j) 732, (k) 775 (l) 832, (m) 875 and (n) 898 K .*

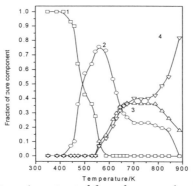

Figure 4 *Evolution of Pt species extracted from the near edge analysis as Pt/Ba/Al$_2$O$_3$ as heated to 873 K in a stoichiometric C$_3$H$_6$/air mix.*

PCA was applied to the Pt L_{III} edge for a sample which was exposed to NO_2 following calcination and then heated in a stoichiometric propene/air mixture. Results suggest that four species were present during the temperature ramp. The concentration profiles (Fig. 4), indicate the existence of only one initial oxidized species (species 1) which converts above 400 K to an intermediate (species 2) and finally disappears below 600 K. The intermediate species reaches a maximum concentration around 550 K before being converted simultaneously to species 3 and 4. All 3 (2,3 and 4) species exhibit temperature stability in the reaction mixture between 673 and 800 K, after which species 2 and 3 were simultaneously converted to yield species 4. The shape and edge position of the XANES spectra suggest that species number 1 and 3 correspond to Pt(IV) species in oxidic environments while 4 is Pt^0.[9] Species number 2 is a partially reduced $Pt^{\delta+}$ (δ close to 1,2) intermediate.[9]

4 DISCUSSION

The stability of stored NOx during the warm-up is of relevance to the contribution to total NOx emissions during a test cycle. For $Pt/Ba/Al_2O_3$ catalysts containing 5% baria and treated with NO_2 at room temperature, we have shown that NO_2 may be released between 400-500 K while temperature ramping under stoichiometric conditions, although under air, this stored NOx is much more stable.[10] Despite the fact that NO_2 treatment was performed here at 673 K, a condition under which it is known that bulk carbonate decomposition ocurrs[5], NO_2 was still released as a single peak falling within this range suggesting that the NO_2 pretreatment temperature does not significantly alter the nature in which NOx is stored. Such a scenario would be consistent with a previous suggestion[5] that the majority of NOx is stored as surface or top few layers of the baria, with the rest held on the alumina surface. Significant IR intensity losses at 1618, 1585 and 1359 cm^{-1} were observed between 478-511 K (Fig. 1g-h), consistent with the maximum in NO_2 release at 486 K (Fig. 2). This temperature does not appear to be significant in terms of the evolution of Pt species (Fig. 4) and it can be assumed that Pt plays no direct part in the desorption process. NO release then shadows the release of NO_2 suggesting that the processes may be linked. The ratio of the NO_2/NO signals was 1.42 whereas a similar run conducted in air in the absence of propene gave a ratio of 3.02, which might suggest that the presence of the reductant allowed a degree of reduction of the released NOx to take place. However, the $m/e = 44$ profile quite clearly shows that the onset of propene conversion occurred after the majority of the NOx had been released, suggesting that the NO_2/NO ratio had a greater dependence on the temperature of NOx release or the relative oxygen concentration than on catalytic reduction. It is more likely that at the higher temperatures of NOx release in air alone[9], rapid oxidation of NO to NO_2 occurred.

Propene combustion by oxygen was initiated around 590 K reaching 50% conversion by 613 K (Fig. 2). Note that Nova *et al*[11] using similar catalysts report that propene consumption was always observed between 473 and 503 K. As the addition of baria to Pt/Al_2O_3 is thought to inhibit the adsorption of propene on Pt[12], it might be suggested that the partial pressure of propene is significant in determining the onset temperature. However, similar experiments with 0.5 times the current propene concentration (lean mixture) only increased the 50% temperature by 5 K.[9] Our values for the onset of combustion correspond to a temperature range beyond the temperature at which Pt species 2 (Fig. 4) exhibited its maximum and species 3 and 4 were being rapidly evolved, suggesting a role for reduced Pt. However, the onset of propene oxidation also occurred at *ca.* 590 K over alumina alone and over Ba/Al_2O_3[9] which would negate a role of Pt in this process. As the presence of Pt did not result in improved onset temperatures relative to the support alone, this would strongly suggest that inhibition took place, most likely attributable to readsorption on Pt of released NOx. The presence of the oxidized Pt(IV) species number 3 detected by XANES is consistent with this interpretation. Given that the addition of baria may enhance the strength of NO adsorption but weaken propene adsorption on platinum[12] it is likely that a significant increase in propene concentration is required before the activation of the hydrocarbon may occur in the presence of significant concentrations of released NOx. However, once initiated, the 50% conversion temperatures were 650 and 661 K for Al_2O_3 and Ba/Al_2O_3, respectively, but 613 K in the presence of Pt. This, and the initial sharp light-off following ignition indicate that following the onset, Pt was active in the oxidation reaction. In addition to the detection of the mass peak at m/e=44, formation of CO_2 was indicated by the increased intensity at 1432 cm^{-1} in FTIR spectra recorded above 601 K (Fig. 1 j) confirming the assignment to carbonate species.[9] This release of CO_2 also led to the formation of bulk carbonate phases as indicated by the increased line intensity at 15.9° (2θ) in XRD patterns with the sample temperature around 640 K (Fig. 3 h).

5 CONCLUSIONS

During temperature programmed reaction under stoichiometric conditions, all of the stored NOx is released as nitrogen oxides between 373 and 600 K. As this occurs prior to the onset temperature of propene activation (600 K), this released NOx is liberated from the system rather that being reduced in reaction by propene. The activation of propene takes place at a temperature above that at which reduced Pt metal is detected, suggesting that the availability of active metal sites is not the limiting feature. The oxidation of propene above the temperature at which NOx is released conflicts with the NOx release mechanism suggested by Aberntsson *et al*[13] where it is proposed that it is the combustion of propene leading to the reduction in oxygen levels which destabilizes the stored NO, leading to its release. It is likely that oxygen levels do dictate the decomposition temperature of stored NOx and if this occurs at a sufficiently low temperature, the released NOx may be readsorbed on Pt sites thus inhibiting the activation of propene.

References

1   M. Fernández-Garcia, C. Márquez-Alvarez, and G.L Haller, *J. Phys. Chem.*, 1995, **99**, 12565.
2   M. Fernández-Garcia, J.A. Anderson, and G.L. Haller, *J. Phys. Chem.*, 1996, **100**, 16247
3   J.A.Anderson, M. Fernández-Garcia, and G.L. Haller, *J. Catal.*, 1996, **164**, 477.
4   M. Fernández-Garcia, F.K. Chong, J.A. Anderson, C.H. Rochester, and G.L. Haller, *J. Catal.*, 1999, **182**, 199.
5   J.A.Anderson and M. Fernández-Garcia, *Trans IChemE.*, 2000, **78**, 935.
6   B.S.Clausen, L. Grabaek, G. Steffensen, P.L. Hansen, and H. Topsoe, *Catal. Letts.*, 1993, **20**, 23.
7   G. Sankar and J.M. Thomas, *Topics in Catal.*, 1999, **8**, 1
8   E.R. Malinoswki, "Factor Analysis in Chemistry" (Wiley, New York, 1991).
9   M. Fernández-Garcia, B. Bachiller-Baeza and J.A. Anderson, *submitted to PCCP*.
10 J.A.Anderson, A.J. Paterson and M. Fernández-Garcia, *Stud. Surf. Sci and Catal.*, 2000, **130**, 1331
11 I. Nova, L. Castoldi, L. Lietti, E. Tronconi and P. Forzatti, *Catal. Today*, 2002, **75**, 431.
12 M. Konsolakis and I.V. Yentekakis, *J. Catal.*, 2001, **198**, 142.
13 A. Amberntsson, H. Persson, P. Engström and B. Kasemo, *Appl. Catal., B.*, 2001, **31**, 27.

CATALYTIC UTILIZATION OF LOW-MOLECULAR ALKANES

S.I. Abasov, S.B. Agayeva, D.B. Tagiyev

Institute of Petrochemical Processes of National Academy of Sciences of Azerbaijan
30, N.Rafiyev str., 370025 Baku, Azerbaijan, E-mail: dtagiyev@hotmail.com

1 INTRODUCTION

Lower alkanes C_2-C_4 being chemically low active are mainly used as fuel material. However, in some oil extracting and oil refining zones the accumulation and desirable consumption of such hydrocarbons is technically hindered. Therefore, they are destroyed by burning on torches. In this case, not only hydrocarbon feedstock is uselessly lost, but the environment in the regions with the torches is badly damaged. Thus, the rational use of the hydrocarbon feedstock is an important problem not only from the viewpoint of economy, but also from the ecological consideration.

Taking into account the complexity of C_2-C_4 alkanes functionalization, one of methods of solution of this problem is the conversion of these alkanes into liquid hydrocarbons. The investigations in this direction showed the possibility of one-stage catalytic conversion of lower alkanes into aromatic hydrocarbons [1-3]. Taking into account the reactivity differences of C_2-C_4 alkanes, it is interesting to elucidate the reactivity of gas mixtures consisted of these molecules and their aromatic products. The present report is devoted to the results of studies in this direction.

2 METHODS AND RESULTS

2.1 Preparation of Catalysts

The catalysts used in studies were prepared on the basis of ZSM-5 zeolite and industrial catalyst of selective hydrocracking (SHC-1) prepared on the basis of pentasyl type zeolite [4]. During catalysts synthesis on the basis of ZSM-5, the initial Na-form of zeolite (Si/Al=34) was fully converted into the ammonium form by 1 N NH_4Cl, washed, dried and calcinated. The H-form obtained was shaped with the alumogel binder, and repeatedly dried and calcinated at 650^0 for 4h. The resulting material was then used for ion exchange with Zn, Ga or Cr salts, respectively. Then the samples were repeatedly dried and calcinated.

The catalysts on the basis of SHC-1 were modified similarly.

The ion content in the prepared catalysts (as oxides) was changed from 1% to 5 wt %.

Pt, Re-Al_2O_3 catalyst (PRAC) was prepared by the method described by [4-5]. Platinum and rhenium content in the prepared catalysts was 0.5 wt % Pt and 1.0 wt % Re, respectively.

2.2 Carrying Out of Reactions

Reactions were carried out with a continuous flow-reactor operating at atmospheric pressure. Before the reaction the catalyst was pretreated in a stream of air at 550° for 1.5h.

The torch gas from Baku refinery plant (C_1-13.5, C_2-18.6, C_3-37.1, C_4-13.8, C_5-3.5 wt %), C_2-C_4 alkanes mixture isolated off from associated oil gases (C_1-2.8, C_2-3.9, C_3-45.7, C_4-39.8, C_5-9.3), individual lower alkanes C_2-C_4, and also the mixtures of C_3H_6:C_3H_8 (4:1 mol ratio) and C_6H_6:C_3H_8 were used as a reactant for the catalytic reactions.

2.3 Conversion Of C_2-C_4 Alkanes Mixture

The reaction of C_2-C_4 alkanes mixture on modified Zn, Ga or Cr zeolite and SHC-1 catalysts was carried out, and its total conversion, products' yield and selectivity are summarized in Table 1.

Table 1 *Conversion of lower C_2-C_4 alkanes mixture*

Ion modifier	Temperature °C	GHSV h^{-1}	Conversion %	ArH yield %	Coke %	Selectivity %
ZSM-5						
0	550	500	90.8	24.5	2.4	26.9
1 wt % Zn	550	500	89.6	41.2	3.2	45.1
2.5 wt % Zn	500	500	84.7	30.5	1.7	34.8
		750	84.3	34.0	1.5	40.3
	550	250	91.4	34.9	5.2	38.2
		500	88.4	44.5	3.1	50.7
	600	500	90.3	50.1	3.9	55.4
5 wt % Zn	550	500	90.1	43.47	3.3	48.5
2.5 wt % Ga	550	500	69.5	32.5	3.5	46.9
2.5 wt % Cr	550	500	74.2	41.7	2.1	52.1
SHC-1						
2.5 wt % Zn	550	250	90.7	38.6	3.4	42.5

Concerning the reaction of C_2-C_4 alkanes mixture, the yield and selectivity of aromatics (ArH) appreciably increase with the introduction of zinc or gallium or chromium cations into ZSM-5. At the same time, the selectivity of undesirable products of methane and ethane, decreased by introducing these cations. The aromatic hydrocarbons produced were mainly benzene, toluene and xylene. Zn-ZSM-5 was a little more effective than other ZSM-5 catalysts in the aromatization of mixture and much more selective than H-ZSM-5.

The data of Table 1 show the dependence of lower alkanes mixture conversion over Zn-ZSM-5 from promoter concentration and he reaction conditions. The data demonstrate that the selectivity of aromatic hydrocarbons increase and coke deposites decrease with increasing of space velocity. The temperature increase leads to both the activity increase of catalyst and the increase of selectivity and aromatic hydrocarbons yield.

In Table 1 the data obtained were compared from which it is seen that the main changes of the catalytic properties of the samples occur at increasing the introduced zinc concentrations up to 1 wt %. Zinc concentration increase from 2.5 wt % to 5.0 wt %, practically does not influence the properties of Zn-ZSM-5 catalysts.

Nevertheless, the sample containing 2.5 wt % Zn turns out to be the most optimal among the studied Zn-ZSM-5 catalysts of lower alkanes aromatization.

2.4 Conversion of Individual Lower Alkanes

Comparative results of studies of conversions of individual C_1-C_4 alkanes carried out over 2.5 wt % Zn-ZSM-5 catalyst are presented in table 2.

Table 2 *Individual conversion of C_1-C_4 alkanes over 2.5 wt % Zn-ZSM-5.*
   T=600°, GHCV=500h^{-1}

Hydrocarbon	Conversion %	Hydropcarbon product yield %		Coke %	Selectivity %
		gaseous	aromatic		
CH_4	-	-		-	-
C_2H_6	-	-		-	-
C_3H_8	76.3	59.5	33.2	7.3	43.5
C_4H_{10}	94.8	57.9	38.7	1.1	40.8
i-C_4H_{10}	99.1	51.3	45.6	-	46.1

Table 2 shows that only propane and butanes undergo aromatization at the conversion conditions of lower alkanes mixture, presented above. However, the conversion of propane is lower by 20-25% than the conversion of butanes. At the same time, it should be noted that availability of unsaturated hydrocarbons leads to increase of propane conversion and aromatic hydrocarbons yield. For instance, the conversion of C_3H_8:C_3H_6=4:1 mixture reaches 95%, and the selectivity of the aromatic hydrocarbons formation makes 60% at the conditions pointed in Table 2.

2.5 Conversion of Torch Gas

Conversion of torch gas was carried out on Zn-ZSM-5 catalysts. It should be noted that torch gas contains lower amount of C_4 hydrocarbons, but it contains up to 20% of unsaturated hydrocarbons. The results of conversion of this gas mixture are summarized in Table 3.

Table 3 *Conversion of torch gas (T=550°)*

Ion modifier wt %	GHSV h^{-1}	Conversion %	Yield of liquid products, %	Coke %	Selectivity %
ZSM-5					
0	750	76.6	35.0	2.8	45.7
1	500	66.2	41.1	4.2	61.9
	750	70.4	45.1	2.5	63.9
	1000	81.7	54.1	1.4	66.1
2.5	750	77.4	46.2	2.4	59.7
5.0	500	80.4	47.2	2.4	58.7
SHC-1					
0	750	89.1	34.2	3.5	48.4
2.5	750	68.6	48.7	1.6	70.8
5.0	750	70.9	55.2	1.4	77.5

These data show that Zn-ZSM-5 catalysts possess high catalytic activity for torch gas conversion into liquid hydrocarbons. The yield of these products depends on the conditions the reaction carried out in and the catalyst composition. With the increase of space velocity from 500h^{-1} to 1000h^{-1} the conversion of gas mixture and the yield of liquid products increase by 15% and 13%, respectively. For all that, coke deposit decreases more than 2 times. The promoting effect of zinc on HZSM-5 furthers, in main, an appreciable increase of desirable products yield. The sample containing 10 wt % Zn (as oxide) possesses the best catalytic properties at torch gas conversion.

The results of conversion of C_2-C_4 alkanes mixture and torch gas over the catalysts prepared on the basis of SCH-1, in Table 1 and 3 show that their catalytic properties are very close to that of Zn-ZSM-5. This conclusion allows to simplify the problem of catalytic preparation for the utilization of lower alkanes.

2.6 Conversion of C_6H_6:C_3H_8 Mixtures

Taking into account the investigation results of propane interaction with one of aromatization reaction products, i.e. with benzene [6-7], the study of C_6H_6:C_3H_8 mixture conversion over mixed catalysts (MC) containing zeolite H-form and PRAC was carried out. Table 4 shows the influence of mole ratio on the conversion of C_6H_6:C_3H_8 mixtures.

Table 4 *Conversion of C_6H_6:C_3H_8 mixtures. Influence of mole ratio.*
$T=350^\circ$, $GHSV=500h^{-1}$

Content, mol %		Conversion, %		Product yield*, % C**		
C_6H_6	C_3H_8	C_6H_6	C_3H_8	IPB	C_3H_6	ArH
0	100	-	0	-	-	-
10	90	12	1.9	3.6	0.5	-
25	75	14	7.2	6.2	0.7	2.2
50	50	23	7.7	6.4	0.4	2.6
100	0	0	-	-	-	-

* the quantity of the formed hydrogen is close to that of the converted propane
** total amount of C-atoms contained in a mixture

Benzene introduced in reaction leads to appreciable activation of propane. As a result of the conversion of these mixtures, isopropylbenzene (IPB), propylene, alkylaromatic hydrocarbons (ArH) and hydrogen are formed.

The dependence of reaction products yield on temperature is shown in Table 5.

Table 5 *Conversion of C_6H_6:C_3H_8 mixture. Influence of temperature.*
C_6H_6:C_3H_8=1:9; $GHSV=500h^{-1}$

Temperature, °C	Conversion, %		Products yield, % C			
	C_6H_6	C_3H_8	IPB	C_3H_6	ArH	C_1-C_2
180	1.0	-	0.3	0	0	0
200	5.0	-	1.5	0	0	0
250	12.0	1.9	3.6	0.5	0	0
300	34.2	6.3	9.0	2.2	0.6	0
320	62.1	12.6	12.5	5.0	6.1	0
375	16.1	19.3	0.3	15.5	4.5	0
400	7.9	7.9	0.1	6.3	2.3	t race
450	10.5	7.1	t race	2.0	3.1	2.5

ArH formation is a result of IPB disproportionation. Thus, IPB and propane are the main products of propane activation. These products yields are also dependent on GHSV.

By GHSV value increased from $125h^{-1}$ to $1000h^{-1}$, IPB yield is decreased, and propylene yield is increased.

The conversions of benzene and propane increase with the temperature growth and pass through maxima at $320°$ and $375°$. However, the relatively sharp benzene conversion decrease within this temperature range is characterized by analogous increase of the propane conversion and the propylene yield.

2.7 Action Mechanism

Comparing the C_2-C_4 alkanes mixture and torch gas conversions (Table 1 and 3) with the conversion of the individual alkanes (Table 2) one can note the higher value of propane conversion in the mixture than in a pure form. At the same time, conversion of propane in the mixture with propylene also leads to the increase of total propane conversion and the formation of aromatic hydrocarbons.

The noted increase of propane conversion and selectivity can be explained in the framework of the accepted mechanism of the lower alkanes aromatization. As is known carbenium ion formed on a zeolite is then either disintegrated or added to the formed unsaturated hydrocarbons [2, 3]. Due to the higher concentration of unsaturated hydrocarbons in the used mixtures in composition with pure propane, the probability of addition products formation grows, and hence, the first reversible stage of reaction moves towards carbenium ion formation.

Thus, the availability of the unsaturated hydrocarbons or hydrocarbons capable of easily dehydrogenizing in a mixture containing propane leads to the increase of the aromatization activity and selectivity and the decrease of the formation of undesirable products, C_1-C_2 alkanes.

It should be noted that aromatic hydrocarbons formed during the reaction can not accept the formed carbenium ions. Contrary to alkanes, they are able to accept the products of C_2-C_4 alkanes destruction and form the alkyl aromatic substances. Dehydroalkylation of aromatic hydrocarbons by carbenium ions can proceed at lower temperatures [7].

The data of Table 5 show that propane activation with the formation of aromatic hydrocarbons and propylene is possible at appreciably lower temperatures. The possible mechanism of benzene dehydroalkylation by propane with the formation of IPB and propylene occurs with the carbenium ion formation by the scheme:

$$\text{(zeolite) } BH + C_6H_6 \rightleftarrows B^- + C_6H_7^+$$

$$Pt\text{-}O\text{-}ReO_x + C_3H_8 \longrightarrow [Pt \cdot \overset{OH}{\underset{C_3H_7}{ReO_x}}] \xrightarrow[H_2O]{+C_6H_7^+} [Pt \cdot ReO_x] + C_3H_7^+$$

$$C_3H_7^+ + C_6H_6 \longrightarrow [C_6H_6 \cdot C_3H_7]^+ \xrightarrow{B^-} \begin{array}{c} C_6H_5C_3H_7 \\ C_6H_6 + C_3H_6 \end{array}$$

$$[Pt \cdot ReO_x] + H_2O \longrightarrow [Pt \cdot O \cdot ReO_x] + H_2$$

The development of the investigation in this direction permits to define the supposed mechanism of low-temperature activation of low-molecular alkanes.

3 CONCLUSION

i) The aromatization of lower C_2-C_4 alkanes mixture and torch gas over ZSM-5 modified by Zn, Ga, Cr has been studied and the high activity and selectivity of Zn-ZSM-5 in this process has been established.

ii) The availability of the unsaturated hydrocarbons or hydrocarbons capable of easily dehydrogenizing in a mixture containing propane leads to the increase of conversion of this hydrocarbons into aromatic hydrocarbons.

iii) A supposition has been made about low-temperature propane activation over mixed catalyst consisted of zeolite in H-form and nonreduced Pt, Re-Al_2O_3 catalyst.

References

1. S.M. Csicsery, *J.Catal* 1970,**17** ,205
2. Yoshio Ono, Hiroyoshi Kitagava and Yoko Sendoda. *Sekiyu Gakkaishi* 1987, **30**, 77
3. Kh. M. Minachev and A.A. Dergachev, *Izv. Akad. Nauk SSSR Ser.Khim.*, 1993, **6**, 1018
4. B.K. Nefedov. *Khimiya i tekhnologiya topliv i masel.* 1985, **9**, 18
5. S.I. Abasov, F.A. Babayeva and B.A. Dadashev, USSR A.C. 1608180 (1990); Rus. Pat. 1811159 (1992)
6. S.I. Abasov, F.A. Babayeva and B.A. Dadashev, *Kinetika i kataliz* 1995, **3**, 428
7. S.I. Abasov, T.V. Vasina and O.V. Bragin, *Izv. Acad. Nauk SSSR, Ser.Khim.*, 1991, **10**, 2228
8. A.V. Smirnov, E.V. Mazin, V.V. Yushenko, E.E. Knyazeva, S.V. Nesterenko, I.I. Ivanova, L. Galperin and R. Iensen, *J.Catal.* 2000, **194**, 266

STRUCTURE-ACTIVITY RELATIONSHIPS IN N_2O CONVERSION OVER FeMFI ZEOLITES. PREPARATION OF CATALYSTS WITH DIFFERENT DISTRIBUTION OF IRON SPECIES

Javier Pérez-Ramírez,[1] Angelika Brückner,[2] Santosh Kumar[2] and Freek Kapteijn[3]

[1] Nitric Acid Technology, Hydro Agri Research Centre, P.O. Box 2560, N-3907, Porsgrunn, Norway, *email: javier.perez.ramirez@hydro.com
[2] Institute for Applied Chemistry Berlin-Adlershof, Richard-Willstätter-Str.12, D-12484, Berlin, Germany
[3] Reactor & Catalysis Engineering, DelftChemTech, Delft University of Technology, Julianalaan 136, 2628 BL, Delft, The Netherlands

1 INTRODUCTION

The o rigin o f t he c atalytic a ctivity i n t he v arious r eactions c atalyzed b y F e-zeolites h as been intensively debated over the last decade. In spite of the considerable efforts to characterize these materials, available data are not yet sufficient to conclude exclusively on the structure of the active iron. Binuclear iron species in FeZSM-5 have been designated as the active site in various reactions, including direct N_2O decomposition, selective oxidation of benzene to phenol with N_2O, and selective catalytic reduction (SCR) of NO_x with hydrocarbons.[1-3] However, in the last reaction, small oligonuclear species of composition Fe_4O_4 and isolated iron ions have also been proposed as active sites.[4,5]

The unification of the various interpretations with respect to the active sites is extremely complicated due to the intrinsic heterogeneous nature of iron species in the catalyst. Particularly challenging in practice is suppressing clustering of iron species into large inactive iron oxide particles. A further complicating aspect for a rational unification is the application of Fe-zeolites in a wide range of catalytic reactions with a different mechanism.

In this paper, steam-activated FeMFI zeolites with a different distribution of iron species have been prepared and characterized by UV-Vis/DRS. Syntheses were aimed at suppressing clustering or association of iron species in the final catalyst. The catalysts were tested in various N_2O-related reactions, including N_2O decomposition and N_2O reduction using C_3H_8 or CO. From the characterization and testing, it is concluded that isolated iron species are essential in N_2O conversions involving reducing agents, while some degree of iron clustering appears favourable in direct N_2O decomposition. The intrinsic reaction mechanism determines the required structure of the active sites in FeMFI zeolites.

2 METHODS

Details on the hydrothermal synthesis of FeMFI zeolites (with Fe-Al-Si and Fe-Si frameworks), and subsequent post-synthesis treatments have been described elsewhere.[6,7] The isomorphously substituted zeolites were activated in steam (30 vol.% H_2O in 30 ml (STP) min^{-1} of N_2 flow) at two different temperatures (873 and 1173 K) during 5 h. Hereafter, the catalysts a re denoted followed by the steam temperature, $e.g.$ Fe-silicalite

(873 K). UV-Vis/DRS measurements were carried out in a Cary 400 spectrometer (Varian) equipped with a diffuse reflectance accessory (Harrick).

Catalytic activity was measured in a parallel reactor system,[8] using 50 mg of catalyst (125-200 μm) and a gas-hourly space velocity (GHSV) of 60,000 h^{-1} at atmospheric pressure. Three reactions were investigated: (a) direct N_2O decomposition (1.5 mbar N_2O in He), (b) reduction of N_2O by CO (1.5 mbar N_2O + 1.0 mbar CO in He), and (c) selective catalytic reduction of N_2O with C_3H_8 in the presence of excess oxygen (1.5 mbar N_2O + 1.5 mbar C_3H_8 + 50 mbar O_2 in He). A gas chromatograph equipped with TCD and FID detectors was used to analyze reactant and product gases.

3 RESULTS AND DISCUSSION

3.1. Nature of iron species in the catalysts

A detailed characterization of the as-synthesized FeMFI zeolites upon calcination and steam treatment has been reported elsewhere.[6,7,9,10] The iron content in FeZSM-5 (Si/Al = 31 and 0.67 wt.%) and Fe-silicalite (Si/Al ~ ∞ and 0.68 wt.% Fe) catalysts was very similar. The visual appearance of the steamed Fe-Al-Si and Fe-Si catalysts at different temperatures already suggested a different catalyst constitution with respect to iron. FeZSM-5 (873 K) and Fe-silicalite (1173 K) are light brownish, which evidences a certain accumulation of iron oxide/hydroxide in the zeolite. FeZSM-5 (1173 K) presents a darker brownish color. Fe-silicalite (873 K) was nearly white, suggesting the more isolated nature of the iron species in the catalyst.

The UV-Vis/DRS spectra of the steam-activated catalysts were essential to conclude on the degree of association of the iron species in the catalysts (Fig. 1). Two intense $Fe^{3+} \leftarrow O$ charge-transfer bands are observed in all the samples except for Fe-silicalite (873 K), with a single band at low wavelengths. Bands between 200 and 300 nm are typically assigned to isolated Fe^{3+} species, either tetrahedrally coordinated in the zeolite framework or with higher coordination,[11,12] while bands between 300 and 450 nm are attributed to small oligonuclear iron species, (FeO)$_n$. Extraction of tetrahedral iron to extraframework positions in FeZSM-5 (873 K) was virtually complete, as shown by ^{57}Fe Mössbauer spectroscopy and voltammetric response techniques.[6,7,9,10] Smaller extraframework iron species, including isolated iron ions and oligonuclear species in the zeolite channels have been identified in the UV-Vis spectra and supports previous characterizations.[7,9] Relatively small iron oxide nanoparticles of 1-2 nm were observed by HRTEM, which clearly indicates the significant degree of clustering during steam activation of these samples. This corresponds to the absorption around 500 nm in the UV-Vis/DRS spectrum of FeZSM-5 (873 K), which is typical for absorption of Fe(III) ions in this aggregate form. FeZSM-5 (1173 K) shows a more severe degree of iron clustering, as concluded from the reduced intensity of the band at 200-300 nm, the significant broadening of the band at 300-450 nm, and the appearance of a clear shoulder at 500 nm. This suggests the presence of large iron oxide particles in the sample.

The complete absence of the band in the range of 300-450 nm in Fe-silicalite (873 K) indicates that the majority of Fe^{3+} species in this catalyst is uniform and well isolated. Previous characterization showed that a fraction of iron in this sample remains in the framework, but this fraction is small compared to the extraframework species.[7] The absence of iron clustering in Fe-silicalite (873 K) has been corroborated by voltammetry and HRTEM, where no iron oxide nanoparticles were observed. Increasing the steam-activation temperature of Fe-silicalite to 1173 K leads to the appearance of the band at 300-450 nm, more similar to the spectrum of FeZSM-5 (873 K).

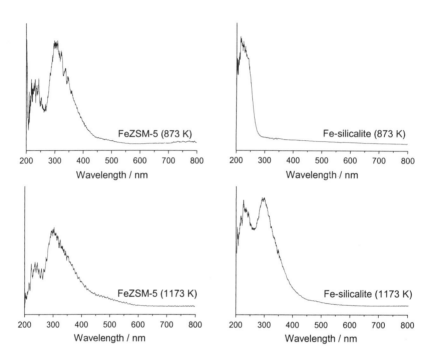

Figure 1   *UV-Vis/DRS spectra of steam-activated FeMFI zeolites.*

3.2. Catalytic activity in N₂O conversions

The different nature of iron species in the catalysts has a strong influence on the catalytic performance in various N_2O-related reactions. N_2O conversion *vs.* temperature in different feed mixtures is shown in Fig. 2. The conversion curves of FeZSM-5 (873 K) and Fe-silicalite (1173 K) in direct N_2O decomposition are very similar, which correlates with the similar nature of iron species in the catalysts. For the sake of clarity, only FeZSM-5 (873 K) is shown in the figure. The similar light-off temperature and activation energy (~140 kJ·mol⁻¹) nicely correlates with the identical nature of iron species in these catalysts. The activity of these samples is higher than that of Fe-silicalite (873 K), which is shifted ~40 K to higher temperatures. In this case, a slightly higher activation energy was obtained (155 kJ·mol⁻¹), suggesting the participation of different iron species in the process.

Addition of reducing agents (C_3H_8 or CO) to the feed mixture lowers the operation temperature of the catalysts by ~150 K. The N_2O conversion over Fe-silicalite (873 K) in $N_2O + C_3H_8 + O_2$ and $N_2O + CO$ mixtures was significantly higher as compared to the other catalysts. The N_2O conversion over this catalyst starts at the same temperature using either C_3H_8 or CO and separates above 575 K. The higher activity of Fe-silicalite (873 K) in the low-temperature range of these processes (500-600 K) compared to the other samples is of special significance. Above 873 K, the activity curves of the catalysts converge. Not only the N_2O conversion is enhanced over Fe-silicalite (873 K), but also the conversion of C_3H_8 and CO to CO_2. This indicates that conversions of N_2O and reducing agent are activated and effectively coupled. In the N_2O + CO mixture a transition is visible. When the limiting reactant CO becomes exhausted (~66% N_2O conversion), the N_2O conversion curve shifts to that of the pure N_2O decomposition. The conversion over FeZSM-5 (1173 K), with a severe degree of iron clustering, is shifted to higher temperatures in the different reactions. In direct N_2O decomposition, significant conversions were obtained above 750-775 K over this catalyst.

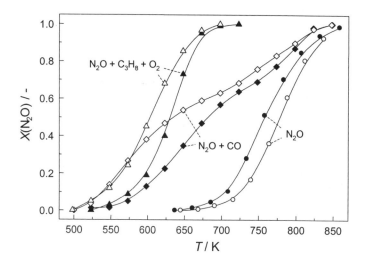

Figure 2 *N_2O conversion vs. temperature over Fe-silicalite (873 K) (open symbols) and FeZSM-5 (873 K) (solid symbols) in different feed mixtures. Experimental conditions as described in Methods.*

The activation of N_2O to produce atomic O* species is an essential step in the reactions investigated here and is controlled by the presence of extraframework iron species. In an earlier work,[13] we demonstrated that extraframework Al or Ga species, as well as Brønsted sites in the catalysts play a minor role in N_2O activation. Taking this into account, the activity differences between FeZSM-5 (873 K) and Fe-silicalite (873 K) in Fig. 2 are strictly related to the distinct nature of the (extraframework) iron species in the catalysts. The similarity between the iron species and activities of FeZSM-5 (873 K) and Fe-silicalite (1173 K) further supports our previous conclusion.

The preparation method of iron-zeolites has been recognized as critical in order to obtain reproducible catalysts with a desired performance.[4,7,14] A distribution of iron species is normally obtained upon activation of catalysts by available methods. Suppressing clustering of iron species into iron oxide is convenient, since these species are proven inactive at low temperatures in the various reactions catalyzed by Fe-zeolites.[1,3,4,7,15,16] Steam activation of isomorphously substituted FeMFI zeolites enables a certain control of the degree of iron clustering, and thus on the relative amount of certain species in the final catalyst, as compared to other methods. A rather unique achievement has been attained here with Fe-silicalite (873 K), in view of the remarkable uniform nature of extraframework species in isolated positions. A minor association of iron species is present in this sample. The presence of Al in such a structure destabilizes framework iron, leading to extraction and clustering of extraframework species, like also happens by steam activation at higher temperatures with Fe-silicalite (1173 K).[7,12]

The activity order of the different catalysts in Fig. 2 strongly depends on the reaction investigated. Isolated extraframework iron sites, present in the largest relative proportion in Fe-silicalite (873 K) appear essential in N_2O processes involving reducing agents, showing higher N_2O and C_3H_8 (or CO) conversions (or the same conversion at lower temperatures) than catalysts with more extensive degree of iron clustering. The small remaining fraction of framework iron in Fe-silicalite (873 K) is inactive for N_2O activation,[13] so their controlled extraction as isolated sites could even lead to a higher activity differences. Oppositely, samples with a larger clustering of iron, either by a different framework composition or higher steam activation temperature, are more active in direct N_2O decomposition. This activation is not caused by the presence of iron oxide nanoparticles observed in TEM, but oligonuclear species in extraframework positions. The low activity of the highly-clustered FeZSM-5 (1173 K) in direct N_2O decomposition supports this statement. The difference in activities between Fe-silicalite catalysts steam-activated at 873 K or 1173 K was in principle assigned to the suboptimal extraction of inactive framework iron in the first sample.[7] But a very pronounced extraction of iron was observed in the first sample, so it is rather due to a different constitution of extraframework species. The apparent activation energy of both samples differs, indicating that different species are involved in the process.

Figure 3   *Schematic mechanisms of the reactions investigated: (a) direct N_2O decomposition, (b) N_2O reduction with CO or C_3H_8. Atomic "O" results from the activation of N_2O over the extraframework iron site.*

Every reaction determines its optimal active species. This is schematically shown in Fig. 3. In direct N_2O decomposition, oligonuclear species are preferred over isolated ions in view of the easier oxygen recombination (rate determining step in the process) of two iron centers that are close together. In N_2O conversions involving reducing agents (CO or hydrocarbons), atomic oxygen deposited on an isolated iron site can be cleaned-off. This implies that the 'activation' of the reducing agent is in fact done by the O* deposited by the N_2O.

In N_2O-related conversions, an iron site can be considered as being active when it easily desorbs molecular oxygen (in N_2O decomposition) or transfer atomic oxygen (in reduction and/or oxidation reactions), starting from Fe-O. This ability should in principle not be exclusive of a certain structure of the site. The operation temperature is often used for labelling a species as "active". Following the previous definition, one can say that in N_2O conversions involving reducing agents, isolated iron sites are the active sites, in view of the higher activity of Fe-silicalite (873 K). It is however, more appropriate to state that isolated iron ions are more reactive towards atomic oxygen transfer to CO or C_3H_8 at a lower temperature. At a somewhat higher temperature, other iron species, *e.g.* oligonuclear iron species, may become operative too. Analogously, in direct N_2O decomposition, oligonuclear iron species appear to facilitate molecular oxygen desorption at the lowest temperature, not excluding other species to contribute at higher temperature, like the isolated iron.

4 CONCLUSIONS

The results presented evidence possibilities of tailoring uniform iron sites in FeMFI zeolites, under specific synthesis and activation conditions. Preparation of steam-activated Fe-silicalite containing mainly isolated iron species in extraframework positions is essential to derive structure-activity relationships in various N_2O conversion reactions over iron zeolite catalysts. The activity of the cluster-free Fe-silicalite was significantly higher in N_2O reduction with C_3H_8 and CO. However, some level of association of iron species leads to higher activities in direct N_2O decomposition. Due to the intrinsic reaction mechanism, this result demonstrates the sensitivity of reactions for the form of the iron species in Fe-zeolites, rather than the existence of a *unique* active site.

References

1 H-Y. Chen and W.M.H. Sachtler, *Catal. Today*, 1998, **42**, 73.
2 El-M. El-Malki, R.A. van Santen and W.M.H. Sachtler, *J. Catal.*, 2000, **196**, 212.
3 K.A. Dubkov, N.S. Ovanesyan, A.A. Shteinman, E.V. Stakoron and G.I. Panov, *J. Catal.*, 2002, **207**, 341.
4 R. Joyner and M. Stockenhuber, *J. Phys. Chem. B.*, 1999, **103**, 5963.
5 F. Heinrich, C. Schmidt, E. Löffler and W. Grünert, *Catal. Commun.*, 2001, **2**, 317.
6 J. Pérez-Ramírez, G. Mul, F. Kapteijn, J.A. Moulijn, A.R. Overweg, A. Doménech, A. Ribera and I.W.C.E. Arends, *J. Catal.*, 2002, **207**, 113.
7 J. Pérez-Ramírez, F. Kapteijn, J.C. Groen, A. Doménech, G. Mul and J.A. Moulijn, *J. Catal.*, 2003, in press.
8 J. Pérez-Ramírez, R.J. Berger, G. Mul, F. Kapteijn and J.A. Moulijn, *Catal. Today*, 2000, **60**, 93.
9 A. Doménech, J. Pérez-Ramírez, A. Ribera, F. Kapteijn, G. Mul and J.A. Moulijn, *Catal. Lett.*, 2002, **78**, 303.
10 A. Doménech, J. Pérez-Ramírez, A. Ribera, G. Mul, F. Kapteijn and I.W.C.E. Arends, *J. Electroanal. Chem.*, 2002, **519**, 72.
11 S. Bordiga, R. Buzzoni, F. Geobaldo, C. Lamberti, E. Giamello, A. Zecchina, G. Leofanti, G. Petrini, G. Tozzolo and G. Vlaic, *J. Catal.*, 1996, **158**, 486.
12 A. Brückner, R. Lück, W. Wieker, B. Fahlke and H. Mehner, *Zeolites*, 1992, **12**, 380.
13 J. Pérez-Ramírez, F. Kapteijn, G. Mul and J.A. Moulijn, *Catal. Commun.*, 2002, **3**, 19.
14 P. Marturano, L. Drozdová, A. Kogelbauer and R. Prins, *J. Catal.*, 2000, **190**, 460.
15 M. Rauscher, K. Kesore, R. Mönnig, W. Schwieger, A. Tißler and T. Turek, *Appl. Catal. A.*, 1999, **184**, 249.
16 P. Marturano, L. Drozdová, A. Kogelbauer and R. Prins, *J. Catal.*, 2000, **192**, 236.

Subject Index